D0700153

CRC SERIES ON CATIONS OF BIOLOGIC SIGNIFICANCE

Editor-in-Chief

Jerry K. Aikawa
Professor of Medicine
University of Colorado
School of Medicine
and
Director
Laboratory Services
University of Colorado
Health Sciences Center
Denver, Colorado

MAGNESIUM: ITS BIOLOGIC SIGNIFICANCE

Author
Jerry K. Aikawa
University of Colorado
School of Medicine
Denver, Colorado

SODIUM: ITS BIOLOGIC SIGNIFICANCE

Editor
Solomon Papper
University of Oklahoma
College of Medicine
and
Veterans Administration Hospital
Oklahoma City, Oklahoma

POTASSIUM: ITS BIOLOGIC SIGNIFICANCE

Editor
Robert Whang
Veterans Administration Hospital
Oklahoma City, Oklahoma

Additional topics to be covered in this series include calcium, zinc, copper, iron, nickel, arsenic, thallium, manganese, molybdenum, chromium, and beryllium.

Sodium: Its Biologic Significance

Editor

Solomon Papper, M.D.
Distinguished Professor and Head
Department of Medicine
University of Oklahoma College of Medicine, and
Staff Physician
Veterans Administration Medical Center
Oklahoma City, Oklahoma

Editor-in-Chief

Jerry K. Aikawa
CRC Series on
Cations of Biologic Significance

CRC Press, Inc.
Boca Raton, Florida

Library of Congress Cataloging in Publication Data

Sodium, its biologic significance.

(Cations of biologic significance, ISSN 8493-5870)
Bibliography: p.
Includes index.
Contents: The normal state: Sodium transport and metabolism/Sidney Solomon and William R. Galey.
Renal regulation of sodium excretion/Mark H.
Gardenswartz and Robert R. Schrier — Sodium excess.
An introductory overview/Solomon Papper.
1. Sodium metabolism disorders. 2. Sodium in the body. 3. Sodium metabolism disorders—Complications and sequelae. I. Papper, Solomon, 1922- .
II. Series: Cations of biologic significance.
[DNLM: 1. Sodium—Physiology. 2. Sodium—Adverse effects. 3. Hyponatremia. 4. Hypernatremia. QV 275 S679]
RC632.S6S66 616.3′99 81-3804
ISBN 0-8493-5873-6 AACR2

Direct all inquiries to CRC Press, Inc., 2000 N.W. 24th Street, Boca Raton, Florida 33431.

International Standard Book Number 0-8493-5873-6

Library of Congress Card Number 81-3804
Printed in the United States

EDITOR-IN-CHIEF

Dr. Jerry K. Aikawa is Professor of Medicine and Biometrics and Acting Head of the Division of Laboratory Medicine and Clinical Pathology at the University of Colorado School of Medicine. He is Director of Laboratory Services at the University of Colorado Health Sciences Center in Denver, Colorado.

Dr. Aikawa received his M.D. degree from the Bowman Gray School of Medicine of Wake Forest University. His graduate medical education includes completion of residency in internal medicine followed by research fellowships sponsored by the National Research Council, the U.S. Atomic Energy Commission, and the American Heart Association. He is author of numerous scientific articles and five books on myxedema, Rocky Mountain spotted fever, clinical data processing, and magnesium. Dr. Aikawa began his research in magnesium in the late 1950s when the U.S. Atomic Energy Commission was encouraging the application of radioactive isotopes to biomedical research, and has maintained his interest in this subject ever since.

THE EDITOR

Solomon Papper, M.D. is a Distinguished Professor and Head of the Department of Medicine at the University of Oklahoma at Oklahoma City Health Sciences Center. He is also a staff physician at the Oklahoma City Veterans Administration Medical Center.

Dr. Papper received his A.B. from Columbia College and his M.D. at New York University in 1944. After completing his internship and residency and a stint in the Army Medical Corps, Dr. Papper became a fellow at Thorndike Memorial Laboratory and Harvard Medical School where he began his research on the regulation of sodium excretion and the kidney in liver disease. He is especially known for his work on the Hepatorenal Syndrome.

In 1962 he became Professor and Chairman of a newly created Department of Medicine at the new medical school of the University of New Mexico. In 1968 he moved to the University of Miami, where he was Professor and Co-Chairman of the Department of Medicine. Before moving to the University of Oklahoma in March 1973, Dr. Papper was Professor of Medicine and Chairman of the Department of Medicine at General Rose Memorial Hospital at the University of Colorado. In April 1974, the Veterans Administration selected him for the designation "Distinguished Physician". He relinquished that title when he accepted the position of Head of the University of Oklahoma Department of Medicine in July 1977.

Dr. Papper has written or co-authored more than 100 published articles in addition to writing a book, *Clinical Nephrology,* whose second edition was published in 1978. He edited *The Kidney,* a publication of the National Kidney Foundation, for ten years. In addition, he has served on the editorial boards of *Journal of Laboratory and Clinical Medicine, Clinical Nephrology,* and *The Forum on Medicine,* and is currently an associate editor of *American Journal of Nephrology.* Dr. Papper is a member of the Association of American Physicians, the American Society for Clinical Investigation, and the American Federation for Clinical Research. He is a Fellow of the American College of Physicians, and at present is Governor of the Oklahoma Region of the American College of Physicians.

PREFACE

Sodium is concerned with the physiology, pathophysiology, and clinical consequences of altered physiology involving the sodium ion.

The first section focuses on the presence and handling of sodium in the normal state. In chapter one, Drs. Solomon and Galey deal with the fundamentals of transport and energy metabolism as they relate to sodium. This is followed by a chapter in which Drs. Gardenswartz and Schrier consider in detail the normal body economy of sodium, and especially the factors (particularly extracellular fluid volume) that regulate the renal handling of sodium and the responses of the various portions of the nephron to these influences.

The rest of the book emphasizes departures from normal physiology. For convenience we have divided these into three broad categories: sodium excess, sodium deficit, and alterations in serum sodium concentration.

The section on sodium excess includes edematous conditions and hypertensive disorders. In the former, Drs. Bernard and Alexander, Epstein, Czerwinski and Llach, and Kem delineate the mechanisms of enhanced renal reabsorption of sodium in the common edematous states. That sodium excess is somehow involved in the causation of some hypertensive states is a fascinating and still incomplete story. Drs. Frohlich and Messerli explore the evidence for the role of sodium and consider the mechanisms whereby excess sodium may be a causative and preventive mechanism in the genesis of hypertension. Finally, diuretic agents are considered by Drs. Whitsett and Chrysant primarily from the pharmacological viewpoint.

Sodium deficit, i.e., extracellular fluid volume deficit, is presented by Dr. Vaamonde in systematic fashion.

The book ends with detailed discourses on hypo- and hypernatremia by Drs. Llach and Czerwinski, and Finberg, respectively.

In all sections there is a commitment to expounding the knowns and exploring the unknowns, and the two are distinguished from each other carefully. There is repetition in the book. Our efforts to maintain each section and chapter as a discrete entity led in several instances to the judgment that repeating material found elsewhere made for a more complete free-standing story.

I thank each author for sharing with skill his knowledge and expertise. As editor I deeply valued and appreciated their cooperative and good natured spirit. Ms. Beverly Clarke's editorial assistance is gratefully acknowledged.

S.P.

CONTRIBUTORS

Edward A. Alexander, M.D.
Professor of Medicine and Professor of Physiology
Boston University School of Medicine
Chief, Renal Section and Associate Director of Medicine
Boston City Hospital
Boston, Massachusetts

David B. Bernard, M.D.
Assistant Professor of Medicine
Boston University School of Medicine
Director, Clinical Nephrology
Evans Memorial Department of Clinical Research
University Hospital
Boston, Massachusetts

Steven G. Chrysant, M.D.
Associate Professor of Medicine
University of Oklahoma College of Medicine
Oklahoma City, Oklahoma

Anthony W. Czerwinski, M.D.
Professor of Medicine
University of Oklahoma College of Medicine
Associate Chief, Nephrology Section
Oklahoma City Veterans Administration Medical Center
Oklahoma City, Oklahoma

Murray Epstein, M.D.
Professor of Medicine
University of Miami School of Medicine
Miami Veterans Administration Medical Center
Miami, Florida

Laurence Finberg, M.D.
Professor
Department of Pediatrics
Montefiore Hospital and Medical Center
Albert Einstein College of Medicine
Yeshiva University
New York, New York

Edward D. Frohlich, M.D.
Vice President, Education and Research
Alton Ochsner Medical Foundation and Division of Hypertensive Diseases
Ochsner Clinic
New Orleans, Louisiana

William R. Galey, Ph.D.
Associate Professor of Physiology
Department of Physiology
University of New Mexico School of Medicine
Albuquerque, New Mexico

Mark H. Gardenswartz, M.D.
Assistant Chief of Nephrology
Lenox Hill Hospital
New York, New York

David C. Kem, M.D.
Professor of Medicine
Chief, Endocrinology, Metabolism and Hypertension Section
Department of Medicine
University of Oklahoma College of Medicine
Oklahoma City, Oklahoma

Francisco Llach, M.D.
Associate Professor of Medicine
Chief, Nephrology Section
Department of Medicine
University of Oklahoma College of Medicine
Oklahoma City Veterans Administration Medical Center
Oklahoma City, Oklahoma

Franz H. Messerli, M.D.
Staff Member
Division of Hypertensive Diseases
Ochsner Clinic and
Division of Research
Alton Ochsner Medical Foundation
New Orleans, Louisiana

Robert W. Schrier, M.D.
Professor and Chairman
Department of Medicine
University of Colorado Health Sciences Center
Denver, Colorado

Sidney Solomon, Ph.D.
Professor of Physiology
University of New Mexico School of Medicine
Albuquerque, New Mexico

Carlos A. Vaamonde, M.D.
Professor of Medicine
University of Miami School of Medicine
Chief, Nephrology Section
Miami Veterans Administration Medical Center
Miami, Florida

Thomas L. Whitsett, M.D.
Professor of Medicine
Chief, Clinical Pharmacology Section
Department of Medicine
University of Oklahoma College of Medicine
Oklahoma City Veterans Administration Medical Center
Oklahoma City, Oklahoma

DEDICATION

To Dr. Maurice B. Strauss
whose work underlies much that is contained
in this book, and who introduced the editor to
Sodium — as well as to the pleasures of asking
questions and trying to answer them.

TABLE OF CONTENTS

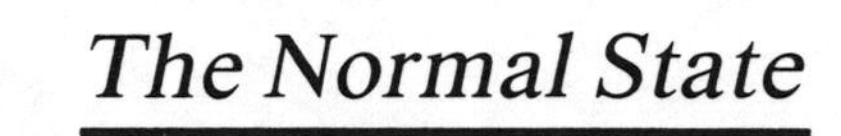

Chapter 1

SODIUM TRANSPORT AND METABOLISM

Sidney Solomon and William R. Galey

TABLE OF CONTENTS

I. INTRODUCTION

During the evolutionary development of cells, one of the early manifestations of these biological units appears to have been the establishment of ionic gradients. Cellular accumulation of potassium and extrusion of sodium became established. The basis of biological actions depending on ionic gradients, i.e., action potential generation, was thereby created. With the evolution of metazoan species, changes in marine environments, invasion of fresh water, and adaptation to a semiarid environment (land), new problems had to be solved to allow for development and diversification of species. A variety of adaptive measures developed. Most species excluded the external environment and maintained the old surrounding for cells—the "internal environment" of Claude Bernard. The most common solution to the homeostasis problem was accomplished through transport of electrolytes, and in particular, sodium. Sodium transport systems serve to maintain ionic balance and cellular transmembrane ionic gradients. On the other hand, some species responded by providing for osmotic equilibration through either the intracellular accumulation of amino acids,[1] or by accumulating extracellular urea[2] or trimethyl amine.[3] In order to maintain functional integrity, however, it remained necessary to keep ionic gradients and this need still persists.

Whenever compartments exist with differing ionic activities or capability for exchange of materials, energy has to be expended to maintain these gradients. Such is the case, whether one is interested in single cells, metazoan cells in contact with the internal environment, or in the interaction of the internal environment of metazoan organisms with the external environment. Current evidence would indicate that transport of sodium is one of the most ubiquitous mechanisms used for maintenance of ionic gradients. As a result, the link between cellular metabolism as a source of energy and sodium transport is an extremely intriguing question. Another consequence of the maintenance of the sodium gradient is that the potential energy of the gradients can be utilized for providing the driving force for movement of organic substrates. The energetic events involved in establishing the sodium gradient, as well as the utilization of the energy the gradient provides, form the basis of this chapter.

II. IONIC MOVEMENTS AND THE NEED FOR ENERGY

Simple passive permeation across membranes is a movement down an electrochemical gradient for the solute being considered and requires no cellular energy. The electrochemical gradient across a membrane has been defined by investigators[4,5] and is given formally by:

$$\mu = \mathrm{RT\,n} \cdot \mathrm{C_1/C_2} + \mathrm{ZFE} \qquad (1)$$

where R and T have their usual definitions of gas constant and absolute temperatures; C_1 and C_2 are the concentrations of the solute under consideration on the two sides of the membrane; Z is the valence of the ion in question (which in the case of sodium has a value of one); F is the Faraday constant; and E is the potential difference between the two compartments which the membrane separates. A simple expression relating potential to concentration gradient is given by the Nernst equation:

$$\mathrm{E} = \frac{\mathrm{RT}}{\mathrm{ZF}} \ln \frac{\mathrm{C_2}}{\mathrm{C_1}} \qquad (2)$$

It is easily seen that Equation 2 can be derived from Equation 1, when μ is set equal to zero.

When solutes permeate a membrane, there is movement in both directions. It was shown by Ussing[6] that the relationships of passive fluxes (M) in each direction are given by the following equation:

$$\frac{M_{1\to 2}}{M_{2\to 1}} = \frac{C_1}{C_2} e^{\frac{ZFE}{RT}} \qquad (3)$$

where e is the base of the natural logarithms and all other symbols have the same meanings as given. One can define a passive movement of a solute as one where the ratio of fluxes $M_{1\to 2}/M_{2\to 1}$ is predicted by Equation 3. If such a relationship is not found, as a first assumption, active transport can be postulated. Active transport, however, is not always required. A movement has been described wherein one ion exchanges for another with no net permeation. This form of membrane transfer has been called exchange diffusion.[7] If exchange diffusion is appreciable, the mathematical ratio will not be observed and although the flux of the solute is not active, an error of interpretation will be made by considering Equation 3 only.

From these considerations it becomes clear that additional criteria are needed for definition when a solute is actively transported. Many current hypotheses hold that for a solute to be transported actively, there must be combination of the solute with a carrier. Metabolic energy is transformed to a mechanical energy so that the carrier either moves or undergoes a conformational change, with the result that the solute is carried to the opposite side of the membrane and then released. Hypothesized models* of the transport process are shown in Figure 1.

Obviously, if the living cell is to maintain ionic disequilibrium between the environment and its internal milieu, energy must be expended. Hence, cells must utilize some of their metabolic energy to accomplish the thermodynamically unfavorable "uphill" transport of the ions against their electrochemical gradients. This, then, is one of the two requirements for showing that the transport of a molecular species is "active". Specifically, the cell must use directly cellular metabolic energy to accomplish the transport of the actively transported solute. As is well known, this energy source is in the form adenosine triphosphate (ATP). The second requirement for active transport has already been alluded to but not directly stated. The process for active transport of a solute must be capable of moving the solute up its electrochemical gradient. These two requirements,

1. The direct expenditure of cellular metabolic energy, and
2. The capability of moving the solute against its electrochemical gradient,

are the necessary and sufficient prerequisites for stating that the solute in question is actively transported. This transport has been called "primary active transport" for reasons that will be discussed later.

The fact that the movement of a solute, be it an electrolyte or nonelectrolyte, is against its electrochemical gradient is not sufficient evidence to prove that it is actively transported. It may be that a solute is being transported at the expense of the ionic gradient that exists across the cell membrane as a result of a previous active transport process. In this case cellular metabolic energy is not being used directly to accomplish the transport and thus should not be considered "active" transport.

The frog skin has been extensively used for studies of sodium transport. In fact, it

* Some investigators do not believe that any cellular accumulation or extrusion of ions requires an active process such as the one to be described (see Ling,[8] for example).

FIGURE 1. Models of a conformational and mobile transport carrier. In the conformational system, the carrier is fixed in place and transports sodium uphill only through a change in shape of the transporting molecule (left panel). In the case of a mobile carrier, the transporting molecule becomes associated with sodium on the side of low concentration, moves across the membrane, and then becomes dissociated from the sodium (right panel). In both models ATP provides energy for doing the transport work.

was Huf who first coined the term "active transport" from an investigation on this tissue.[9] He was first to show that the skin moved sodium chloride against a concentration and that the movement of the sodium ion was also against an electrical gradient. He and others further established that these movements were suppressed by inhibitors of both aerobic and anaerobic metabolism.[10,11] Ussing and Zerahn carried out their first "short circuit" studies on this tissue and were able to establish that sodium was the ionic species being actively moved.[12] Subsequent work involved studies of the quantitative relationship between metabolism (i.e., oxygen consumption) and active sodium transport.[13] Experiments with this tissue have been extensively used as a prototype for investigation of sodium transport in other tissues.

Energy may be put directly into a system involved directly in the movement of a solute — primary active transport — or a solute may be moved, even uphill, in conjunction with the flow of another solute or of solvent. The movement of one solute can be coupled to the flow of another solute either by having the solutes move in opposite directions (exchange processes or antiport) or by the flux of solutes being in the same direction (symport). An example of exchange is the movement of K^+ into a cell secondary to the active extrusion of the Na^+ out of the cell. These types of exchange reactions are very common, occurring in cells (red cells[14] and muscle[15]) and epithelial sheets of cells (kidney[16] and intestine[17]). Another common form of antiport is that of hydrogen exchanging for sodium.[18,19]

Within the past decade or so, the significance of symport has been increasingly recognized as a mechanism for moving solutes. At a much earlier date, however, the basis of the symport phenomenon was established when it was found that Na was required for optimal transport of glucose by the intestine and that the presence of glucose also increased the transport of sodium.[20] Current evidence indicates a linkage between sodium movement and that of a variety of organic solutes. This evidence will be reviewed later in this chapter to indicate that energy is put into the system to create a sodium gradient and that energy for transport of the organic solute is derived in whole or in part through dissipation of this gradient.

In consideration of cell transport, it becomes important to define the localization of the energetic steps. Two problems are encountered: the first is the question of what transport process is involved in maintaining electrolyte gradients between a cell and its environment. Since all cells maintain such gradients, it is likely that this process is relatively primitive and ubiquitous. The second problem relates to cell sheets and

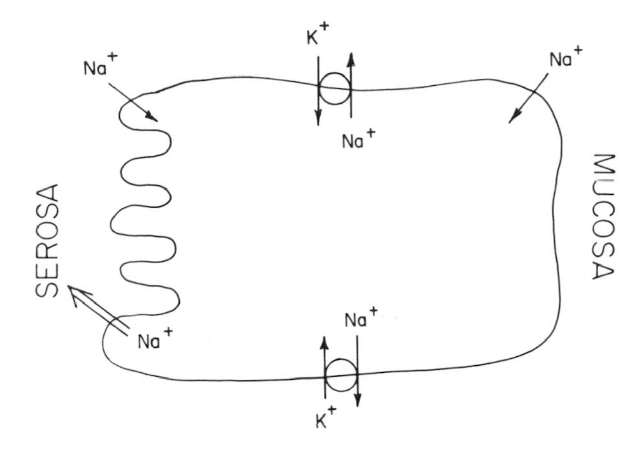

FIGURE 2. Schematic diagram of different paths for sodium movement. Inward passive diffusion of extracellular sodium is indicated by the arrows pointing downward at the upper corners of the cell. Na—K exchange is represented by the loop showing the movement of these two ions in opposite directions. Both of these pathways may undergo membrane-localized changes in density, but probably exist on all cell surfaces. The outward active transport step for sodium is indicated by the double arrow pointing upward and located in the serosal side of the cell.

glands wherein unidirectional transport processes occur. The question is one of how unidirectional transport across an epithelium can be accomplished. It seems likely that the unidirectional processes arose as a modification of the mechanism responsible for maintaining the cellular electrolyte gradients.

Figure 2 shows an idealized cell which illustrates both kinds of transport. It has been shown for a variety of single cells (e.g., nerve,[21] muscle,[22] and kidney[23]) that sodium enters the cell by passive permeation, i.e., down its electrical and chemical gradient. A membrane-localized process then actively extrudes sodium from the cell. Evidence has accumulated that the extrusion of sodium can also be accompanied by the entrance of potassium. When such an exchange process occurs, i.e., one ion of K replacing one ion of Na, no net electrical charge is moved, and the process is considered to be nonelectrogenic. Consequently, only concentrative work is required to move the ions up their concentration gradients. No electrical work is done. However, in most systems, the ratio is not a one-for-one exchange (e.g., 3 Na for 2 K in the red cell[24]). Additional electrical energy is required for the fraction of sodium moving against the electrical gradient, unaccompanied by another ion of like sign moving in the opposite direction.

Transcellular transport involves the same steps, i.e., a passive leak of Na into the cell followed by active extrusion. If, as seems to be the case, the same processes involved in vectorial transport are involved in maintaining cellular electrolyte balance, the question arises as to how directionality is achieved. Two mechanisms which have been suggested are:

1. Infoldings of the membrane on the side to which the solute (usually Na) is to be moved (usually the serosal side of cell sheets, the peritubular side of the renal tubules) could provide for more transport sites on this side.
2. More ATP for energy may be made available on the serosal side through localization of mitochondria, the major ATP source. This kind of localization is well demonstrated by the nephron.[25]

Another possibility which should be investigated is simply that there are more transport sites on the serosal or peritubular membrane independent of membrane area. Finally, the basic assumption may be incorrect and a separate pathway may exist for vectorial transport unrelated to maintenance of the ionic balance of cells.

III. THE ROLE OF METABOLISM IN ACTIVE TRANSPORT

There are several ways by which cellular metabolism may affect the transport of ions across cell membranes. Primarily, cellular metabolism is necessary to provide energy for the active transport of the solute. Often, however, it is simply assumed that if the transmembrane movement of a solute is dependent upon metabolism, that movement of the solute must be active. Although the direct utilization of cellular energy is a necessary condition, it is in itself not sufficient evidence for active transport. There are other ways by which metabolism can influence solute movement without implicating active transport. The fact that the passive movement of ions down their electrochemical gradients requires specific structures within the cell membrane, whether they be hydrophilic conductive channels or specific carriers, demands that such structures must by necessity be produced and maintained functional through processes which utilize energy.[26] Thus, although metabolism is a requisite for the transport of these solutes, it merely plays a facilitory role in providing the appropriate structure for transport and is not directly involved in the translocation of the solute itself.

Another way by which metabolism may effect transmembrane solute movement is through the active transport of some species other than the solute of interest. The other species provides a concentration gradient (i.e., potential energy) for the carrier mediated but passive transport of the solute of interest. Although cellular metabolic energy is being utilized, it does not directly participate in the translocation of the solute. This process is generally not thought of as being active transport. In some cases it is called "secondary active transport" since the movement of the solute is secondary to the active transport of the driving solute which provides the energy gradient for the "uphill transport".

Although one must be cautious not to couple nonactive transport processes erroneously with cellular metabolism, the necessary coupling between metabolism and active transport has proven to be useful in understanding the energetics of the active transport of Na^+ and K^+.[27-29]

In cells such as erythrocytes, which utilize glycolysis as their main metabolic source of energy, numerous authors[27,30-32] have shown that glucose utilization and lactate production are coupled to active transport of sodium and potassium. Furthermore, as early as 1928, Lund[33] perceived the dependence of action transport on oxygen consumption by frog skin. Since that time there have been numerous studies of the role and stoichiometry of oxygen consumption with active ion transport.[13,34,35] Clearly, for several tissues it has been shown that the rate of oxygen consumption is related to the rate of active sodium transport.[29,34,36] When care is taken to eliminate the contributions of bacterial contamination to oxygen consumption, as well as the required "basal metabolism" of cells, studies have shown not only that oxygen consumption by the various tissues parallels active transport, but also that there is an extremely tight coupling stoichiometry between the utilization of oxygen and the amount of sodium transported across the tissue.[37,38] While studies of the stoichiometry of catabolic metabolism to active transport have been useful and have led to several unique and potentially useful tools for studying active transport,[37,39] their success is the direct result of the tight coupling of both anaerobic and aerobic metabolism to the production of adenosine triphosphate (ATP).

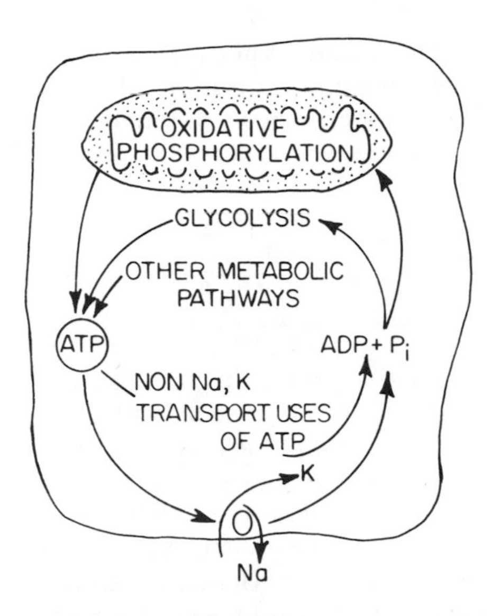

FIGURE 3. Schematic representation of pathways for generation of ATP and for its utilization in providing energy for sodium transport.

IV. THE ROLE OF ATP

As is seen in Figure 3, whether the main metabolic source of energy is glycolysis, oxidative phosphorylation, or even another less important metabolic pathway,[40] the currency for energy is ATP. This ubiquitous molecule is found in virtually every mammalian cell which transports sodium and potassium. Clearly the transport machinery of the cell cannot tell where the ATP it will utilize to bring about the translocation of the ionic species came from. Thus, the production of ATP by whatever cellular metabolic process and its hydrolysis by the ionic transporting mechanism are of paramount importance in understanding the nature of active transport.

V. THE MOLECULAR BASIS FOR ACTIVE TRANSPORT — Na-K-ATPase

It is clear that transport of sodium out of cells and the simultaneous transport of potassium into cells is an active transport process. Not only are both ions moved against their electrochemical gradients but also these movements require cellular energy in the form of ATP.

Within the last 20 years numerous studies have provided overwhelming evidence that the active transport of Na^+ and K^+ are associated with the K^+- Na^+-, Mg^{++}-dependent ATPase enzyme commonly called Na-K-ATPase (see Glynn and Karlish[41] for a review of this subject). From the earliest work of Skou[42] on this subject to the most recent papers,[41,43] parallels have been established between active Na^+ and K^+ transport and this enzyme system. Not only are both the enzyme system and the active transport "pump" associated with cell membranes, but also both require the same substrates, Na^+, K^+, Mg^{++}, and ATP,[44] and they are both poisoned by the same agents, ouabain,[45] and Ca^{++},[46] as well as others.[47] Furthermore, both systems have high activities in cells and tissues[48] where the active Na and K transport is high.

In a recent excellent review, Glynn and Karlish[41] survey much of the evidence for the association of the sodium pump to Na-K-ATPase and discuss the mechanism by which the pump is believed to operate. While the mechanism by which the pump truly operates must await detailed structural knowledge which does not yet exist, numerous general mechanistic schemes have been proposed, including internal translocation, gated channel, and carrier mechanisms.[41] The events in the hydrolysis of ATP seem to be represented by:

$$E_1 \xrightarrow[\text{ATP}]{(\text{Mg, Na})} E_1\sim P \xrightarrow{(\text{Mg})} E_2\sim P \xrightarrow{\text{K}} E_2 + \text{Pi} \quad (E_2 \rightarrow E_1)$$

The enzyme in one of its stable forms (E_1) which has a high affinity for Na appears to be phosphorylated by ATP in the presence of Na and Mg. This new unstable form $E_1 \sim P$ then is proposed to undergo, in the presence of Mg, a change to a new lower energy form $E_2\sim P$.[49] The next step in the sequence involves the dephosphorylation of the enzyme which requires the presence of K^+ to produce the unphosphorylated E_2 enzyme state, a complex of lower energy than $E_2\sim P$. Finally, the E_2 enzyme form returns to the high Na affinity E_1 state. Hence, a stepwise cycle is completed which can easily be related to the translocation of the ionic species. It is particularly tempting to associate the translocation of Na with the steps between E_1 to $E_2\sim P$ and the translocation of potassium with the steps between $E_2\sim P$ to E_1. The exact mechanisms must, however, await more knowledge about the structure and dynamic processes occurring in the enzyme system.

Additional support for the ATPase hypothesis exists in the parallel asymmetries of the transport and ATPase systems. In particular, an increased concentration of Na must be present on the inside of cells and of K on the outside in order to activate significant ATPase activity, as well as being required for the active transport of both Na and K.[50] Furthermore, ouabain and other cardiac glycosides inhibit both active transport and the Na-K-ATPase only when they are present on the outside of the cells.[45] Inhibition can be overcome by increasing the extracellular potassium levels.[51] This is entirely consistent with inhibition resulting from the competition of the active transport inhibitor with potassium for the outer surface binding sites of the ATPase-transport system.

Within recent years, it has been shown that if the transmembrane ion gradients are adjusted so as to "run the pump in reverse", cells will produce ATP from ADP and inorganic phosphate, rather than hydrolyze ATP to Pi and ADP, as happens when active transport is taking place.[52] This strongly indicates that the two systems are closely linked, if not one and the same.

Probably the most conclusive evidence showing the Na-K-ATPase to be responsible for active Na and K transport comes from reconstitution of the enzyme system in lipid membrane vesicles. In these experiments, the Na-K-ATPase enzyme system was purified by conventional biochemical techniques. The enzyme complex was then incorporated into small lipid bilayer membrane-bounded vesicles. If ionic and ATP concentration are appropriately set, the membrane-bound ATPase enzyme actively transports Na and K across the vesicle membrane in opposite directions.[53] While the enzyme complex obtained is not absolutely pure, it is difficult to explain how such transport could be accomplished if the Na-K-ATPase were not the active transport system.

VI. THE CHEMIOSMOTIC THEORY

The involvement of Na-K-ATPase has provided the major biochemical basis for movement of sodium into cells and across cells. Another scheme has been presented which directly links metabolism to a variety of ionic movements. In 1961, Peter Mitch-

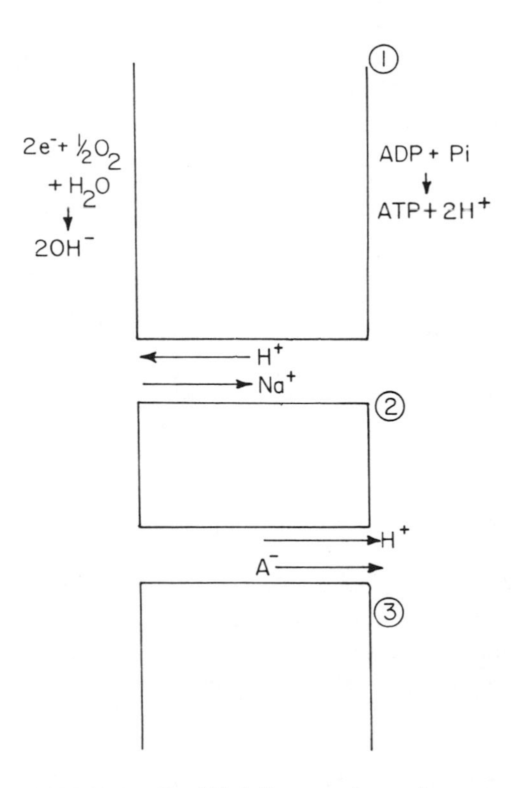

FIGURE 4. Simplified diagram of overall events of the Mitchell chemiosmotic hypothesis. Path 1 represents the reaction leading to the establishment of a pH gradient. Path 2 indicates how passive movement of H^+ can be linked to sodium extrusion Path 3 shows a possible mechanism for extrusion of anions.

ell suggested that a flow of electrons could drive protons across membranes, specifically across chloroplasts, mitochondria, and bacterial cells.[54] This hypothesis was developed into the "chemiosmotic theory" for which Mitchell received the Nobel Prize in 1979. The proposed mechanism could result in a linked movement of sodium by some additional assumptions and modifications of the original scheme. It will, therefore, be considered. At present the chemiosmotic theory has supportive evidence that it is applicable only to select organelles and microorganisms.

The fundamental principle that underlies the Mitchell chemiosmotic theory is that during the process of respiration, charge separation occurs and thereby creates an electrical and chemical gradient. Figure 4 shows a schematic presentation of the primary steps involved. Within the membrane (or in the right compartment) respiratory substrates provide two protons and two electrons. The electrons are used to oxidize a reduced electron carrier of the membrane separating the two phases. The oxidized carrier accompanied by the H^+ is reduced in two steps. At the right side the hydrogen ions are extruded, while at the left side the carrier gives up its electrons and the reaction $e^- + \frac{1}{2} O_2 + H_2O \rightarrow 2\ OH^-$ proceeds. As a result, a pH (hydrogen-ion) gradient is developed. Because of the separation of charge, an electrical gradient is also developed. In the case of the mitochondrion and microorganism, the left compartment would be inside the limiting membrane so that the inside of the organelle would become alkaline and negative to the outside. (It should be pointed out that this schematic presentation is highly oversimplified but presents the salient features of the system. Many of the intermediate details are still controversial, although the overall scheme is fairly well accepted. A more complete analysis can be found in several reviews.)[55,56]

A direct way in which the proton and electrical gradients can be effective in moving Na^+ is by assuming that the limiting membrane is permeable to this cation. The pH gradient would drive sodium out of the left compartment by exchange of protons for the cation. In this case, if Na^+-H^+ exchange were one for one, there would be no electrical work done and the electrical gradient would have no effect. It is conceivable, however, that the electrical gradient itself could serve as a driving force for moving Na^+ out of the inner compartment. It would appear that Na extrusion by mitochondria is only pH gradient-dependent (quoted in Reference 56), while in *E. coli,* it also depends on the electrical gradient (quoted in Reference 56).

The chemiosmotic theory was first conceived in order to explain oxidative phosphorylation. The essential reaction is ADP + Pi → $2H^+$ + ATP. The major pathway for dissipation of the proton gradient is through the ATP generating system. Since ATP is the substrate for membrane Na-K-ATPase action, the chemiosmotic mechanism may play an indirect role in regulating cellular sodium transport by determining the amount of ATP available.

It should be pointed out that the generation of the electron, proton gradient could also act as a driving force for movement of anions. If an anion permeant pathway existed across the limiting membrane, the extrusion of the proton could be accompanied by anion movements. It is also obvious that the movement of the anion would be accelerated by the potential difference.

VII. THE ASSOCIATION — INDUCTION HYPOTHESIS

Although the ATPase theory for active transport is well supported by experimental observations and has been widely accepted, it has been criticized by some investigators. The primary criticism is based on the belief that energetic requirements for the ATPase system exceed the total cellular energy output.[8,57] The alternative theory which has been proposed by G. N. Ling is called the association-induction hypothesis.[8] This theory states that specific cations (potassium) are bound to negatively charged sites on intracellular macromolecules. Furthermore, the high concentration of intracellular proteins perturb the solvent properties of the intracellular water so that it is unable to dissolve ions (sodium). Thusly, the high intracellular potassium levels and the low intracellular sodium levels are explained. It is further postulated that ATP plays an important role in maintaining the state of the intracellular proteins so as to selectively bind potassium over sodium. In this way, the metabolic dependence of the ionic gradients across the cell membrane is explained. While we shall leave the decision as to the validity of these hypotheses to the reader, we should like to leave this subject by stating that the association-induction hypothesis has been seriously attacked on many fronts.[58] More important, even the originator of the hypothesis does not suggest that it is responsible for the active transport of sodium across tissue sheets.

VIII. CELL AND ORGAN METABOLISM AND TRANSPORT—SOME EXPERIMENTAL RESULTS

Some of the earliest evidence that active transport of sodium requires metabolic energy came from studies with metabolic inhibitors. It was established that substances which inhibited both aerobic and anaerobic respiration were effective also in inhibiting sodium movement by frog skin.[10] When anaerobic metabolism was inhibited by bromoacetate, Huf was able to show that lactate addition could provide adequate energy through the aerobic pathway so as to support transport. This result was significant since it showed that the effects of the inhibitor resulted only from an alteration of metabolism rather than from a change in the transport system itself. Since these pioneering studies, other investigators have increased the number of effective inhibitors to show that interruption of the supply of available energy in the form of energy-rich compounds results in a reduction in sodium transport (see Kinne[59] for a listing of these inhibitors).

The results are not always clear-cut, however. For example, dinitrophenol (DNP), an uncoupler of oxidative phosphorylation, does not reduce sodium transport by the kidney when injected into the renal artery.[60] Originally, this result was interpreted to indicate that the cell contained enough energy-rich compounds other than ATP to support transport.[60,61] With micropuncture studies, it was shown that intralumenal application of DNP did produce inhibition of sodium transport.[62] It is, therefore, obvious that with complex organs such as the kidney, interpretation of metabolic inhibitor results can be most difficult. The effectiveness of the inhibitor will depend not only on its concentration but also upon the route of administration. Since transporting segments exist in the kidney which can and do have somewhat differing transport properties, one would expect that the site of application of an inhibitor would also be significant in partially determining its action. Lack of effect of metabolic inhibitors on Na transport in addition to DNP could be cited, but since the topic has recently been reviewed,[59] it will not be considered further.

It has been pointed out previously that oxygen transport increases linearly with increases in active sodium transport. This observation has been made for simple epithelial sheets as well as for more complex organs such as the kidney.[63] The general aim of these studies has been to determine if a fixed relationship exists between the amount of sodium transported and oxygen consumption. The Na/O_2 ratio at first glance appears to be variable since in toad bladder it has been found to be about 16,[64] while in the kidney, it was between 25 and 30.[65] In the kidney, however, a large fraction of sodium moves by other than active processes. It has been estimated that in the proximal tubule only about 40% of the transport is mediated by active processes.[66] If O_2 consumption is then corrected for the passive component, the Na/O_2 ratio is reduced to about 15 on 16 as found in other systems. Further, if one assumes the production of 6 mol of ATP per mole of O_2 consumed, the sodium transported per ATP is between 2.5 and 3.0. Such a calculation shows that adequate energy is available from ATP hydrolysis in aerobic systems to account for all of the sodium transport. Further indirect evidence that ATP hydrolysis is the major energy source for sodium transport is also found in anaerobic systems. It has been calculated (see Kinne[59]) that in the red cell 2.8 to 3.0 sodium ions are extruded per mole of ATP utilized.

With respect to overall renal metabolism, it has been popular at times to look for a single substrate which is preferentially utilized as a source of energy for supporting sodium transport. From consideration of the fact that ATP has a central role as the energy source for maintaining transport and can be generated from numerous different substrates, it would seem that such a search is really a non-problem. Nevertheless, reports have appeared at times indicating that lactate is a primary substrate. Such reports have usually used results from measurement of renal arterial-venous differences[67] and whole kidney perfusion studies[68] as the basis for this conclusion. In our opinion, it is unlikely that a single substrate plays a central role in providing energy. Although lactate seems to be a substrate capable of increasing oxygen consumption more than others studied,[69] this does not mean that it is a substrate directly linked to any function. It may merely reflect the possibility that lactate can permeate cells more readily than other exogenous substrates. This explanation, as well as other alternatives for explaining the ability for lactate to preferentially support function requires evaluation.

IX. COUPLING OF Na MOVEMENT TO THE TRANSPORT OF OTHER SOLUTES

It appears that in some tissues the ability to actively exchange sodium for the hydrogen ion exists. While this system is of great importance in the HCL-secreting oxyntic cells of the stomach[70] as well as the nephron, it is but one of a number of coupled transport systems.

By far the most quantitatively important transport process involving Na, aside from Na and K active transport, is the coupled passive transport of sodium with a large variety of organic solutes. These transport systems are said to be carrier mediated since they show characteristics which are consistent with the presence of a finite number of transporter molecules located in the cell membrane. Specifically these "carriers" show: (a) specific binding of a single species of solute or those having particular structural similarities, (b) saturation transport kinetics which are characteristic of site-mediated processes, (c) poisonability by specific carrier poisons such as phloretin, and (d) reversibility, meaning that if solute gradients are oriented in a direction opposite to that usually found, the transport process can be run in reverse. The sole distinguishing characteristic which separates such carriers from active transport is that carriers are passive in that they do not utilize cellular metabolic energy. In particular, the driving force for the transport of a solute is provided by an electrochemical gradient of itself or another species. In some cases this gradient is that of a hydrogen ion,[71] as we have previously mentioned in our discussion of the Mitchell hypothesis. However, in many cells, the sodium gradient, from the high sodium content extracellular environment to the low intracellular sodium concentration established by the active transport of sodium from the cell, is used to drive this process. Although counter transport or antiport in which the "driving solute" supplying the energy for the transport moves in a direction opposite to the carried solute is possible in some systems,[72] the sodium gradient is usually found to drive co-transport or symport in which both the driving solute (in this case sodium) and the "carried solute" are transported at the expense of and secondary to the primary active transport of sodium. The passive carrier-mediated process is sometimes called secondary active transport even though the process is not at all active.

There are many examples of sodium dependent co-transport of solutes in a vast number of tissues. For example, amino acid absorption,[73] organic acid secretion,[74] and sugar absorption[75] in the proximal tubule of the kidney, as well as amino acid[76] and sugar absorption[77] from the small intestine, are transported by Na-dependent symport. It should be made clear that these carriers are rather specific. As an example of this specificity, the carrier in the small intestine which is responsible for the absorption of the monosaccharides glucose and galactose does not carry the monosaccharide fructose.[78] Similarly, the carrier responsible for neutral amino acid reabsorption from the intestinal lumen does not transport dibasic, acidic, or imido-amino acids. Each of these amino acid classes has its own sodium-dependent carrier system.[79]

Once the solutes have been carried into the absorbing cells, they are either metabolized[80] or diffused from the cells, usually through the basal or lateral cell membranes. Hence, the transcellular movement of such solutes is energized by the existing sodium gradient ranging from the solution from which the transport is to take place to the cell interior. Subsequently, the movement of the carrier solute from this cell is usually by passive diffusion. The net effect is that by utilizing the existing transmembrane sodium gradient resulting from active extrusion of sodium from a cell, cells are able to accomplish an asymmetric net transport of a second carried solute. The cost of such transport is the dissipation of the sodium gradient which can be easily reestablished by the active sodium transport process.

The exact mechanism by which synport is achieved has yet to be explained in detail. A reasonable working model has been proposed by Christiansen.[21] In this model, Na^+ and the co-transported species are both considered to be located on a carrier at sites close to each other. The carrier then moves or is moved across the membrane. In the case of amino acids, the Na binding site is about the distance of four carbon atoms away from the carboxyl group. This theory would predict that amino acids having an intrinsic positive charge close to this position (lysine, arginine, citrulline) might not require sodium for co-transport. This prediction is, in fact, born out by experimental results.

X. PROBLEMS REMAINING

If a summary were made of the substantiated knowledge that relates active sodium transport to metabolism, it would be very small. It seems clear that a linear relationship exists between transport and energy production by cells, and quantitatively, the energy produced is adequate to account for sodium transport. ATP has a central role as the substance with adequate chemical energy able to link directly to the transport system to provide it with the necessary driving force. The evidence is becoming stronger that Na-K-ATPase is not only the energy transducer of ATP to transport, but it is probably also an integral part of the transport system itself. Finally, it is certain that sodium transport can result in the movement of other ions by symport and by antiport mechanisms.

Multiple problems as yet remain unsolved. For example, although evidence is increasing that a mobile carrier system is unlikely, it has yet to be established if, in fact, a conformational carrier system is the only possible mechanical shuttle. In fact, the detailed nature of the molecular events remain to be clarified.

At a somewhat higher order of organization, a significant amount of information exists about the way ATP is generated by mitochondria, as well as other energy-generating systems. It is not yet known by what pathway ATP is directed to the Na-K-ATPase. Instinctively, one might expect that simple diffusion may not be an efficient process. If ATP movement is directed to the Na-K-ATPase, how is it directed?

Probably one of the most fascinating problems yet to be solved is how the transport systems are regulated. Concentration of Na and K per se seems to be adequate to explain maintenance of cellular ionic gradients through activation of Na-K-ATPase. In the case of epithelial sheets and glands, however, sodium transport is turned on and off to meet the needs of the whole organism. It would seem that this is one area in which we are most ignorant.

One can assess our current state of knowledge with the observation that great strides have been made in our understanding of Na transport processes and metabolism in the past two or three decades. The next several decades promise to be equally as exciting and, hopefully, as productive.

ACKNOWLEDGMENTS

The large number of publications in the area covered by the chapter has precluded our listing of all of the significant papers. We believe those references cited are representative and can lead the interested reader to explore more of the relevant data. If we have omitted critical publications, we apologize and accept responsibility for all errors. Finally, we would like to thank Ms. Arlene Martinez for the extra effort expended by her and for her patience during preparation of this manuscript.

REFERENCES

1. **Florkin, M.**, La régulation isosmotique intracellulaire chez les invertébres marins euryhalins, *Bull. Acad. Royale de Belg.*, 48, 687, 1962.
2. **Smith, H. W.**, The composition of the body fluids of the elasmobranchs, *J. Biol. Chem.*, 81, 407, 1929.
3. **Lange, R. and Fugelli, K.**, The osmotic adjustments in the euryhaline teleosts, the flounder, *Pleunectes flesus L.* and the three spined stickleback *Gesterosteus acrilatus* L., *Comp. Biochem. Physiol.* 15, 283, 1965.
4. **Spanner, P. C.**, *Introduction to Thermodynamics,* Academic Press, New York, 1964, chap. 10.
5. **Katchalsky, A. and Curran, P. F.**, *Nonequilibrium Thermodynamics in Biophysics,* Harvard University Press, Cambridge, Mass., 1965, chap. 7.
6. **Ussing, H. H.**, The distinction by means of tracers between active transport and diffusion, *Acta Physiol. Scand.*, 23, 110, 1949.
7. **Ussing, H. H.**, Interpretation of the exchange of radiosodium in isolated muscle, *Nature,* 160, 262, 1947.
8. **Ling, G. N.**, *A Physiological Theory of the Living State,* Blaisdell Publishing, New York, 1962.
9. **Huf, E.**, Über aktiven Wasser-und Salztransport durch die Froschhaut, *Pflüg. Arch. Ges. Physiol.*, 237, 143, 1936.
10. **Huf, E.**, Versuche über den Zuzammenhang zwischen Stoffwechsel, Potentialbildung and Funktion der Froschhaut, *Pflüg. Arch. Ges. Physiol.*, 235, 655, 1935.
11. **Huf, E.**, Über den Anteil vitaler Krafte bei der Resorption von Flüssigkeit durch die Froschhaut, *Pflüg. Arch. Ges. Physiol.*, 236, 1, 1935.
12. **Ussing, H. H. and Zerahn, K.**, Active transport of sodium as the source of electric current in the short-circuited isolated frog skin, *Acta Physiol. Scand.*, 23, 110, 1951.
13. **Zerahn, K.**, Oxygen consumption and active sodium transport in the isolated and short-circuited frog skin, *Acta. Physiol. Scand.*, 36, 300, 1956.
14. **Glynn, I. M.**, Sodium and potassium movements in human cells, *J. Physiol.*, 134, 278, 1956.
15. **Steinback, H. B.**, On the Na and K balance of isolated frog muscle, *Proc. Natl. Acad. Sci. U.S.A.*, 38, 451, 1952.
16. **Berliner, R. W., Kennedy, T. J., and Hilton, J. G.**, Renal mechanisms for excretion of potassium, *Am. J. Physiol.*, 162, 348, 1952.
17. **Nellans, H. N. and Schultz, S. G.**, Relations among transepithelial sodium transport, potassium exchange and cell volume in rabbit ileum, *J. Gen. Physiol.*, 68, 441, 1967.
18. **Murer, H., Hopfer, U., and Kinne, R.**, Sodium/proton antiport in brush-border membrane vesicles isolated from rat small intestine and kidney, *Biochem. J.*, 154, 597, 1976.
19. **Pitts, R. F. and Alexander, R. S.**, The nature of the renal tubular mechanisms for acidifying the urine, *Am. J. Physiol.*, 144, 239, 1945.
20. **Riklis, E. and Quastel, J. H.**, Effects of cations on sugar absorption by isolated surviving guinea pig intestine, *Can. J. Biochem. Physiol.*, 108, 37, 1949.
21. **Hodgkin, A. L. and Katz, B.**, The effect of sodium ions on the electrical activity of the giant axon of the squid, *J. Physiol.*, 108, 37, 1949.
22. **Hodgkin, H. L. and Horowicz, P.**, The effect of sudden changes in ionic concentrations on the membrane potential of single muscle fibers, *J. Physiol.*, 148, 127, 1959.
23. **Giebisch, G.**, Measurements of electrical potential differences on single nephrons of the perfused *Necturus* kidney, *J. Gen. Physiol.*, 44, 659, 1961.
24. **Post, R. L. and Jolly, P.**, The linkage of sodium, potassium and ammonium active transport across the human erythrocyte membrane, *Biochem. Biophys. Acta,* 24, 118, 1957.
25. **Sjøstrand, F. S. and Rhodin, J.**, The ultrastructure of the proximal tubule of the mouse kidney as revealed by high resolution electron microscopy, *Exp. Cell Res.*, 4, 426, 1953.
26. **Politoff, A. L., Socolar, S. J., and Lowenstein, W. R.**, Permeability of a cell membrane junction. Dependence on energy metabolism, *J. Gen. Physiol.*, 53, 498, 1969.
27. **Harris, J. E.**, The influence of the metabolism of human erythrocytes on their potassium content, *J. Biol. Chem.*, 141, 579, 1941.
28. **Gordon, E. E. and De Hartog, M.**, The relationship between cell membrane potassium ion transport and glycolysis, *J. Gen. Physiol.*, 54, 650, 1969.
29. **Maffly, R. H. and Edelman, I. S.**, The coupling of the short-circuit current to metabolism in the urinary bladder of the toad, *J. Gen. Physiol.*, 46, 733, 1963.
30. **Maizels, M.**, Factors in the active transport of cation, *J. Physiol.*, 112, 59, 1951.
31. **Whittam, R. and Agar, M. E.**, The connection between active transport and metabolism in erythrocytes, *Biochem. J.*, 97, 214, 1965.
32. **Klahr, S. and Bricker, N. S.**, Energetic of anaerobic sodium transport by fresh water turtle bladder, *J. Gen. Physiol.*, 48, 571, 1965.

33. **Lund, E. J.,** Relationship between continuous bioelectric currents and cell respiration. III. Effects of concentration of oxygen on cell polarity in the frog skin, *J. Exp. Zool.*, 51, 291, 1928.
34. **Rottenberg, H., Caplan, S. R., and Essig, A.,** Stoichiometry and coupling: theories of oxidative phosphorylation, *Nature*, 216, 610, 1967.
35. **Elshave, A. and Van Rossum, G. D. V.,** Net movements of sodium and potassium and their relation to respiration in slices of rat liver incubated *in vitro*, *J. Physiol.*, 168, 531, 1963.
36. **Sharp, G. W. and Leaf, A.,** The central role of pyruvate in the stimulation of sodium transport by aldosterone, *Proc. Natl. Acad. Sci. U.S.A.*, 52, 114, 1964.
37. **Viera, F. L., Caplan, S. R., and Essig, A.,** Energetics of sodium transport in frog skin. I. Oxygen consumption in the short-circuited state, *J. Gen. Physiol.*, 59, 60, 1972.
38. **Saito, T., Essig, A., and Caplan, S. R.,** The effect of aldosterone on the energetics of sodium transport in the frog skin, *Biochem. Biophys. Acta*, 318, 371, 1973.
39. **Caplan, N. S., Steele, R. F., and Maffly, R. H.,** Interrelationships of sodium transport and carbon dioxide production by the toad bladder. Response to changes in mucosal sodium concentration to vasopressin and to availability of metabolic substrate, *J. Membr. Biol.*, 34, 289, 1977.
40. **Randall, H. M., Jr. and Cohen, J. J.,** Anaerobic CO_2 production by dog kidney in vitro, *Am. J. Physiol.*, 211, 493, 1966.
41. **Glynn, J. M. and Karlish, S. J. D.,** The sodium pump, *Ann. Rev. Physiol.*, 37, 13, 1975.
42. **Skou, J. C.,** Influence of some cations on adenenosine triphosphatase from peripheral nerves, *Biochim. Biophys. Acta*, 23, 399, 1957.
43. **Koepsell, H.,** Conformational change of membrane bound ($Na^+ - K^+$) ATPase as revealed by antibody inhibition, *J. Membr. Biol.*, 45, 1, 1979.
44. **Skou, J. C.,** Further investigations on Mg^{++} and Na^+ activated adenosine triphosphatase, possibly related to active, linked transport of Na^+ and K^+ across cell membranes, *Biochim. Biophys. Acta*, 42, 6, 1960.
45. **Glynn, I. M.,** Action of cardiac glycosides on ion movements, *Pharmacol. Rev.*, 16, 381, 1964.
46. **Epstein, F. H. and Whittam, R.,** Mode of inhibition by calcium of cell membrane adenosine triphosphatase activity, *Biochem. J.*, 99, 232, 1966.
47. **Fahn, S., Hurley, M. R., Koval, G. J., and Albers, R. W.,** Sodium-potassium-activated adenosine triphosphatase of *Electrophorus* electric organ. II. Effects of *N*-ethylmaleimide and other sulfhydryl reagents, *J. Biol. Chem.*, 241, 1890, 1966.
48. **Katz, A. I. and Epstein, R. H.,** Physiologic role of sodium-potassium activated adenosine triphosphatase in the transport of cations across biologic membranes, *N. Engl. J. Med.*, 278, 253, 1968.
49. **Fahn, S., Koval, G. J., and Albers, R. W.,** Sodium-potassium-activated adenosine triphosphatase of *Electrophorus* electric organ. I. An associated sodium-activated transphosphorylation, *J. Biol. Chem.*, 241, 1882, 1966.
50. **Post, R. L., Merritt, C. R., Kinsolving, C. R., and Albright, C. D.,** Membrane adenosine triphosphatase as participant in active transport of sodium and potassium in human erythrocyte, *J. Biol. Chem.*, 235, 1796, 1960.
51. **Charnock, J. S. and Post, R. L.,** Mechanism of ouabain inhibition of cation activated adenosine triphosphatase, *Nature*, 199, 910, 1963.
52. **Glynn, J. M. and Lew, V. L.,** Synthesis of adenosine triphosphate at the expense of downhill cation movements in intact human red cells, *J. Physiol.*, 207, 393, 1970.
53. **Hilden, S., Rhee, H. M., and Hopkin, L. E.,** Sodium transport by phospholipid vesicles containing purified sodium and potassium ion-activated adenosine triphosphatase, *J. Biol. Chem.*, 249, 7432, 1974.
54. **Mitchell, P.,** Coupling of phosphorylation to electron and hydrogen transfer by a chemi-osmotic type of mechanism, *Nature*, 191, 144, 1961.
55. **Robertson, R. N.,** *Protons, Electrons, Phosphorylation and Active Transport*, Cambridge University Press, Cambridge, England, 1968.
56. **Hinkle, P. C. and McCarty, R. E.,** How cells make ATP, *Sci. Amer.*, 238, 104, 1978.
57. **Ling, G. N. and Cope, F. W.,** Potassium ion: is the bulk of intracellular K^+ absorbed?, *Science*, 163, 1335, 1969.
58. **Hazlewood, C. F.,** Pumps or no pumps, *Science*, 177, 815, 1972.
59. **Kinne, R.,** Metabolic correlates of tubular transport, in *Membrane Transport in Biology*, Vol. 4B, Giebisch, G., Tosteson, D. C., and Ussing, H. H., Eds., Springer-Verlag, New York, 1979.
60. **Fujimoto, M., Nash, F. C., and Kessler, R. H.,** Effects of cyanide, Q_{O_2} and dinitrophenol on renal sodium reabsorption and oxygen consumption, *Am. J. Physiol.*, 206, 1327, 1964.
61. **Strickler, J. C. and Kessler, R. H.,** Effects of of certain inhibitors on renal excretion of salt and water, *Am. J. Physiol.*, 205, 117, 1963.
62. **Györy, A. Z. and Kinne, R.,** Energy source for transepithelial sodium transport in rat proximal tubules, *Pflüg. Arch. Ges. Physiol.*, 327, 234, 1971.

63. **Deetjen, P. and Kramer, K.**, Die Abhängigkeit des O_2 -Verbrauchs der Niere von der Na-Ruckresorption, *Pflüg. Arch. Ges. Physiol.*, 273, 636, 1961.
64. **Leaf, P., Page, L. B., and Anderson, J.** Respiration and active transport of sodium, *J. Biol. Chem.*, 234, 1625, 1959.
65. **Frömter, E., Rumrich, G., and Ullrich, K. J.**, Phenomenologic description of Na^+, Cl^- and HCO_3^- absorption from proximal tubules of rat kidney, *Pflüg. Arch. Ges. Physiol.*, 343, 189, 1973.
66. **Toretti, J. E., Hendler, E., Weinstein, E., Longnecker, A. E., and Epstein, F. H.**, Functional significance of Na-K-ATPase in the kidney and effects of ouabain inhibition, *Am. J. Physiol.*, 22, 1398, 1972.
67. **Dieś, F. G., Ramos, G., Avelar, E., and Lennhoff, M.**, Renal excretion of lactic acid in the dog, *Am. J. Physiol.*, 116, 106, 1969.
68. **Cohen, J. J., Kook, Y. J., and Little, J. R.**, Substrate-limited function and metabolism of the isolated perfused rat kidney: effects of lactate and glucose, *J. Physiol.*, 266, 103, 1977.
69. **Caldwell, T. and Solomon, S.**, Changes in oxygen consumption of kidney during maturation, *Biol. Neonate*, 25, 1, 1975.
70. **Hunter, M. J.**, A qualitative estimate of the non-exchange-restricted chloride permeability of the human red cell, *J. Physiol.*, 218, 49P, 1971.
71. **Slayman, C. L.**, Proton pumping and generalized energetics of transport: a review, in *Membrane Transport in Plants*, Zimmerman, E. and Dainty, J., Eds., Springer-Verlag, New York, 1974.
72. **Blaustein, R. P.**, The interrelationship between sodium and calcium fluxes across cell membranes, *Rev. Physiol. Biochem. Pharmacol.*, 70, 33, 1974.
73. **Ullrich, K. J., Rumrich, G., and Kloss, S.**, Sodium dependence of the amino acid transport in the proximal convolution of the rat kidney, *Pflüg. Arch. Ges. Physiol.*, 351, 49, 1974.
74. **Vogel, G. and Kroger, W.**, Die Bedeutung des Transportes der Konzentration and der Darbientungsrichtung von Na^+ für den tubulärer Glucose-und PAH-Transport, *Pflüg. Arch. Ges. Physiol.*, 288, 342, 1966.
75. **Ullrich, K. J., Rumrich, G., and Kloss, S.**, Specifically and sodium dependence of the active sugar transport in the proximal convolution of the rat kidney, *Pflüg. Arch. Ges. Physiol.*, 351, 49, 1974.
76. **Curran, P. F., Schultz, S. G., Chez, R. A., Fuisz, R. E.**, Kinetic relations of the Na-amino acid interaction at the mucosal border of intestine, *J. Gen. Physiol.*, 50, 1261, 1967.
77. **Frizzell, R. A., Nellans, H. Y., and Schultz, S. G.**, Effects of sugars and amino acids on sodium and potassium influx in rabbit ileum, *J. Clin. Invest.*, 52, 215, 1973.
78. **Holdsworth, D. C. and Dawson, A. M.**, The absorption of monosaccharides in man, *Clin. Sci.*, 27, 371, 1964.
79. **Levitan, R. and Wilson, D. E.**, Absorption of water soluble substances, in *Gastrointestinal Physiology*, Jacobson, E. E. and Shanbour, L. L., Eds., University Park Press, Baltimore, 1974.
80. **White, L. W. and Landau, B. R.**, Sugar transport and fructose metabolism in human intestine *in vitro*, *J. Clin. Invest.*, 44, 1200, 1965.
81. **Thomas, E. L. and Christensen, H. N.**, Nature of the cosubstrate action of Na^+ and neutral amino acids in a transport system, *J. Biol. Chem.*, 246, 1682, 1971.

Chapter 2

RENAL REGULATION OF SODIUM EXCRETION

Mark H. Gardenswartz and Robert W. Schrier

TABLE OF CONTENTS

I. INTRODUCTION

The exquisite precision of animal biology is nowhere more evident than in the physiology of normal sodium homeostasis. As the most abundant extracellular cation, sodium is the prime determinant of the extracellular fluid volume (ECFV), the perturbations of which may gravely threaten the health and life of the organism. The critical importance of sodium is reflected in the extraordinary complexity of the apparatus, only partially understood, which governs its metabolism. This discussion furnishes an overview of total body sodium economy and its relationship to the ECFV. It then focuses upon the component mechanisms which regulate sodium balance, and specifically the contributions of the intrinsic regulation by the kidney itself, and of neural and hormonal processes. Because subsequent chapters consider derangements of these regulatory mechanisms in common disease states this chapter concentrates primarily upon normal physiology.

II. RELATIONSHIP OF SODIUM TO THE EXTRACELLULAR FLUID VOLUME

The well-established relationship between the state of external sodium balance and the ECFV derives from the distribution of the cation within the body. Sodium is largely confined to the extracellular fluid (EFC) compartment, where its salts constitute virtually all of the ECFV. The concentration of these salts normally is regulated very tightly through the appropriate intake of water and renal regulation of water excretion, processes which demand an intact thirst mechanism and the modulation of vasopressin release and its hydro-osmotic effect, respectively. Thus, alterations in total body sodium balance perturb the ECFV, but not the fluid osmolality.

The effect of changes in external sodium balance on the ECFV as reflected in body weight is well-illustrated in Figure 1.[1] This balance study in a normal individual illustrates the relationship between sodium intake and ECFV. Initially the subject is in steady state on a sodium intake of 10 mEq/day. When dietary sodium intake is increased to 150 mEq/day, the urinary excretion of sodium transiently lags behind, and the subject is in positive sodium balance for several days. During this time the patient maintains normal plasma tonicity by ingesting and conserving an appropriate quantity of water. When he achieves sodium balance (i.e., intake equals output) on his augmented daily sodium intake, he is in a new steady state, with an increased body weight reflecting the state of ECFV expansion. A return to the prior low sodium diet reverses this sequence of events, with a period of transient negative sodium balance accompanied by a decline to the initial body weight, and a normalization of the ECFV.

The kidney, as the major excretor of sodium, is thus the guarantor of the integrity of the ECFV. To discharge this enormous responsibility the kidney is endowed with a spectacular wisdom, partially intrinsic and partially the result of extrarenal influences. This "wisdom" was clearly demonstrated decades ago, in elegant studies by McCance.[2] In a series of balance studies upon himself, he achieved a net loss of 758 mEq of sodium, primarily through induced sweating. His urine became free of sodium, with daily excretion of less than 1 mEq/day. When he began sodium repletion, his daily urine sodium remained at 1 mEq/day, until the third day of recovery. On that day he ingested enough sodium to make the 3-day total intake 732 mEq, and the urinary sodium content increased to 16 mEq/day.

Subsequent studies by Strauss et al.[3] underscored the remarkable capacity of the normal kidney to protect the ECFV through appropriate adjustments in urinary sodium excretion. Sodium-restricted subjects, in sodium balance on a daily intake of 5 mEq, were given a 15-min hypotonic saline infusion calculated to increase the total

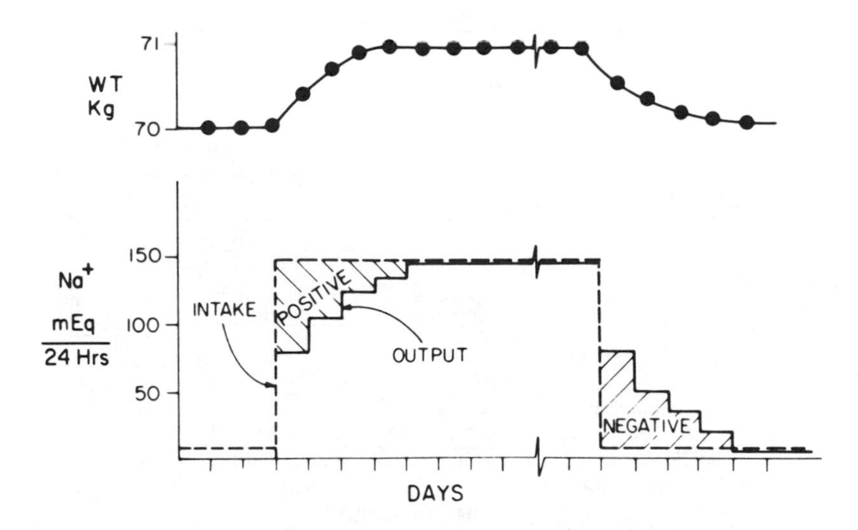

FIGURE 1. Normal sodium balance. An abrupt increase in dietary sodium intake results in a period of positive sodium balance and concomitant weight gain lasting 3 to 5 days. Thereafter urinary sodium equals sodium intake and weight stabilizes. When dietary sodium is decreased, negative sodium balance occurs and body weight decreases. After 3 to 5 days, a steady state is again achieved. (From Reineck, H. J. and Stein, J. H., in *Clinical Disorders of Fluid and Electrolyte Metabolism,* 3rd ed, Maxwell, M. and Kleeman, C., Eds. Copyright (c) 1980, McGraw-Hill, New York. Used with the permission of the McGraw-Hill Book Company.)

ECF sodium content by less than 1%. Within the time course of the infusion, the patients began to demonstrate a prompt natriuresis, and ultimately a threefold increase in the rate of sodium excretion was observed (Figure 2). By contrast, subjects who, through dietary sodium restriction and induced sweating, were prepared by moderate and mild extrarenal sodium depletion, demonstrated no natriuresis and a delayed, blunted natriuresis, respectively (Figure 3). The demonstration of such an exquisitely sensitive natriuretic apparatus in the sodium-restricted subjects, and its appropriate modification after sodium depletion, led the authors to the conceptual framework illustrated in Figure 4.

This figure depicts the relationship between daily renal sodium excretion and the state of the ECFV. Most individuals, who eat usual quantities of salt, exist at point B, representing a state in which a fluctuating, mild volume expansion after dietary intake of sodium results in the excretion of the surfeit. Point A is the state of sodium-restricted individuals, whose urine contains very little sodium, and in whom a minute degree of ECFV expansion causes an immediate natriuresis. Point C represents sodium-depleted subjects, who excrete no sodium until the deficit has been repaired. The erect figure in this illustration and its parallel regression line portray the importance of posture in the distribution of the ECFV. The quantity of sodium or volume of extracellular fluid which is surplus in recumbency and subject to excretion may represent a deficit in the erect position.

It is clear, then, that the ECFV and the state of sodium balance are inextricably linked. The precise way in which alterations in external sodium balance are perceived by the body's volume-sensing apparatus is not clear. However, this afferent pathway is thought to exist anatomically within the chest and/or great vessels. Considerable investigative effort has been devoted to the study of both the afferent and efferent limbs of the control of renal sodium handling.

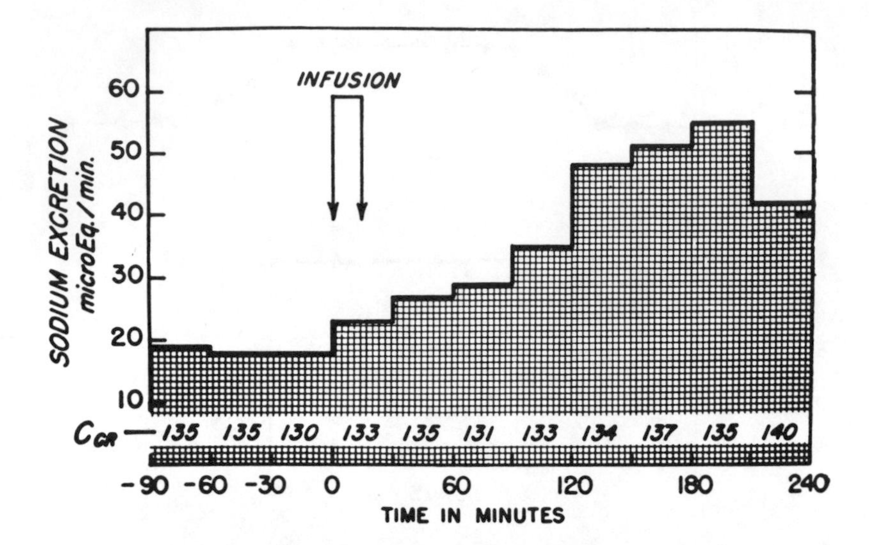

FIGURE 2. Mean renal sodium excretion in ten recumbent sodium-restricted but not sodium-depleted subjects before and after the intravenous infusion of 15 mM of NaCl in hypotonic solution. Endogenous creatinine clearance is shown in lower figures. (From Strauss, M. B., Lamdin, E., Smith, W. P., and Bleifer, D. J., *Arch. Intern. Med.*, 102, 527, 1958. Copyright, 1958, American Medical Association.)

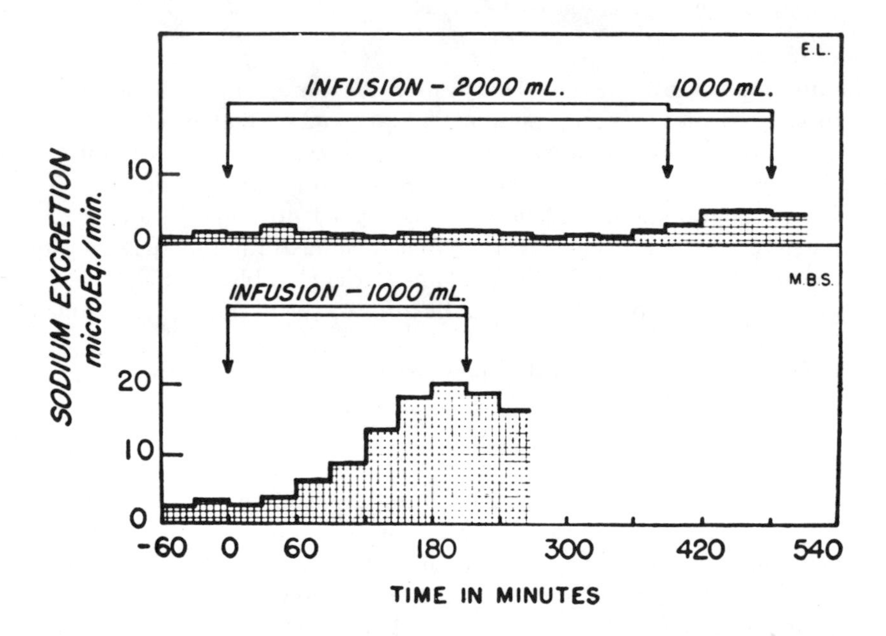

FIGURE 3. Lack of response of a subject (E.L.) considerably depleted of sodium, to the intravenous infusion of 200 mM of NaCl in hypotonic solution and minimal response to the additional infusion of 100 mM (upper chart). Delayed and slight response of a mildly depleted subject (M.B.S.) to the infusion of 100 mM of NaCl in hypotonic solution. (From Strauss, M. B., Lamdin, E., Smith, W. P., and Bleifer, D. J., *Arch. Intern. Med.*, 102, 527, 1958. Copyright, 1958, American Medical Association.)

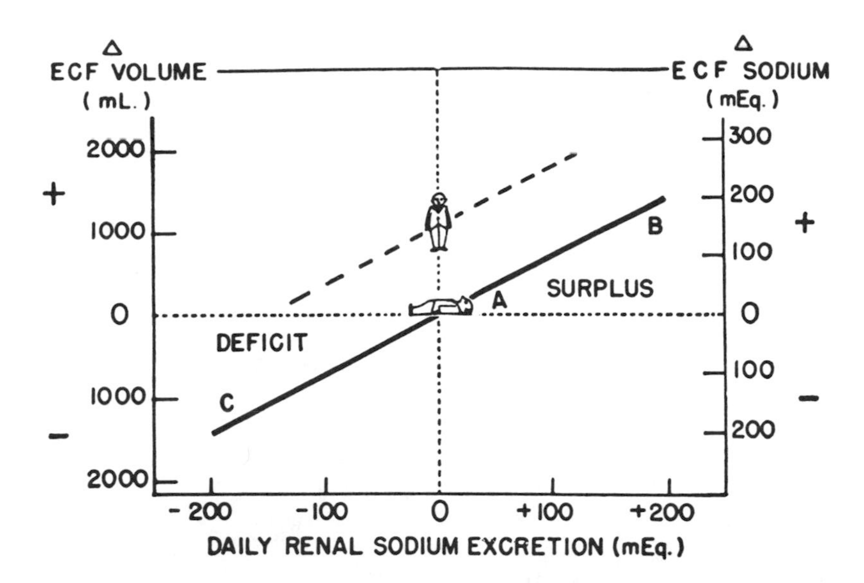

FIGURE 4. A diagram expressing the very close relationship between renal sodium excretion and extracellular fluid volume. (From Strauss, M. B., Lamdin, E., Smith, W. P., and Bleifer, D. J., *Arch. Intern. Med.*, 102, 527, 1958. Copyright, 1958, American Medical Association.)

III. REGULATION OF SODIUM EXCRETION

A. Glomerular Filtration Rate

1. Regulation of Renal Blood Flow — Autoregulation

The initial step in the excretion of sodium is the formation of an isotonic plasma ultrafiltrate at the glomerulus. Hence, normal control of sodium excretion is linked to the relative constancy of the glomerular filtration rate (GFR). The process whereby the kidney is able to maintain a constant renal blood flow (RBF) and GFR over a wide range of renal perfusion pressures is known as renal autoregulation. This phenomenon occurs within seconds after changes in perfusion pressure, whether or not the kidney is innervated, and in the presence or absence of filtration.[4] A number of "active" and "passive" mechanisms have been proposed to explain renal autoregulation, but detailed consideration of these theories is beyond the scope of this chapter. Suffice it to say that the most likely explanation for renal autoregulation is a vascular myogenic response, which dictates an appropriate arteriolar vasoconstriction or vasodilatation. Thus, RBF or GFR are preserved by alterations in renal vascular resistance in response to changes in renal perfusion pressure. Whether this vascular myogenic mechanism is the result of properties intrinsic to the vascular smooth muscle or a reflex humoral system is not clear.

2. Tubuloglomerular Feedback Mechanism

Whatever the mechanism, renal autoregulation is the initial safeguard of GFR and, thus, filtered load of sodium. The next mechanism which assures relative constancy of sodium delivery into the proximal tubule is called tubuloglomerular feedback. This term describes the reflex whereby the macula densa somehow senses changes in the delivery of some solute and initiates changes in afferent arteriolar tone which dampens the initial alteration. This reflex is represented graphically in Figure 5, in which an increase in the microperfusion rate to the macula densa reduces the stop-flow pressure in the early part of the proximal tubule of the same nephron. This suggests a reflex decrease in filtration rate in response to increased distal delivery. Though the very

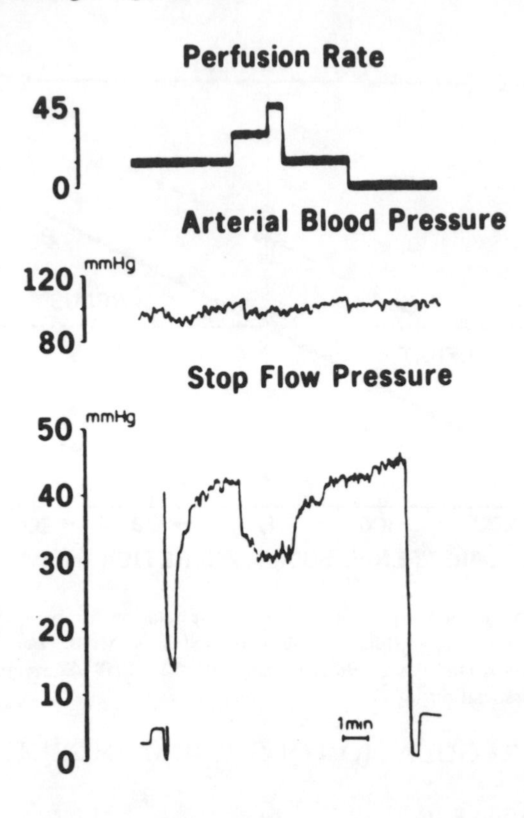

FIGURE 5. Tubuloglomerular feedback. Effect on stop-flow pressure in early proximal tubule of varying the rate of microperfusion of loop of Henle with modified Ringer's solution. (From Schnermann, J., Hermle, M., Schmidmeier, E., and Dalheim, H., *Pfluegers Arch.*, 358, 325, 1975. With permission.)

existence of this reflex has been doubted by some investigators who have not been able to reproduce the phenomenon, work from a number of laboratories has demonstrated some relationship between distal delivery and change in afferent or efferent arteriolar tone.

The mechanism underlying this tubuloglomerular feedback remains unclear. While tubular fluid sodium and osmolality have been proposed as the species which trigger the response, it now appears that changes in macula densa chloride concentration may provide the afferent stimulus, with possibly angiotensin constituting the efferent mediator of the phenomenon. Thus, according to this latter hypothesis, an increase in chloride delivery to the macula densa would increase angiotensin production and thereby mediate afferent arteriolar vasoconstriction. Conversely, a macula densa which had been deprived of sodium chloride would decrease renin activity with consequent diminished afferent arteriolar tone. The latter scenario was supported by the study of Morgan and Gillies[5] in which the venous effluent from proximal tubules which were blocked for several minutes contained a lowered renin content.

Several lines of evidence suggest that the decrease in stop-flow pressure depicted in Figure 5 may result from efferent arteriolar vasodilatation rather than afferent vasoconstriction. These include the controversial anatomic data that the distal tubule has a closer contact with the efferent arteriole than with the afferent arteriole.[6,7] The evidence also includes physiologic data regarding changes in renin release that would imply a vasodilatory role of the efferent arteriole. In vitro work on isolated, superfused glomeruli has shown that, while a sustained high sodium chloride concentration increases renin production, a sudden rise in sodium chloride concentration diminishes renin production.[8,9] This latter observation correlates well with the known decrease in renin release from the whole kidney after saline loading. It also fits with Vander and Miller's findings[10] that increased sodium chloride delivery to the juxtaglomerular apparatus, achieved by a diuretic, reverses the increased renin production from a whole kidney with experimental renal artery constriction. Thus, in order for the decrease in stop-flow pressure which results from increased perfusion (Figure 5) to be due to a diminution in renin release, vasodilatation of the efferent arteriole must occur.

3. Glomerulotubular Balance

The rate of tubular reabsorption of sodium in the proximal tubule is known to change in a direction to buffer against alterations in GFR and, thus, filtered sodium, causing large changes in urinary excretion. This phenomenon is known as glomerulotubular balance (GTB) and it constitutes the next line of defense when renal autoregulation and tubuloglomerular feedback mechanisms fail to maintain the constancy of sodium filtration. Through this regulatory process, as GFR rises, absolute proximal tubular reabsorption increases, and when GFR falls, absolute proximal tubular reabsorption decreases. Thus, the fractional reabsorption of filtrate remains relatively constant, thereby buffering against large changes in urinary excretion of water and electrolytes. The importance of this phenomenon is emphasized by the fact that 99% of the normal daily glomerular ultrafiltrate is reabsorbed, and only 1% is excreted. Therefore, adaptive changes in tubular function which are relatively small may substantially affect the absolute excretion of sodium and chloride. Conversely, if tubular reabsorption were constant, changes in GFR would be a major determinant of the rate of sodium excretion.

Though the entire tubular apparatus is involved in the regulation of sodium reabsorption, the term GTB is popularly applied to the contribution of the proximal tubule to this regulatory process. The means whereby the proximal nephron performs this homeostatic function remains unresolved, in spite of an adequate number of plausible hypotheses over the past two decades and intense investigation designed to prove them. These experiments have utilized a variety of maneuvers to alter the GFR and thus the tubular load of sodium, including aortic and renal artery constriction, ureteral and renal venous obstruction, and volume expansion. The effects on tubular handling of sodium have been assessed with clearance and micropuncture techniques. The considerable imagination and industry which have been devoted to the understanding of GTB have resulted in some understanding of the phenomenon and the identification of areas of controversy. (The reader is referred to extensive reviews of the early work in the area[11-16] and to a more recent review of areas of current contention.[9])

Most theories of GTB have been formulated along two broad concepts:

1. That GTB depends upon changes in the peritubular environment, largely as the result of changes in peritubular capillary protein concentration
2. That GTB results from intraluminal effects arising from physical or chemical properties of glomerular filtrate

The first hypothesis arose from the observation that GTB does not occur[17-20] or is limited[21] when the changes in sodium load are produced artificially by tubular microperfusion with saline or Ringer's solution. As tubular microperfusion does not alter the peritubular environment, it has been suggested that this lack of peritubular effect accounts for the absence of GTB. A number of studies designed to confirm or refute this hypothesis have yielded conflicting results, which are discussed in greater detail in the subsequent section on the importance of physical factors in sodium reabsorption. Suffice it to say that, while numerous observations in whole animals and some microperfusion experiments support the importance of the proximal tubular pericapillary oncotic force in determining sodium reabsorption, many lines of evidence (vide infra) negate this conclusion and leave the question, at best, unresolved.

The other proposal which has received serious consideration is the one linking intraluminal pressures or constituents to the regulatory process. Initially, Kelman[22] suggested that transtubular sodium transport was limited by the rate of radial diffusion of sodium from the proximal tubular lumen across the apical cell membrane. This hypothesis associates a decreased GFR with decreased intraluminal turbulence and thus decreased diffusion of luminal sodium into the cells. This concept may be erroneous, since turbulent intraluminal flow is as improbable as the notion that radial diffusion of sodium limits its reabsorption. The proposal by Gertz[23] that the degree of distention of the proximal tubules regulates the proximal reabsorptive rate through an effect on surface area is also unlikely. Though in some conditions the intratubular volume correlates with reabsorptive rate, such a correlation is lacking when GFR is decreased from ureteral or renal venous obstruction. A decrease in tubular reabsorptive activity in these latter settings in spite of intratubular distention argues compellingly against Gertz's hypothesis.

Leyssac[24] proposed that GTB results from an effect of the renin-angiotensin system upon the rate of tubular reabsorption of sodium. A decrease in GFR would decrease distal delivery of tubular fluid, leading to an increased release of renin and formation of angiotensin. Angiotensin would then depress the rate of tubular sodium reabsorption so as to return the rate of distal fluid and sodium to normal. However, this proposal was not supported by microperfusion studies using angiotensin[25] or by the more recent demonstration[26] that angiotensin II increases rather than decreases renal sodium reabsorption. Also, in rats subject to chronic sodium depletion, angiotensin II blockade with the angiotensin competitive inhibitor, saralasin, caused a decrease in absolute proximal reabsorption in spite of a concurrent increase in single nephron GFR.[27]

The inability to demonstrate GTB consistently, utilizing tubular microperfusion with artificial solutions, suggested that previous negative studies with respect to the intraluminal theory of GTB may have been artifactual. Bartoli and Earley[28] therefore meticulously examined the methodology of microperfusion and demonstrated that such small, but critical, pitfalls as errors in perfusion rate, incomplete collection, and contamination of perfusion fluid by interstitial fluid are avoidable. They also illustrated the importance of considering the length of tubule through which the microperfusion flows in assessing the relationship of tubule flow rate to reabsorption. Of major importance was the finding that microperfusion with artificial solutions was associated with a spontaneous decrease in the tubular reabsorptive rate, whereas microperfusion with a protein-free plasma ultrafiltrate demonstrated no such decrement. Such a spontaneous fall in reabsorption could have obscured an effect of increased tubular flow to enhance reabsorption. Since these results suggested that the stability of the tubular reabsorptive rate might depend upon the presence of unidentified plasma constituents not contained in artificial perfusate, the earlier, negative microperfusion experiments with artificial solutions were open to obvious question.

Bartoli et al.[29] extended these earlier findings by demonstrating that a reduction in the delivery of normal filtrate was associated with a reduced reabsorptive rate in spite of constancy of the filtration fraction and renal hemodynamics. The investigators utilized an elegant dual collection technique depicted in Figure 6. Through the placement of separate pipettes in the earlier and more distal proximal tubule, some filtrate was aspirated proximally, thereby decreasing flow and delivery of filtrate to the late proximal tubule. This diversion of filtrate was accomplished without change in nephron

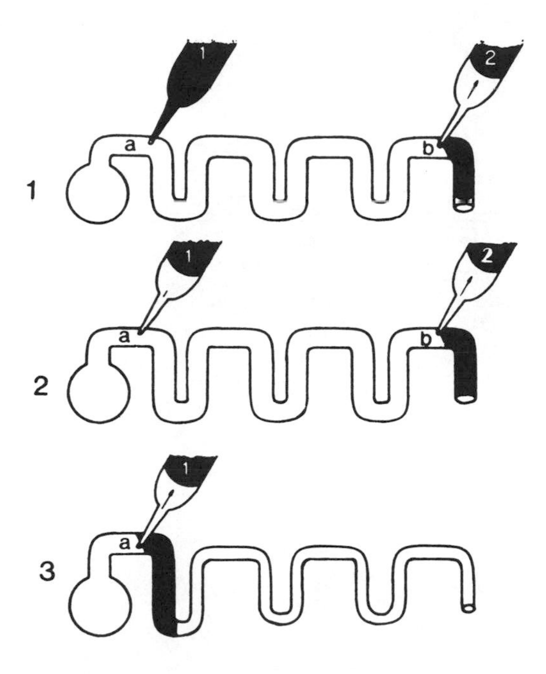

FIGURE 6. Technique for reducing intratubular flow rate by dual collection. (1) Pipette number 1 was placed in an early proximal convolution and (a) an oil droplet was injected to identify the last accessible proximal tubular convolution (b) where second pipette number 2 was inserted. A total collection then was made at site b only (collection 1). (2) Dual collections were then made by aspirating part of the tubular fluid at site a and simultaneously collecting the remaining tubular fluid at site b (collections 2a and 2b). Finally, a total free-flow collection was made at site a (collection 3). (From Bartoli, E. and Earley, L. E., *Kidney Int.*, 3, 142, 1973. With permission.)

filtration rate. The rate of tubular fluid flow was reduced an average of 45%, and tubular reabsorptive rate was decreased by an average of 29%. This GTB balance of 63% between tubular fluid delivery and reabsorptive rate, in spite of the lack of change in the peritubular environment, supported the importance of the intraluminal load per se in the phenomenon of GTB. This observation has since been confirmed by Habeler and Shiigai.[9] The notion that certain ingredients of glomerular filtrate somehow promoted the reabsorption of the filtered fluid has been developed further.

As reviewed by de Wardener,[9] microperfusion experiments by Habeler and Shiigai have confirmed the absence of GTB when the perfusion fluid was either lactated Ringer's solution or plasma ultrafiltered to exclude substances with molecular weight greater than 500 daltons (Figure 7). However, when the perfusate was previously harvested proximal tubular fluid, GTB resembled that seen with spontaneous changes in GFR (Figure 8). These findings suggest that the filtrate contains substances which stimulate its own reabsorption. The effects were not reproduced with glucose, and amino acid content was added to artificial perfusate. Some evidence in other transport systems has fostered speculation that the effect of glomerular filtrate on the tubular reabsorptive rate may be secondary to effects on the motility of the proximal tubular brush border.[9]

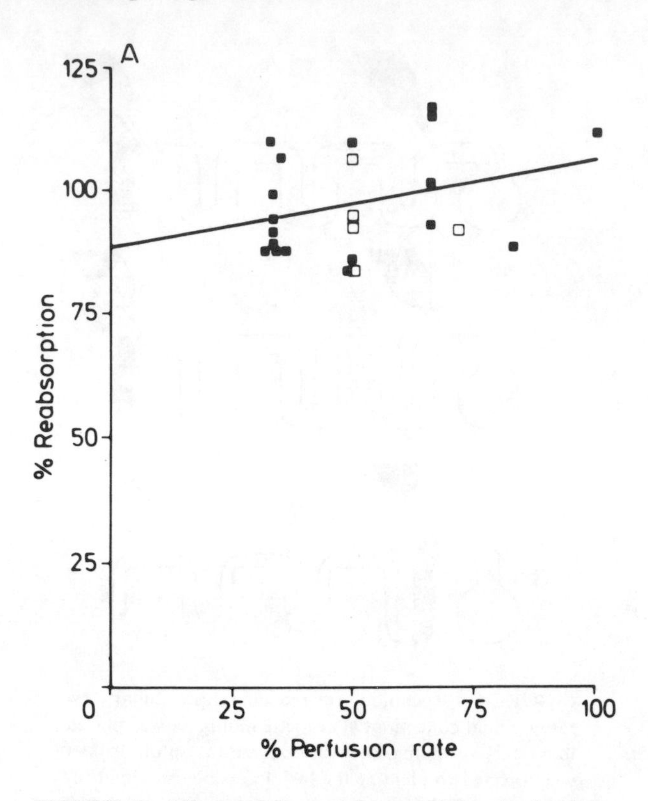

FIGURE 7. Microperfusion of lumen of proximal tubule with either Ringer solution (closed squares) or ultrafiltrate of rat plasma ultrafiltered through an Amicon UM05 filter (open squares). In each experiment the tubule was perfused at two rates of flow, one between 10 and 20 nℓ/min and the other between 20 and 30 nℓ/min. Perfusion rates are expressed as a percentage of highest perfusion rate used in each experiment. Reabsorption of highest rate in that experiment. Each repuncture value is corrected for differences in length of tubule perfused. (From Häbeler, D. A. and Shiigai, T., *Proc. 6th Hoechst Workshop Conf.*, Excerpta Medica, Amsterdam, 1978. With permission.)

This discussion of GTB illustrates the extent to which its mechanisms remain in question. As support exists for both peritubular and intraluminal effects, and as no single experimental model or maneuver has explained the entirety of the phenomenon, it is probably reasonable to consider GTB the result of forces acting in concert.

B. Aldosterone

The sodium retaining influence of aldosterone has been recognized for nearly three decades.[30] Brief summaries of the early work with the hormone and the classic review articles which pertain to it are found in earlier reviews of sodium metabolism.[30-34] This mineralocorticoid hormone increases active sodium transport in a number of epithelial tissues, including sweat glands, gastrointestinal tract, and, of greatest importance, the renal tubule. Aldosterone is secreted by the zona glomerulosa of the adrenal gland in response to a rise in renin-angiotensin activity, plasma potassium concentration, and adrenocorticotropic hormone (ACTH).[35] Its importance in total body sodium homeostasis is suggested by the observation that adrenalectomy and adrenal insufficiency lead to marked salt wasting and volume depletion, effects that can be overcome only with extremely high sodium intake. Though some clearance and micropuncture data support an effect of aldosterone on the proximal nephron and loop of Henle, the major

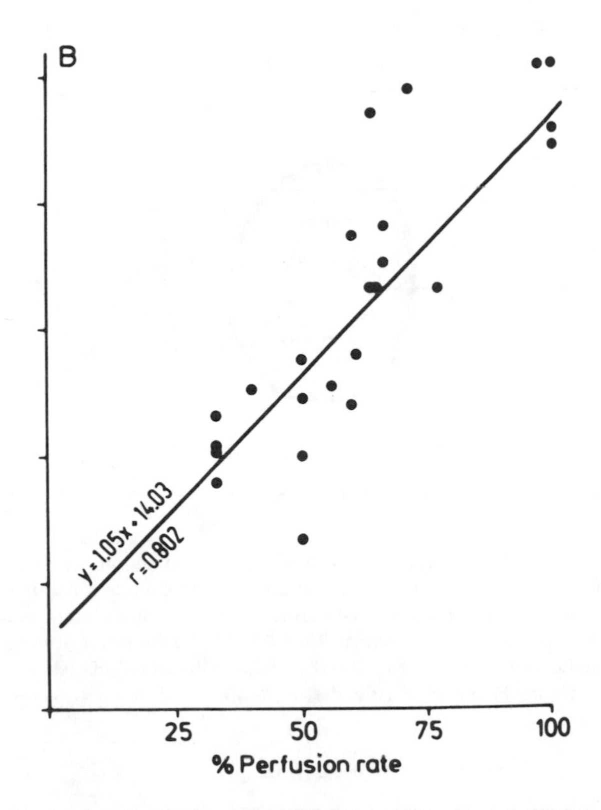

FIGURE 8. Microperfusion of lumen of proximal tubule with proximal tubule fluid harvested from late proximal tubules. (From Häbeler, D. A. and Shiigai, T., *Proc. 6th Hoechst Workshop Conf.*, Excerpta Medica, Amsterdam, 1978. With permission.)

impact of aldosterone on the volume status and urinary composition of the organism derives from its effect upon the distal nephron. Thus, while the quantity of sodium subject to a distal aldosterone effect is relatively small, the strategic location of the aldosterone effect is thought to permit "fine-tuning" in overall sodium balance.

Extensive work in a number of in vivo and in vitro systems has shown that epithelial tissues conserve sodium and lose potassium in response to aldosterone. In addition, aldosterone appears to have a direct effect in stimulating hydrogen ion secretion, as reviewed recently by Al-Awqati.[36] The subcellular mechanisms by which aldosterone exerts its effects have been the subject of some controversy.[36,37] It is widely accepted that the early phases of the cellular effect of aldosterone are those characteristic of steroid hormones in general. That is, the hormone penetrates the cell membrane, probably by diffusion, and binds noncovalently to a specific, cytoplasmic binding protein. This "aporeceptor" and aldosterone form an active complex, which translocates to the nucleus and somehow interacts with chromatin. The result is the new transcription both of messenger ribonucleic acid (mRNA) and ribosomal ribonucleic acid (rRNA), the translation of which yields aldosterone-induced protein (AIP). This series of events is depicted in Figure 9.

The function of AIP is the subject of the debate. Three theories have emerged to explain the effect of this protein in promoting active sodium transport and are depicted in Figure 10:

1. The sodium pump hypothesis — that AIP accelerates the activity of sodium pumps or creates more sodium pumps

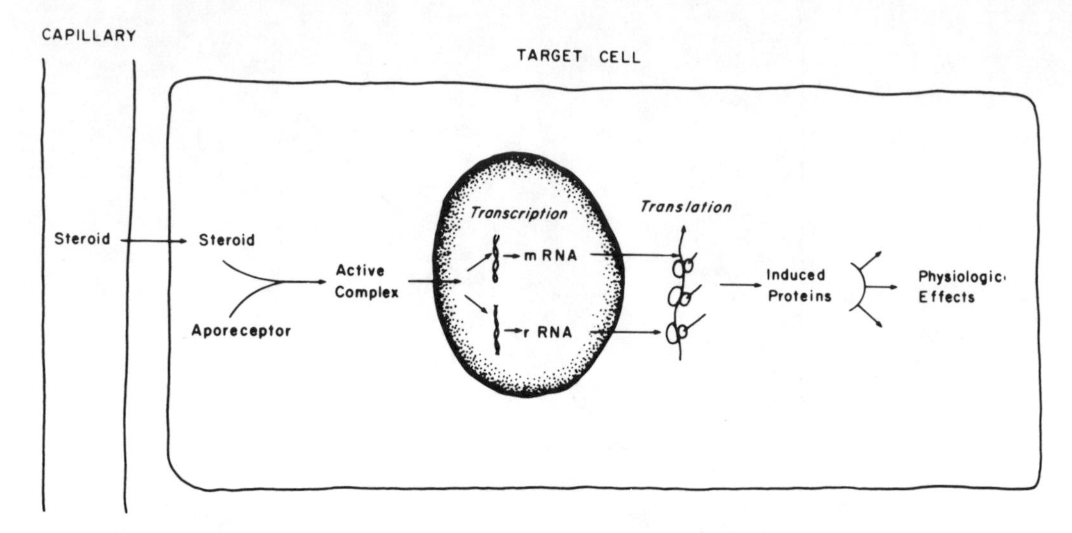

FIGURE 9. General model proposed for the mechanism of steroid action. The steroid enters the cell by diffusion and combines with an aporeceptor in the cytoplasm to generate an active complex which then translocates to the nucleus. Within the nucleus the steroid-receptor complex interacts with the chromatin to enhance messenger RNA (mRNA) and ribosomal RNA (rRNA) transcription which, in turn, leads to increased translation of induced proteins. The synthesis of induced protein ultimately mediates the physiologic effect of the hormone. (From Feldman, E., Funder, J. W., and Edelman, I. S., *Am. J. Med.*, 53, 545, 1972. With permission.)

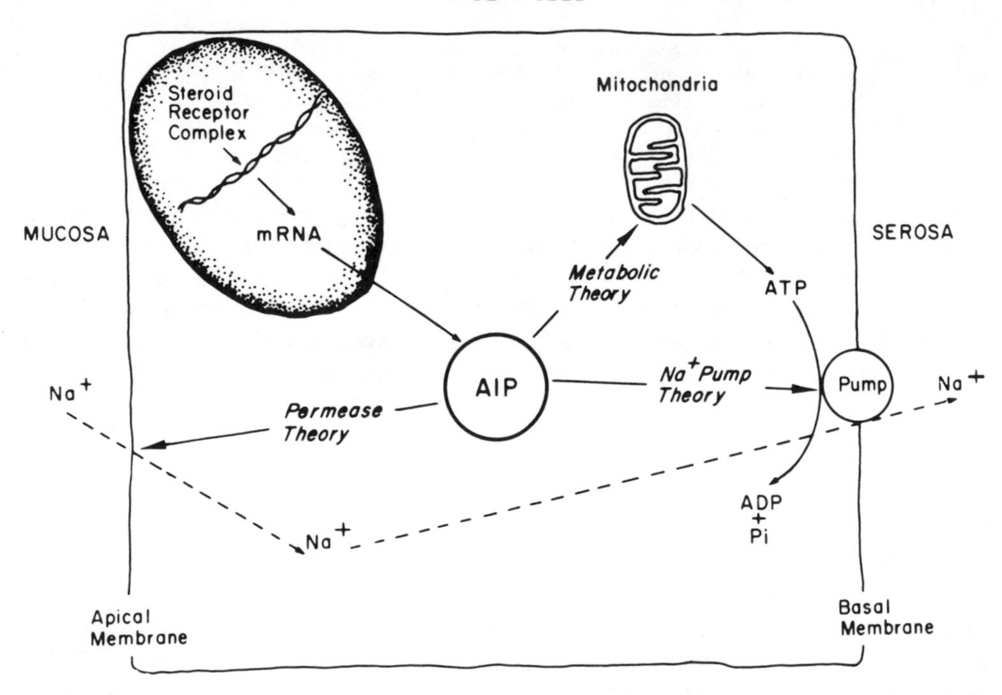

FIGURE 10. Possible sites of action of aldosterone-induced protein (AIP). Na^+ enters the cell at the apical membrane along its electrochemical gradient and is extruded from the cell by a Na^+, K^+ activated ATPase (pump) located in the basal and lateral membranes: AIP has been suggested to exert its effect at three possible sites: (1) at the apical entry step (permease theory), (2) directly on the NaK-ATPase (pump theory), or (3) on the oxidative pathway generating high energy intermediates to fuel the pump (metabolic theory). (From Feldman, E., Funder, J. W., and Edelman, I. S., *Am. J. Med.*, 53, 545, 1972. With permission.)

2. The permease hypothesis — that AIP enables more sodium to enter through the apical membrane and thereby promotes the increased extrusion of sodium via the basal membrane pump
3. The metabolic hypothesis — that AIP stimulates ATP synthesis and makes more energy available for transport process.

For a more detailed consideration of each hypothesis the reader is referred to the review of Feldman et al.[37] For our purpose here it should be noted that experimental evidence for and against each hypothesis exists, but that a definitive concept regarding the subcellular effects of AIP awaits more investigation. Of interest is the conclusion of Al-Awqati[36] based upon toad[38] and turtle[39,40] bladder work, that the effect of aldosterone in stimulating hydrogen ion transport is dissociable from its effect on sodium transport. This in vitro evidence correlates well with the dog studies of Lifschitz et al.,[41] in which actinomycin D, an inhibitor of DNA-dependent RNA synthesis, abolished the antinatriuretic effect of aldosterone, but did not prevent the hormone's kaliuretic or proton-secreting effect. These data suggest that the transport effects of aldosterone on sodium, potassium, and hydrogen ions are multiple, complex, and to some extent, independent of each other.

1. Mineralocorticoid Escape

One observation which has intrigued physiologists is the so-called "escape" phenomenon seen with chronic administration of a mineralocorticoid hormone, such as desoxycorticosterone acetate (DOCA). This escape is depicted in Figure 11, which shows a study in a 70-kg individual initially in sodium balance on a daily sodium intake of 100 mEq. The administration of DOCA results in positive sodium and water balance and an increase in body weight. This positive sodium balance is not progressive, however, and after several days a natriuresis ensues which restores sodium balance, but at a higher ECFV.

The mechanism whereby the mineralocorticoid effect to enhance sodium reabsorption is short-lived is unresolved. Earlier suggestions that the adaptive natriuresis results from tubular insensitivity to mineralocorticoid, or from a DOCA-induced natriuretic factor, have not been substantiated in the laboratory. Currently, controversy exists as to the relative importance of proximal and distal adaptations in this phenomenon. Experiments in both dogs[42] and rats[43] have demonstrated a decrease in proximal tubular reabsorptive rate in DOCA-treated animals, supporting the proposition that DOCA escape is largely secondary to increased distal sodium delivery. This enhanced proximal tubular rejection is probably not a direct effect of aldosterone but rather reflects an adaptation to concomitant ECFV expansion or some consequence thereof.

Although Sonnenberg[44] (using micropuncture methods) has also observed an increase in distal sodium delivery during DOCA escape, he has emphasized the importance of distal events in the natriuresis of DOCA escape. Utilizing microcatheterization techniques,[45] Sonnenberg has observed that the fractional reabsorption of sodium by the medullary collecting duct is reduced when sodium excretion is increased. This finding suggests that the increased sodium excretion of DOCA-treated rats is due to inhibition of sodium transport beyond the distal tubule at the collecting duct level. These data must be reconciled with that of Haas et al.[43] who found that in the volume-expanded rat, DOCA stimulated tubular sodium reabsorption in the cortical and/or outer medullary collecting duct. Our present knowledge does not permit the complete resolution of these data, and thus the locus, as well as the mechanism, of DOCA escape is an unresolved issue. In a recent extensive review of mineralocorticoid escape, Knox and colleagues[46] conclude that the phenomenon is multifactorial in origin, likely representing an adaptive interplay of neural, hemodynamic, and possibly hormonal alterations.

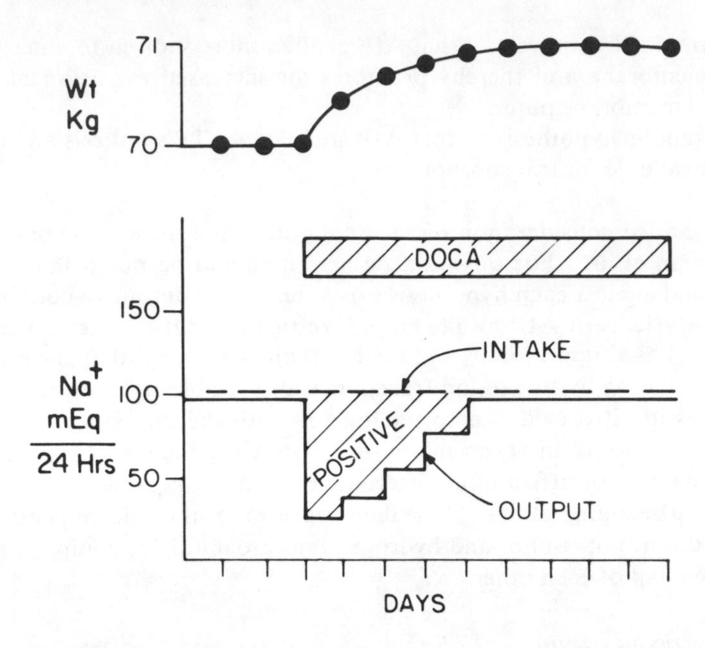

FIGURE 11. "DOCA escape" phenomenon. When an individual on a constant sodium intake receives mineralocorticoid, a period of positive sodium balance and weight gain occurs. After 3 to 5 days, however, a new steady state ensues with sodium excretion again equal to sodium intake and no further weight gain occurs. (From Reineck, H. J. and Stein, J. H., in *Clinical Disorders of Fluid and Electrolyte Metabolism*, 3rd ed., Maxwell, M. and Kleeman, C., Eds. Copyright (c) 1980, McGraw-Hill, New York, Used with permission of the McGraw-Hill Book Company.)

Although there is ample documentation of physiological effects of aldosterone, and the clear demonstration that adrenalectomy or adrenal insufficiency are associated with potentially fatal renal salt wasting, a number of observations have demonstrated that aldosterone is not the sole regulator of tubular sodium reabsorption. These include the aforementioned DOCA escape phenomenon and the fact that, while the subcellular events necessary for the onset of the aldosterone effect take up to an hour, alterations in sodium transport in response to a variety of maneuvers are momentary. In addition, adrenalectomized patients on constant mineralocorticoid replacement can maintain sodium balance and respond appropriately to such maneuvers as changes in posture and constriction of an extremity.[31] These data, taken together, therefore implicate regulatory processes which were independent of aldosterone. However, because of imprecision in measuring GFR, nagging doubts persisted that alterations in sodium excretion could be due to subtle and undetectable changes in GFR. Thus, the first unequivocal evidence that sodium excretion was governed by factors other than GFR and aldosterone was provided by the classical experiment of de Wardener and colleagues[47] and the subsequent work which confirmed these findings.

C. "Third Factor" — The de Wardener Experiment

The seminal observations of de Wardener et al.[47] are illustrated in Figure 12. An anesthetized dog was prepared with supraphysiological amounts of mineralocorticoid hormone and vasopressin, and a balloon was placed in the animal's thoracic aorta. With the inflation of the balloon, abdominal aortic pressure and inulin and PAH clearance all fell. Nevertheless, with the saline infusion, a brisk natriuresis ensued. This natriuresis occurred in spite of a diminished filtered load of sodium and in the presence

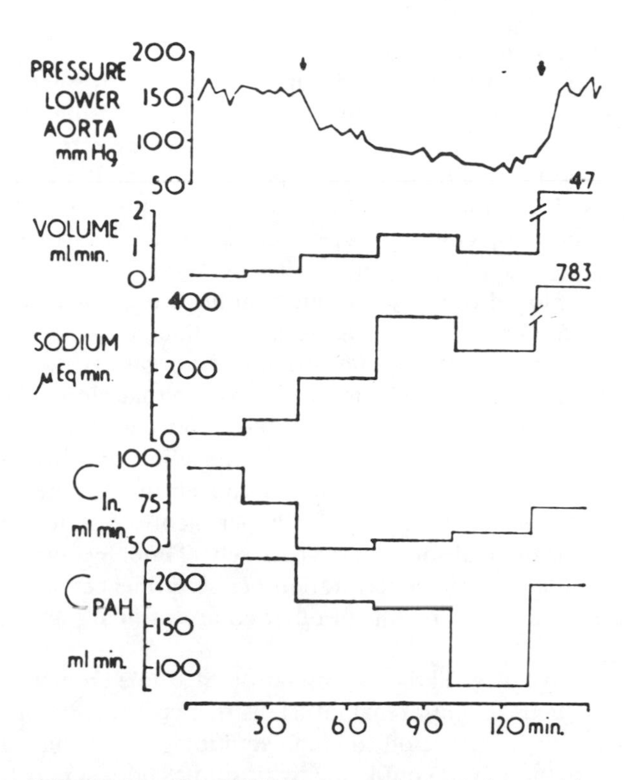

FIGURE 12. The de Wardener experiment. A dog receiving mineralocorticoid was volume expanded with isotonic saline. Despite a fall in renal perfusion pressure, the GFR (C_{In} and RPF (C_{PAH}), that was achieved by C_{In} inflating an aortic ballon proximal to the renal arteries, a large increase in urinary sodium excretion occurred. This group of studies provided unequivocal evidence that factors other than the GFR and mineralocorticoid were involved in determining urinary sodium excretion. (From de Wardener, H. E., Mills, I. H., Clapham, W. F., and Hayter, C. J., *Clin. Sci.*, 21, 249, 1961. With permission.)

of supraphysiological levels of mineralocorticoid hormone. Thus, some unidentified factor or combination of factors must have influenced tubular reabsorption of sodium. Subsequent confirmatory studies[48-51] also underscored the importance of "third factor". This term, which historically described regulatory influences other than GFR ("first factor") and aldosterone ("second factor"), is not used in the current literature. Instead, as our knowledge has expanded, the full panoply of physical, neural, and hormonal influences governing sodium excretion has been examined in detail. The following is a discussion of these factors.

D. Peritubular Physical Forces

In the earlier discussion of GTB, the intraluminal effects of filtrate upon proximal sodium transport were considered. Here we shall examine the other proposed determinant of GTB, i.e., the peritubular Starling forces. The capillary Starling forces, specifically the peritubular hydrostatic and oncotic pressures, have been the subject of much study and discussion. The following summary updates prior reviews of the subject.[52,53]

1. Peritubular Oncotic Forces

Over a century ago, Carl Ludwig[54] postulated that urine formation resulted from plasma ultrafiltration and subsequent tubular modification of the filtrate due to peritubular hydrostatic and oncotic forces. Table 1 lists and categorizes the intensive works devoted to the question of the importance of peritubular capillary oncotic pressure in regulating tubular sodium transport and includes a brief summary of results. In spite of the commitment of considerable energies over half a century by a number of distinguished renal physiologists, the question remains unanswered.

The techniques employed to study this question have been many and varied. Early studies by Starling and Verney[55] utilizing isotonic saline administration into a heart-lung-kidney perfusion preparation demonstrated what was termed a "dilutional diuresis". This work was perhaps the first to demonstrate that blood compositional factors might influence tubular sodium handling. These concepts were extended by studies in the isolated perfused kidney by Craig et al.[56] and Nizet and his colleagues.[57,58] In these experiments, addition of isotonic saline or concentrated protein solutions to the perfusate to make its composition hypo- or hyperoncotic resulted in diminished or augmented tubular sodium reabsorption, respectively. Though compatible with the hypothesis that peritubular protein concentration per se influences sodium reabsorption rate, these data do not exclude the role of other compositional factors, such as hematocrit and viscosity.

Similar problems have plagued the interpretation of whole animal studies (see Table 1) in which either systemic or intrarenal infusions of hyperoncotic fluids were administered to examine their influence on sodium excretion. As demonstrated in Table 1, the outcome of these studies was conflicting, with studies both supporting and refuting a role of peritubular oncotic pressure as a determinant of tubular sodium reabsorption. Though the bulk of the evidence favored an effect of hyperoncotic infusions on tubular transport, in many studies it was difficult to exclude the influence of changes in other blood compositional factors, changes in filtered load, and an effect of the ECFV. Probably the most convincing group of studies were those of Earley and his colleagues.[59-61] These investigators found that when tubular sodium reabsorption was depressed by saline infusion or induced renal vasodilatation, the infusion of concentrated albumin resulted in decreased excretion of sodium. The reversal of this tubular rejection of sodium was not attributable to changes in the GFR. Thus, these studies provided indirect evidence that peritubular protein concentration was a mediator of the rate of tubular sodium transport.

A more direct approach in investigating the role of peritubular protein on sodium reabsorption was permitted with the use of micropuncture techniques. Initial studies utilizing the split oil drop technique of quantifying the transport rate in the proximal tubule demonstrated that an increase in intraluminal or peritubular oncotic pressure had little or no effect on sodium reabsorption. However, subsequent studies in which the contraluminal environment was changed, either through spontaneous fluctuations of filtration fraction[62] or through experimental perfusion of the peritubular capillaries with dextran-containing solutions,[63] did suggest a role of the peritubular oncotic pressure in regulatory sodium excretion. This was further supported by the demonstration by Brenner and associates[64] that suprarenal intraaortic injections of concentrated albumen resulted in a prompt increase in the proximal tubular fractional reabsorption of sodium. However, other unidentified regulatory factors were also implicated in the latter study, as the proximal tubular fractional reabsorption of sodium later fell despite persistent elevation of the peritubular oncotic pressure (Figure 13).

The data from a number of microperfusion experiments also are controversial. While Windhager et al.[65] and Brenner and colleagues[66-70] have been able to demonstrate the importance of the peritubular protein concentration in mediating sodium

Table 1
ROLE OF PERITUBULAR ONCOTIC PRESSURE IN SODIUM REABSORPTION

Authors	Ref.	Preparation	Findings	Comment
Starling, E. H. and Verney, E. B.	*Proc. R. Soc. London, Sec B*, 97, 321, 1925	Isotonic saline perfusion of heart-lung-kidney preparation	"Dilutional diuresis"	Initial evidence that urinary sodium excretion influenced by blood compositional factors
Bresler, E. H. Vander, A. J. et al.	*Am. J. Med. Sci.*, 232, 92, 1956; *Am. J. Physiol.*, 199, 517, 1960; *Am. Heart J.*, 62, 1, 1961 *Am. J. Med.*, 25, 497, 1958			Early proposals that force for proximal tubular reabsorption generated by peritubular Starling forces
Isolated Perfused Kidney				
Craig, G. M. et al. Nizet, A. et al.	*J. Physiol. London*, 186, 113, 1966 *Pfluegers Arch.*, 304, 30, 1968	Perfusion fluid diluted by isotonic saline	Increased urinary sodium excretion	Impossible to ascertain which alteration(s) in compositional factors is responsible
Nizet, A.	*Pfluegers Arch.*, 301, 7, 1968	Concentrated protein solution infused into perfusate	Decreased urinary sodium excretion	Effect of hyperoncotic solutions on sodium excretion appear to be due to intrarenal changes
Whole Organism				
Goodyer, A. V. et al. Welt, L. G. and Orloff, J. Orloff, J. and Blake, W. D. Petersdorf, R. G. and Welt, L. G. Schrier, R. W. et al. Vereerstraeten, P. and DeMyttenaere, M.	*J. Appl. Physiol*,, 1, 671, 1949 *J. Clin. Invest.*, 30, 751, 1951 *Am. J. Physiol.*, 164, 167, 1951 *J. Clin. Invest.*, 32, 283, 1953 *Clin. Sci.*, 34, 57, 1968 *Arch. Ges. Physiol.*, 302, 1, 1968	Concentrated colloid infusions into whole organism	Unlike results with saline infusions, colloid infusions cause little or no increase in sodium excretion	Initial indication that plasma protein concentration, per se, may influence tubular sodium handling; sometimes changes in GFR made results difficult to interpret
Earley, L. E. Vereerstraeten, P. et al. Vereerstraeten, P. and Toussaint, C. Vogel, G. and Heym, E. Vogel, G. et al. Vereerstraeten, P. and DeMyttenaere, M.	*Proc. Soc. Exp. Biol. Med.*, 116, 262, 1964 *Nephron*, 3, 103, 1966 *Nephron*, 2, 355, 1965 *Arch. Ges. Physiol.*, 262, 226, 1956 *Z. Ges. Exp. Med.*, 126, 485, 1955 *Pfluegers Arch.*, 302, 1, 1968	Concentrated protein solutions infused directly into renal circulation of experimental animals	Decreased urinary sodium excretion	Effect of protein, per se, as separated from filtered load, difficult to dissociate; results suggest direct relationship of peritubular protein concentration with sodium reabsorption
Martino, J. A. and Earley, L. E. Daugharty, T. M. et al. Earley, L. E. et al.	*J. Clin. Invest.*, 46, 1963, 1967 *Am. J. Physiol.*, 215, 1442, 1968 *J. Clin. Invest.*, 45, 1668, 1966	Manipulation of tubular sodium reabsorption in dog and utilization of clearance techniques	The depression of tubular sodium reabsorption due to saline infusion or renal vasodilatation is reversed by infusion of concentrated albumin infusion; decrement not attributable to changes in GFR	Suggestion that increased plasma oncotic pressure results in increased tubular reabsorption of sodium; increases in sodium reabsorption both proximal and distal

Table 1 (continued)
ROLE OF PERITUBULAR ONCOTIC PRESSURE IN SODIUM REABSORPTION

Authors	Ref.	Preparation	Findings	Comment
Howards, S. S.	*J. Clin. Invest.*	Intravenous infusions of hyperoncotic albumin in dogs; sodium excretion judged by micropuncture and clearance techniques	Intravenous colloid infusion decreased proximal sodium reabsorption and increased urine flow	Apparent predominant effect of colloid infusion related to some alteration of ECFV expansion
Levy, M. and Levinsky, N. G.	*Am. J. Physiol.*, 220, 415, 1971	Systemic or renal arterial infusions altering renal artery oncotic pressure	"Positive" study; protein affects transport	Further evidence of a correlation between pericapillary oncotic pressure and reabsorptive rate
Morgan, T.	*Circ. Res.*, 26, 27 (Suppl 2), 245, 1970			
Ott, C. E. et al.	*J. Clin. Invest.*, 55, 612, 1975	Systemic or renal arterial infusions altering renal artery oncotic pressure	Effect of oncotic pressure occurs only in the volume expanded state	Raises questions about physiologic relevance
Leyssac, P. P.	*Acta Physiol. Scand.*, Suppl. 442, 1976	Renal artery occlusion followed by measurement of reabsorptive rate	From time of renal artery occlusion until tubular collapse reabsorptive rate constant	Constant reabsorptive rate despite dilution of peritubular protein argues against importance of peritubular protein effect
Leyssac, P. P.	*Acta Physiol. Scand.*, 58, 236, 1963			
Heller, J.	*Pfluegers Arch.*, 329, 115, 1971	Systemic or renal arterial infusions altering renal oncotic pressure	No effect on tubular reabsorptive rate	"Negative" studies; no correlation between oncotic pressure and reabsorptive rate
Knox, F. G. et al.	*Am. J. Physiol.*, 223, 741, 1972			
Kuchinsky, W. et al.	*Pfluegers Arch.*, 321, 102, 1972			
Micropuncture (Split Oil Drop)				
Whittembury, G. et al.	*Am. J. Physiol.*, 197, 1121, 1960	Albumin solutions placed into proximal tubular lumen; sodium reabsorptive rate measured by split oil drop technique	Albumin solutions reabsorbed at same rate or only partially inhibited rate as isotonic saline solutions	Suggests protein oncotic gradient across tubular wall without major effect on sodium excretion; negative study
Kashgarian, M. et al.	*Proc. Soc. Exp. Biol. Med.*, 117, 848, 1964			
Giebish, G. et al.	*J. Genol. Physiol.*, 47, 1175, 1964			
Gyory, A. Z. and Kinne, R.	*Pfluegers Arch.*, 327, 234, 1971	Artificial perfusion of peritubular capillary bed to vary oncotic force; reabsorptive rate measured by split oil drop	No change in reabsorptive rate to parallel oncotic force changes	Negative study
Sato, K.	*Biochemical Aspects of Renal Function*, Hans Huber, Bern, 1975, 175	Artificial perfusion of peritubular capillary bed to vary oncotic force; reabsorptive rate measured by split oil drop	Reabsorptive rate varies with oncotic force	Positive study
Lewy, J. E. and Windhager, E. E.	*Am. J. Physiol.*, 214, 943, 1968	Oncotic forces varied through spontaneous changes in filtration fraction	The higher the filtration fraction the greater the tubular reabsorption	Suggests that oncotic gradient induced by changes on peritubular side affects proximal sodium transport
Spitzer, A. and Windhager, E. E.	*Am. J. Physiol.*, 218, 1188, 1970	Direct perfusion of peritubular capillaries with dextran and saline solutions; sodium reabsorption measured with free flow and split oil drop techniques	Saline decreased reabsorptive rate, dextran without effect; linear correlation between reabsorptive rate and calculated oncotic pressure	Suggests importance of peritubular capillary oncotic pressure in sodium reabsorptive rate, but decreased reabsorptive associated with decreased SNGFR

Falchuk, K. H. et al.	*Am. J. Physiol.*, 220, 1427, 1971	Proximal reabsorption studied during partial occlusion of renal vein, renal artery, carotid occlusion with vagotomy, and during intravenous infusion with 6 or 15 g % albumin solution	Changes in capillary oncotic and hydrostatic pressure were in the direction consistent with regulation of proximal reabsorption by capillary fluid uptake	Positive study
		Micropuncture (Free Flow)		
Brenner, B. et al.	*J. Clin. Invest.*, 48, 1519, 1969	Suprarenal aortic injections of concentrated albumin; micropuncture from peritubular capillaries and proximal tubules	Transient, marked, immediate increase in proximal tubular fractional reabsorption of sodium	Apparent effect of increased peritubular capillary protein concentration in promoting proximal sodium reabsorption; effect appears intrarenal; volume expansion may explain later decreases in fractional reabsorption below preinfusion controls despite persistent elevation of pertubular protein concentration; changes in SNGFR impossible to exclude definitively, due to method
Daugharty, T. M. et al.	*Am. J. Physiol.*, 222, 225, 1972	Volume expansion in rats with iso-oncotic and colloid-free solutions	Ringer's solution caused decrease in postglomerular protein concentration and reduction of proximal tubular reabsorption, unlike plasma-treated group	Supports importance of peritubular oncotic pressure
Brenner, B. M. and Galla, J. H.	*Amer. J. Physiol.*, 220, 148, 1971	Effect of changes in hematocrit on systemic and peritubular protein concentration and proximal sodium reabsorption	Reabsorption correlated directly with changes in postglomerular protein concentrations	Positive study
		Microperfusion		
Windhager, E. E. et al.	*Nephron*, 6, 247, 1969	Microperfusion of peritubular venous capillaries with hyperoncotic dextran; sodium reabsorption measured with split oil drop and free flow sampling	Sodium and water transport directly related to peritubular capillary oncotic pressure	Apparent direct influence of oncotic pressure upon reabsorptive rate without extrarenal or secondary intrarenal mechanism
Brenner, B. M. and Troy, J. L.	*J. Clin. Invest.*, 50, 336, 1971	In hydropenic rats efferent arterioles and peritubular capillaries perfused with colloid-free Ringer's, iso-oncotic and hyperoncotic perfusates; proximal tubular reabsorption measured by free flow micropuncture	Using different perfusates, proximal tubular reabsorption varied in direct linear relationship to postglomerular protein concentration; alterations not explained by changes in filtered load, and reversible minutes after cessation of perfusion	Still considered very strong evidence that capillary oncotic pressure mediates proximal tubular sodium reabsorption and may contribute to normal GT balance

Table 1 (continued)
ROLE OF PERITUBULAR ONCOTIC PRESSURE IN SODIUM REABSORPTION

Authors	Ref.	Preparation	Findings	Comment
Brenner, B. M. et al.	*J. Clin. Invest.*, 50, 1596, 1971	Capillary microperfusion studies	Absolute proximal reabsorption decreased during perfusion with colloidfree Ringer's; not changed from control when 9 to 10 g % albumin added	Positive study
Brenner, B. M. et al.	*J. Clin. Invest.*, 52, 190, 1973	Prosimal reabsorption studied during acute volume expansion; protein concentration in peritubular capillaries then restored to normal by capillary perfusion made iso-oncotic by addition of albumin	Inhibition of proximal tubular reabsorption of volume expansion reversed with restoration of oncotic force to normal	More evidence for importance of postglomerular protein concentration
Green, R. et al.	*Am. J. Physiol.*, 226, 265, 1974	Double perfusion experiments on proximal tubules and peritubular capillaries of rat kidneys in vivo; transport examined under "normal" circumstances and when active transport inhibited by cyanide solutions	When transport inhibited, oncotic forces had little or no effect; protein effect marked when transport intact	Protein effect only modifies the isotonic fluid movement resulting from active transport
Bank, N. et al.	*Kidney Int.*, 1, 397, 1972	Perfusion of peritubular capillaries with high and low protein concentration solutions, at high and low rates	Oncotic effects did not explain all of changes in reabsorptive rate	Reemphasizes the role hydrostatic pressure may play in these experimental settings
Rumrich, G. and Ullrich, K. J.	*J. Physiol. London*, 197, 69, 1968	Capillary microperfusion with hyperoncotic solutions; sodium reabsorption measured with split oil drop	No effect of peritubular plasma oncotic pressure on proximal sodium reabsorption	Negative study
Lowitz, H. D. et al.	*Nephron*, 6, 173, 1969	Micropuncture in rats; peritubular capillaries perfused with whole blood and Ringer's solution	Proximal tubular reabsorption unaffected by composition of perfusate	Negative study
Holzgreve, H. and Schrier, R. W.	*Pfluegers Arch.*, 356, 73, 1975	Micropuncture techniques; proximal tubular capillaries perfused with hyperoncotic Ringer's and by blood during aortic constriction and saline expansion	No difference in response between sets of tubules	Suggests a lack of importance of oncotic forces in determination of reabsorptive rate
Holzgreve H. and Schrier, R. W.	*Pfluegers Arch.*, 356, 57, 1975			
Conger, J. D. et al.	*Clin. Sci. Mol. Med.*, 51, 379, 1976	Capillary perfusion with pooled plasma varying in protein concentration between 1 and 13 mg/100 mℓ	Variations in protein concentration did not affect tubular reabsorption	Negative study; protocol similar to other perfusion studies which were positive; difficult to reconcile
Bartoli, E. et al.	*J. Clin. Invest.*, 52, 843, 1973	Flow of filtrate between early and late proximal tubule partially diverted by microaspiration	Glomerulotubular balance may occur in part without peritubular oncotic changes	Implicates factors other than oncotic force in reabsorptive rate

Isolated Perfused Tubule

Grantham, J. J. et al.	*Kidney Int.*, 2, 66, 1972	Isolated perfused rabbit tubule; bath fluid changed from serum to serum ultrafiltrate and then to hyperoncotic bath	Transtubular fluid transport decreased by change to plasma ultrafiltrate; no effect with hyperoncotic bath	Peritubular protein concentration may influence net absorption at the antiluminal side of epithelial cell
Horster, M. et al.	*Kidney Int.*, 4, 6, 1973	Isolated perfused rabbit tubule; bath fluid changed from serum to serum ultrafiltrate and then to hyperoncotic bath	No effect of oncotic pressure	Negative study
Burg, M. et al.	*Amer. J. Physiol.*, 231, 627, 1976	Increased tubular reabsorption when protein present	Increased tubular reabsorption when protein present	Reversal of previous results; positive study
Imai, M. and Kokko, J. P.	*Kidney Int.*, 6, 138, 1974	Isolated perfused rabbit tubule; bath fluid changed from serum to serum ultrafiltrate and then to hyperoncotic bath	Transtubular fluid transport decreased when bath an ultrafiltrate; increased by hyperoncotic bath	Positive study

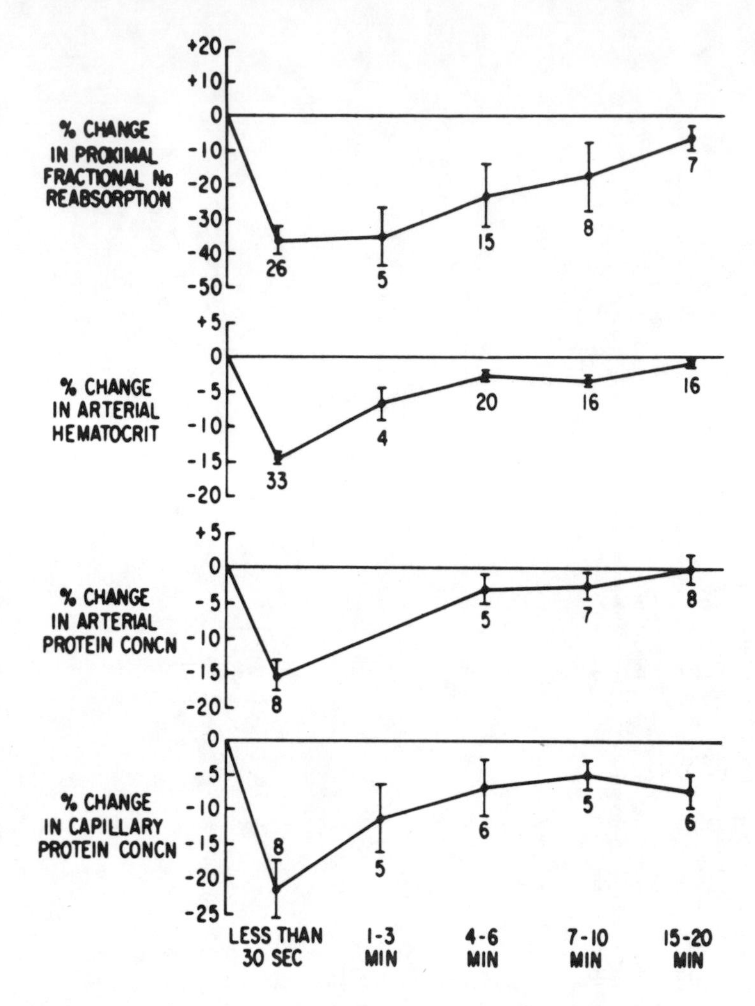

FIGURE 13. Effects of suprarenal arterial injection of hyperoncotic albumin in the rat. Micropuncture samples were collected from peritubular capillaries and from the same site in proximal tubules before and at the time intervals shown, after injecting a 15-g/100 mℓ solution of albumin into the aorta just above the renal arteries. Proximal tubular fractional reabsorption increased immediately, associated with the rise in plasma protein concentration. This increased proximal reabsorption did not persist and actually decreased below control at 20 min despite a persistent elevation of arterial and peritubular plasma protein concentration. The immediate effects of the hyperoncotic solution indicate an effect of protein concentration to increase proximal tubular reabsorption through intrarenal mechanisms. (From Brenner, B. M., Falchuk, K. H., Keimowitz, R. I., and Berliner, R. W., *J. Clin. Invest.*, 48, 1519 1969. With permission.)

transport, a number of other laboratories have not confirmed these results.[25,71] Specifically, the findings of Holzgreve and Schrier[72,73] and Conger et al.[74] are difficult to reconcile with the positive studies. In the latter study, the investigators used techniques very similar to those of the earlier, positive experiments, except for one change. Instead of perfusion with artificial solutions, they utilized rat plasma which was either concentrated or a protein-free ultrafiltrate. Perfusion of the peritubular capillaries with either protein-rich (13 g%) or protein-free solutions did not alter proximal tubular sodium transport.

Thus, the evidence favoring an important role of the peritubular capillary oncotic pressure in the regulation of tubular sodium reabsorption remains controversial. A number of very subtle technical and conceptual issues, largely unidentified, probably account for the disparity of the results in this area. Though many physiologists have concluded that, on balance, there appears to be some effect of peritubular oncotic pressure on sodium transport, de Wardener has eloquently summarized the state of the art and concludes, for a number of reasons, that the peritubular plasma protein concentration has little effect on proximal tubular fluid transport.[9] He suggests that the change in reabsorption by the proximal tubule induced by altering the oncotic pressure of the blood entering the kidney probably relates to some secondary effect beyond that of simply changing oncotic pressure. In addition, the studies of Ott et al.,[75] in which the effects of intrarenal albumin infusion were dependent upon the volume state of the animal, raise some questions about the physiologic importance of peritubular oncotic pressure.

2. Peritubular Hydrostatic Forces

While the importance of peritubular oncotic pressure remains controversial, there exists somewhat stronger evidence for a role of peritubular hydrostatic pressure in governing transepithelial sodium and water transport. The initial evidence for this effect was derived from experiments in a number of different in vitro epithelial preparations, as summarized in a previous review.[53] Though the findings and conclusions were not uniform, and though the relevance of such in vitro systems to mammalian nephrons is uncertain, at least under some circumstances pressure gradients appear to have influenced active sodium transport. Animal studies examining the effects of renal arterial and venous hypertension and renal vasodilatation and constriction have been interpreted as the in vivo extension of the in vitro studies. Thus, the results of alterations in blood pressure and renal vascular resistance have been analyzed as means of altering peritubular hydrostatic pressure.

The initial studies pertaining to the effects of hypertension on sodium excretion did not exclude the possibility that changes in sodium excretion were attributable to parallel changes in GFR. Blake et al.[76] were the first to show that the effects of blood pressure on urinary sodium excretion occurred independently of changes in GFR. Subsequently, the studies of Handley and Moyer[77] and of Earley, Martino, and Friedler[59,78,79] enlarged our understanding of the relationship of hypertension to sodium excretion. Handley and Moyer[77] initially demonstrated that if a vasodilator drug is infused unilaterally into the renal artery, the subsequent intravenous infusion of a vasoconstrictor drug increases sodium excretion in the vasodilated kidney and decreases sodium excretion in the contralateral control kidney. Using the same technique, Earley and Friedler[78] demonstrated that the natriuresis occurred only in the vasodilated kidney despite like changes in renal blood flow and GFR between the two kidneys. The authors suggested that the higher arterial pressure raised the hydrostatic pressure only in the peritubular circulation of the vasodilated kidney and not of the control kidney. They also suggested that the natriuresis following pharmacologic vasodilatation may also be due to a similar increase in peritubular hydrostatic pressure. Unfortunately, however, vasodilatation alters not only the hydrostatic pressure, but also the filtration fraction and hence postglomerular oncotic pressure. Thus, if the oncotic pressure is important, its reduction after vasodilatation may contribute to the observed natriuresis. The above series of experiments therefore supports, but does not prove, the importance of hydrostatic pressure in in vivo sodium transport. Further studies, in which the effects of systemic arterial pressure and renal vascular resistance have been dissociated, are similarly compatible with the importance of peritubular hydrostatic pressure, as reviewed previously.[52] The effects of hypertension on sodium excretion appear to be mediated at various nephron sites, including the proximal nephron, and particularly in a chronic setting, sites beyond the proximal nephron.

a. Experimental Renal Venous Hypertension

As experimental renal vein occlusion with renal venous hypertension increases peritubular hydrostatic pressure, a number of studies have examined the effect of that maneuver on renal sodium excretion. The expectation would be that, as renal venous hypertension increases peritubular hydrostatic pressure, a decrease in tubular sodium reabsorption and a natriuresis would occur. However, in initial studies, renal venous constriction caused an antinatriuresis,[80-84] which in some cases was attributable to a measurable decline in GFR.[81,82] Of interest is the finding that similar degrees of renal venous hypertension in hydropenic and saline-loaded dogs caused a natriuresis and antinatriuresis, respectively. This difference may be explained by the marked decrease in GFR observed in the saline-loaded but not in the hydropenic animals. This decrease in GFR in saline-loaded animals perhaps could occur because saline infusion raises intratubular hydrostatic pressure to a critical level, so that any further increase during elevation of renal venous pressure may substantially diminish net filtration pressure. By contrast, in hydropenic animals the basal intratubular hydrostatic pressure is lower and thus not critically affected by renal venous hypertension.

Thus, studies examining the effect of renal vein constriction on tubular sodium reabsorption must take into account concomitant changes in RBF, GFR, and filtration fraction. As recently reviewed, the interpretation of both clearance and micropuncture studies has been confused by these issues.[52] However, taking together the conclusions from in vitro, clearance, and micropuncture studies, it is probably reasonable to conclude that transepithelial transport can be influenced by hydrostatic pressure. Thus, a natriuresis is induced by increased hydrostatic pressure as long as a significant fall in GFR does not occur.

3. Other Physical Forces/Compositional Factors

a. Hematocrit and Viscosity

Ever since Starling and Verney's[55] demonstration of a "dilutional diuresis", attention has been focused on the gamut of compositional factors which may influence renal sodium handling. The role of protein concentration and oncotic pressure has been discussed earlier. Early investigations into the role of hematocrit were beset with difficulties in manipulating the hematocrit as the single variable.[52] Thus, when changing the hematocrit, the addition of either saline or plasma-like solutions could result in lowering of the protein concentration or in volume expansion with associated changes in blood pressure, renal vascular resistance, and change in the concentration of some pertinent humoral substance(s).

Several investigators have attempted to define the role of hematocrit on sodium excretion, independent of volume expansion or change in plasma protein concentration. In the studies of Bahlmann et al.,[85] the blood of anesthetized dogs was circulated through reservoirs containing 2.5 g/100 mℓ or 5 g/100 mℓ albumin concentration, with resultant reduction of the hematocrit. A natriuresis occurred only during dilution with the 5 g/100 mℓ solution, raising the possibility that the effect was due to a pharmacologic action of the foreign protein. However, studies by Nashat and Portal[86] and Nashat et al.[87] demonstrated increases in urine flow, sodium excretion, and renal plasma flow as the hematocrit was decreased by an isovolemic exchange of blood for iso-oncotic dextran. They did not demonstrate an antinatriuresis with an isovolemic red blood cell exchange with an increase in hematocrit, thus allowing the possibility that the natriuresis observed during the dilution of the hematocrit may have been due to some other effect of the dextran.

Schrier and Earley[88] examined further the effects of increases and decreases in the hematocrit on renal hemodynamics and sodium excretion in the absence of changes in renal perfusion pressure and plasma protein concentration or infusion of foreign pro-

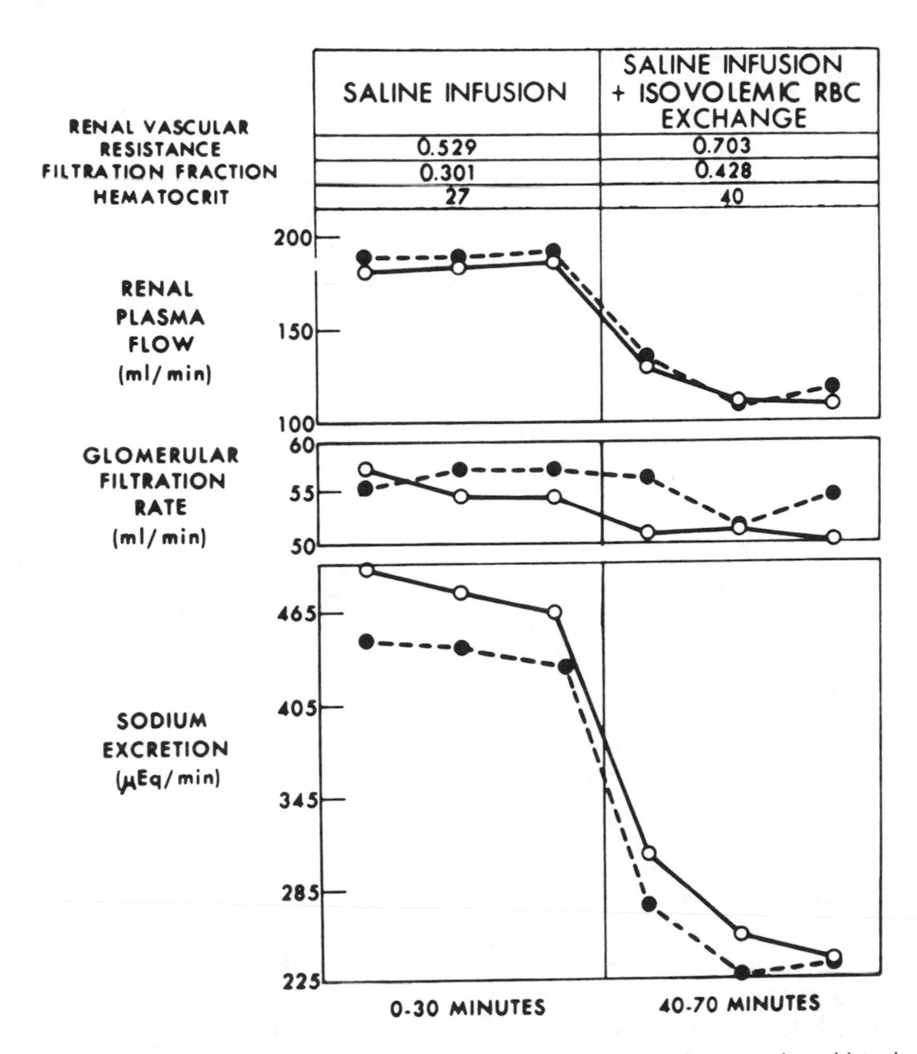

FIGURE 14. Effect of increasing hematocrit during sustained volume expansion with and infusion of isotonic saline in one animal. After stabilization of urine flow, three 10-min control periods were obtained. Then, isovolemic exchange with fresh, washed red blood cells was performed as the saline infusion was continued. The increase in hematocrit diminished sodium excretion and renal plasma flow as renal vascular resistance and filtration fraction increased. Values are for each kidney of a single animal. (From Schrier, R. W. and Earley, L. E., *J. Clin. Invest.*, 49, 1656 1970. With permission.)

tein or dextran. They demonstrated that both increases and decreases in hematocrit significantly affected sodium excretion and renal hemodynamics both in the hydropenic and the saline-loaded dog (Figure 14). The results of this study suggested that the effect of hematocrit on sodium excretion was most probably mediated by altering sodium reabsorption since consistent changes in GFR were not observed. The increase in sodium excretion with decreases in hematocrit were also associated with a concomitant rise in solute-free water reabsorption and potassium excretion. These findings suggested an augmented distal delivery of filtrate, consistent with the hypothesis that changes in hematocrit affect proximal sodium reabsorption. This hypothesis has been supported by findings in micropuncture studies.[70,89,90] A decrease in hematocrit was also consistently associated with a diminution in filtration fraction and renal vascular resistance, whereas an increase in hematocrit was associated with increases in these same parameters. These effects of hematocrit are likely to be mediated through an influence on viscosity. In this regard, Schrier and colleagues[91] have demonstrated that alterations in hematocrit during volume expansion are associated with marked changes in blood viscosity, particularly at low rates of shear found in the microcirculation. The effect of hematocrit on renal sodium handling may be mediated through these changes in viscosity, which by affecting filtration and renal vascular resistance may alter the peritubular oncotic and hydrostatic pressures, respectively.

b. Plasma Sodium Concentration

Another compositional factor which has received attention is the plasma sodium concentration, independent of sodium-associated changes in ECFV or renal and systemic hemodynamics. A more detailed consideration of this factor is beyond the scope of this review and may be found elsewhere.[52] To summarize, however, the initial studies demonstrating a correlation between plasma sodium and sodium excretory rate did not differentiate the effects of plasma sodium concentration from volume expansion. The work of Kamm and Levinsky[92] attempted to address that distinction. The authors performed intrarenal arterial perfusion of hypertonic saline, with maintenance of a constant filtered load of sodium achieved through arterial constriction. This preparation allowed the examination of the effects of sodium concentration in the absence of changes in filtered sodium load. Any natriuresis resulting from ECFV expansion would be manifested also by the contralateral kidney and thus would be subtractable from any additional effect of hypernatremia in the ipsilateral kidney. The results demonstrated an increased sodium excretion in the kidney perfused with hypertonic saline, an effect attributed to diminished tubular reabsorption since the filtered load was maintained constant by lowering GFR with constriction of the arterial perfusion. Interpretation of this study is complicated, however, by the fact that perfusion with a protein-free hypertonic saline solution without red blood cells diminished the plasma protein concentration and hematocrit in the perfused kidney. Therefore, isolating the effect of hypernatremia from dilutional alterations with secondary effects on the peritubular Starling forces is again problematic.

A different approach was taken by Goldsmith et al.[93] and Blythe and Welt[94] who utilized hyponatremic models and maneuvers to ensure constancy of the filtered sodium load. They demonstrated that under their experimental circumstances, hyponatremia was associated with a diminished urinary sodium excretion. These results, taken together, suggest that changes in the plasma sodium concentration may have a direct effect on sodium reabsorption independent of changes in filtered sodium load. Though micropuncture evidence suggests that the locus of this effect is the proximal nephron,[95] the nature of the mechanism of this direct influence of sodium concentration on reabsorptive rate remains unclear.

4. Peritubular Physical Forces — An Overview

The consideration of the above physical factors is not complete without attempting to integrate them into a cohesive, unified view of transport in the proximal nephron. Such a conceptual framework would have to incorporate what is known from electrophysiologic and other studies about transepithelial transport with the in vivo clearance and micropuncture techniques described earlier. The reader is referred to recent thoughtful, articulate reviews on the subject.[1,54,96,97]

The initial view of Carl Ludwig, that tubular transport resulted entirely from the passive effects of peritubular physical forces, was not tenable, in view of the later demonstration of active transport along the nephron and of the ability of the kidney to both concentrate and dilute the urine. However, the participation of active transport in the tubular reabsorptive process would not necessarily preclude a modifying influence of passive physical forces. To clarify the relationship between active transport

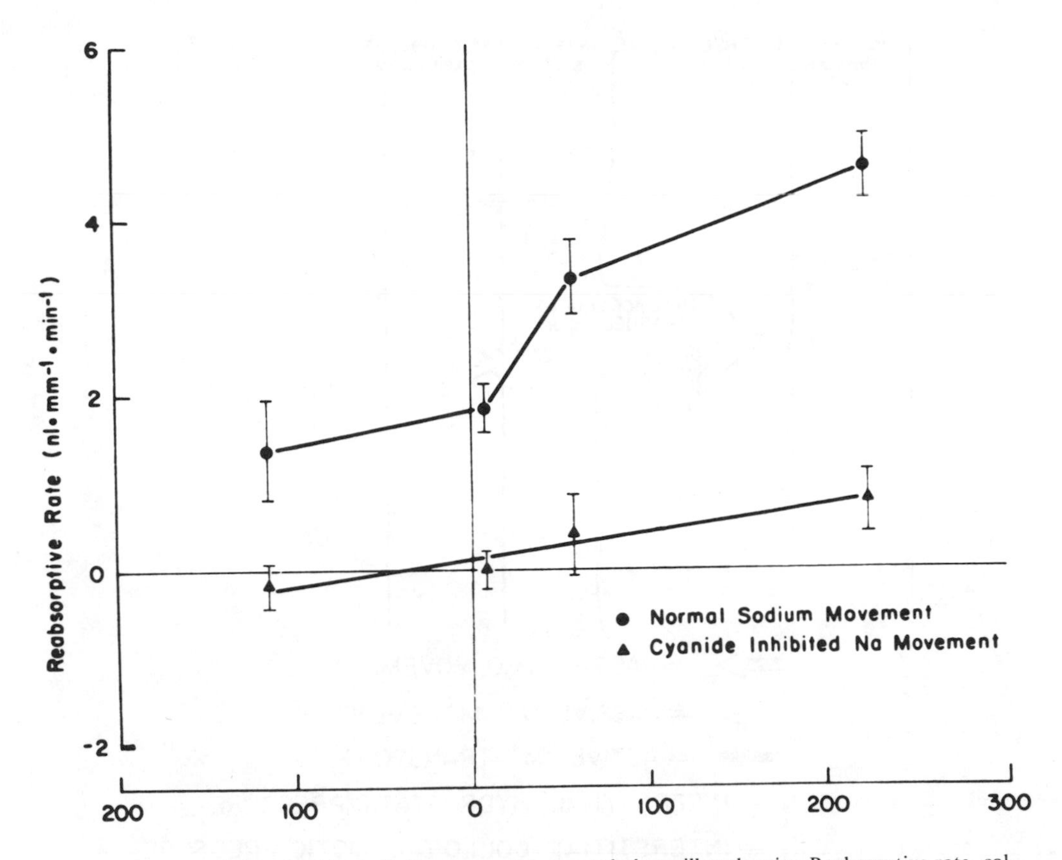

FIGURE 15. Effect of HSA on fluid movement across tubulocapillary barrier. Reabsorptive rate, calculated in $n\ell \cdot mm^{-1} \cdot min^{-1}$, is plotted against net oncotic gradient across tubulocapillary barrier for series 1. In normal sodium movement, series 2, net sodium movement inhibited by poisoning pump with 4 mM cyanide. Apparent "negative" net oncotic gradient was produced with HSA in tubular lumen. Vertical bars are ± SE. (From Green, R., Windhager, E. E., and Giebisch, G., *Am. J. Physiol.*, 226, 265, 1974. With permission.)

and passive physical forces, Green et al. performed an elegant series of studies.[98] Utilizing the split-oil drop technique, and simultaneous perfusion of tubule and peritubular capillaries, the effects of altering intraluminal and peritubular capillary albumin concentrations were examined. The investigators found that intraluminal oncotic pressure did not exert a major effect on fluid reabsorption. Their fascinating results with peritubular capillary addition of albumin are depicted in Figure 15. The investigators altered the peritubular oncotic forces by varying the peritubular capillary albumin concentration over a range from 2.5 g/100 mℓ to 20 g/100 mℓ, with active transport both intact and inhibited by sodium cyanide. As depicted, when net sodium flux was inhibited by cyanide, peritubular capillary albumin concentration had only a minor effect on fluid transfer. By contrast, the effect of an increased transcapillary oncotic gradient was marked when active transport was allowed to proceed normally. These results and their measurements of the hydraulic permeability of the proximal tubular epithelium suggested that albumin has only a small direct oncotic effect across the tubular wall and acts mainly to modify fluid reabsorption which has occurred secondary to active sodium transport.

To visualize the manner in which this occurs, the following model has been proposed. The "pump-leak" model of proximal tubular sodium reabsorption is depicted in Figure 16. This model incorporates the electrophysiologic data that the proximal convoluted tubule is a "leaky" epithelium, with a high conductance pathway enabling

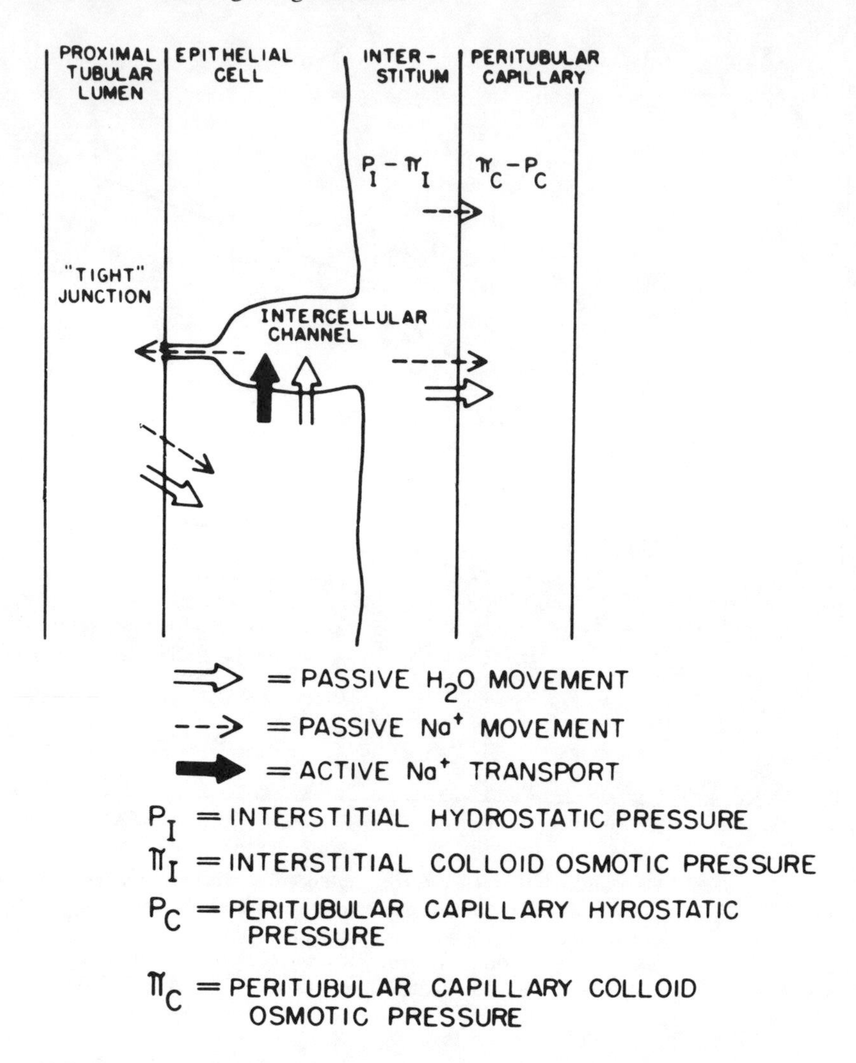

FIGURE 16. Schematic representation of the "pump-leak" model of proximal tubular sodium reabsorption. Filtered sodium is passively reabsorbed from the tubular lumen into the epithelial cell down a concentration gradient. It is then actively transported into the lateral intercellular channel which communicates directly with the interstitial space. The uptake of reabsorbate from the interstitium into the peritubular capillary is determined by Starling forces acting across the capillary wall. If the net force for uptake of reabsorbate into the capillary is decreased, conductive and/or geometric changes in the tight junction-interspace complex may occur which favor increased "back-leak" of reabsorbate into the tubular lumen, diminishing net reabsorption. (From Reineck, H. J. and Stein, J. H., in *Clinical Disorders of Fluid and Electrolyte Metabolism,* 3rd ed., Maxwell, M. and Kleeman, C., Eds. Copyright (c) 1980, McGraw-Hill, New York. Used with permission of the McGraw-Hill Book Company.)

the shunting of an electrical current. It is currently thought that the tight junction and/or lateral intercellular space is the site of this shunt pathway. This model proposes that sodium enters the cell through the luminal membrane along its concentration gradient, and is then actively transported into the lateral intercellular space. Water then moves into the intercellular space because of the osmotic gradient generated by the active sodium transport. Fluid then moves from the intercellular channel into the interstitial space by a hydrostatic pressure generated within the channel. Fluid removal from the interstitium is governed by the Starling forces operating across the peritubular capillary. If the rate of reabsorption exceeds the rate of capillary uptake, interstitial volume will increase. Changes in Starling forces which occur with saline loading therefore could explain a resultant decrease in net sodium reabsorption in the following manner. According to this formulation, the concomitant decrease in capillary oncotic pressure and rise in capillary hydrostatic pressure with ECFV expansion would decrease the egress of fluid from the interstitium into the peritubular capillaries. The attendant increase in interstitial volume could open the tight junction and allow for the increased backleak of reabsorbate from the interstitium into the proximal tubular lumen. Thus, without changing the rate of active sodium transport or unidirectional water flux, net fluid reabsorption is decreased. The converse would occur if the transcapillary Starling forces favored movement of fluid from the interstitium into the peritubular capillaries. If this fluid flux into capillaries were so favored that no backleak occurred, a theoretical maximum in reabsorption could be achieved, not to be augmented by any further Starling forces favoring reabsorption. The data of Green et al.[98] would support this idea. As shown in Figure 15, it appears that with the graded increase in peritubular protein, up to 20 g/100 mℓ albumin solution, the effect of protein on fluid reabsorption diminishes at the higher levels of peritubular protein.

Though the model entails aspects which are still speculative or controversial, it probably is the most reasonable working hypothesis available. These concepts incorporate data using a variety of experimental techniques and provide a framework for the understanding of how peritubular physical forces may influence sodium reabsorption under a variety of conditions.

E. Total and Medullary Blood Flow

Because of the complex interrelationships between peritubular physical factors, it has been difficult to define the influence of renal blood flow on sodium reabsorption and excretion. Though increases in renal blood flow are often associated with a natriuresis, the dissociation of the effects of changes in blood flow per se from the concomitant effects of alterations in peritubular hydrostatic and oncotic pressure has been problematical. For example, an increase in RBF during renal vasodilatation is associated with a decrease in renal vascular resistance, a rise in deep intrarenal venous pressure (presumably an index of peritubular hydrostatic pressure), and a reduction in filtration fraction (presumably an index of initial postglomerular protein concentration).[78,79] Thus, the natriuresis resulting from vasodilatation could derive from changes in physical forces in peritubular capillaries as well as the increased RBF. Willis et al.[99,100] have reported that the natriuresis during renal vasodilatation with acetylcholine and bradykinin is not observed when the renal blood flow is not allowed to increase. However, the decrease in renal perfusion pressure during these experiments may have induced antinatriuretic factors which obscure the natriuresis. These studies thus illustrate the difficulties inherent in attempting to isolate the effects of RBF from changes in physical factors on sodium excretion.

In addition to considering the influence of total RBF, some thought has been devoted to the role of alterations in medullary blood flow as a mediator of changes in sodium excretion. The possible importance of shifting blood flow from outer cortical to inner, more avidly sodium-retaining, juxtamedullary nephrons is considered below. In addition, however, the hemodynamic effect of increased RBF on medullary tonicity has been proposed as a determinant of tubular sodium handling. On the one hand, Pomeranz et al.[101] have suggested that an increase in medullary blood flow may augment reabsorption in the loop of Henle by diminishing medullary hypertonicity, thereby decreasing the passive back diffusion of sodium from the interstitial space into the lumen. By contrast, Earley and Friedler[48] have postulated that the decrease in medullary interstitial tonicity would have the opposite effect. Their theory suggests that the decreased interstitial solute would promote less abstraction of water from the descending limb of the loop of Henle. As a result, an increased volume of luminal fluid with the same sodium content would reach the ascending limb. If there is a minimum sodium chloride concentration which can be achieved by reabsorption along the ascending limb of Henle's loop, less sodium chloride would be reabsorbed and then more sodium will be delivered to more distal sites. To date the controversy as to whether an increased medullary blood flow increases[101] or decreases[48,102-104] or has neither effect on sodium reabsorption in the ascending loop of Henle has not been resolved.

F. Intrarenal Distribution of Blood Flow or Glomerular Filtrate

The notion that renal sodium handling is partially mediated through selective redistribution of glomerular filtration to nephrons with differing functional characteristics is more than two decades old. Key to this proposal is the concept of "nephron heterogeneity", recently the subject of critical review by Lameire et al.[105] These authors ably summarize the anatomic data supporting morphologic heterogeneity in the character of superficial, intermediate, and juxtamedullary nephrons. However, consideration of the functional differences related to this anatomic data is markedly more complex. Utilizing both micropuncture and microdissection techniques, a number of laboratories have found that the juxtamedullary single nephron GFR (SNGFR) is higher than the superficial SNGFR during hydropenia. However, no such consistent difference is observed during acute or chronic alterations in sodium balance or renal perfusion pressure. Thus, the functional importance of morphological differences in nephrons remains to be determined.

The results regarding redistribution of RBF during various circumstances are more consistent, but a relationship between such changes and urinary sodium excretion remains elusive. With initial studies utilizing the inert gas washout method, Barger[106] proposed that the natriuresis with ECFV expansion would be associated with preferential distribution of blood flow to outer cortical nephrons, while salt-retaining states would demonstrate preferential flow to more sodium-avid inner cortical (juxtamedullary) nephrons. However, because the inert gas washout method relies upon several critical assumptions which may not be valid, conclusions derived from these studies are suspect.

More recent work has utilized the radiolabeled microsphere technique of measuring RBF, which, despite some theoretical objections, is generally agreed to be a reasonably accurate method. Utilizing this technique, a number of experimental maneuvers, which have a fall in renal resistance in common, were shown to promote a redistribution of blood flow to inner cortical nephrons. However, there exists no correlation between changes in RBF distribution and urinary sodium excretion. After citing a number of studies in which either a natriuresis or antinatriuresis was associated with inner cortical redistribution, Lamiere et al.[105] concluded that from the existing data, "no strong case can be made for a relationship between the distribution of RBF and the excretion of sodium in the urine."

Though heterogeneity of blood flow or GFR has not been demonstrated to account for predictable changes in urinary sodium excretion, functional differences between nephron populations may still influence renal sodium handling irrespective of hemodynamic considerations. In this regard, differences in transport characteristics of superficial and juxtamedullary nephrons and their different responses to experimental maneuvers which alter the ECFV are considered separately in the later section on segmental sodium transport.

G. Renal Sympathetic Nervous Activity

The effects of the adrenergic nervous system on renal sodium excretion have been studied extensively for decades, yet areas of controversy and confusion remain. In large part, the difficulties in investigating this area relate to the protean physiologic effects of the sympathetic nervous system, which hamper attempts to dissociate its many indirect effects from direct renal influences on tubular sodium reabsorption. That such a direct influence may exist has been suggested by several in vitro studies,[107,108] in which catecholamines were shown to affect active sodium transport directly. However, proving such a direct role in the in vivo circumstance has been more problematic.

Earlier reviews[52,109] summarize previous experiments, in which several techniques of adrenergic manipulation were applied to whole animal studies. These included spinal cord section,[110,111] cardiac denervation,[112,113] interruption of afferent and efferent autonomic pathways,[52,110,114] pharmacologic adrenergic depletion and blockade,[115] hemorrhage and cross-circulation experiments,[116] and renal denervation.[117-119] The results demonstrated that, in the absence of overriding changes in systemic arterial pressure, circumstances with enhanced sympathetic tone were associated with augmented tubular sodium reabsorption. Conversely, states in which renal adrenergic tone is impaired or abolished were associated with an increase in sodium excretion. However, these studies were generally associated with alterations in central blood volume or systemic hemodynamics, with indirect renal effects, any of which could explain the observed differences in sodium handling. Thus, it was concluded then that "in virtually all in vivo experiments demonstrating an effect of alterations in adrenergic tone on sodium excretion, the results do not differentiate between a direct effect on active sodium transport and an indirect effect mediated by some alteration in intrarenal hemodynamics."[109]

Subsequently, DiBona has summarized more recent studies which suggest that there is, indeed, an in vivo effect of catecholamines on sodium excretion, dissociable from concomitant hemodynamic effects.[120] This proposal is supported by anatomical studies demonstrating adrenergic innervation of proximal and distal renal tubules.[120-122]

First, LaGrange and colleagues[123] demonstrated that direct electrical nerve stimulation was associated with a significant decrease in sodium excretion without a change in filtered load of sodium. The measurement of GFR, however, was based on a urineless technique utilizing the Fick principle. In a series of experiments by DiBona and colleagues,[124-126] these results were confirmed and extended. These investigators employed direct stimulation of the left renal nerves of saline-loaded anesthetized dogs. They found that low level direct electrical renal nerve stimulation caused a decrease in urine sodium excretion without altering GFR (inulin clearance) or RBF (*para*-aminohippurate clearance and extraction, and electromagnetic flow meter). Nor was there an associated intrarenal redistribution of blood flow as measured with the microsphere technique. This antinatriuretic effect of renal nerve stimulation was reversible and was prevented by prior alpha-adrenergic receptor blockade. It was not, however, affected by blockade of angiotensin II[127] or by prostaglandin synthesis inhibition.[128] These results therefore suggested an antinatriuretic effect of increased sympathetic neural activity which was not dependent upon either changes in renal hemodynamics or sympathetic-induced increases in angiotensin or prostaglandin activity.

These authors then studied reflex changes in renal tone on sodium excretion by inducing reflex sympathetic nerve stimulation by perfusing the carotid sinus (Figure 17).[129] This experimental maneuver was associated with an antinatriuresis in the absence of measurable changes in renal hemodynamics and was also abolished by pretreatment with adrenergic blocking agents. The demonstration of this antinatriuretic response under circumstances of reflex sympathetic activity suggests that renal adrenergic tone may be an important homeostatic mechanism in sodium regulation. This suggestion is further strengthened by results in which reflex decreases in renal sympathetic nerve activity were induced by left atrial distension and stellate ganglion stimulation.[130] These maneuvers both decreased recorded efferent nervous activity and caused reversible increases in urine flow and sodium excretion in the absence of detectable changes in renal hemodynamics.[130]

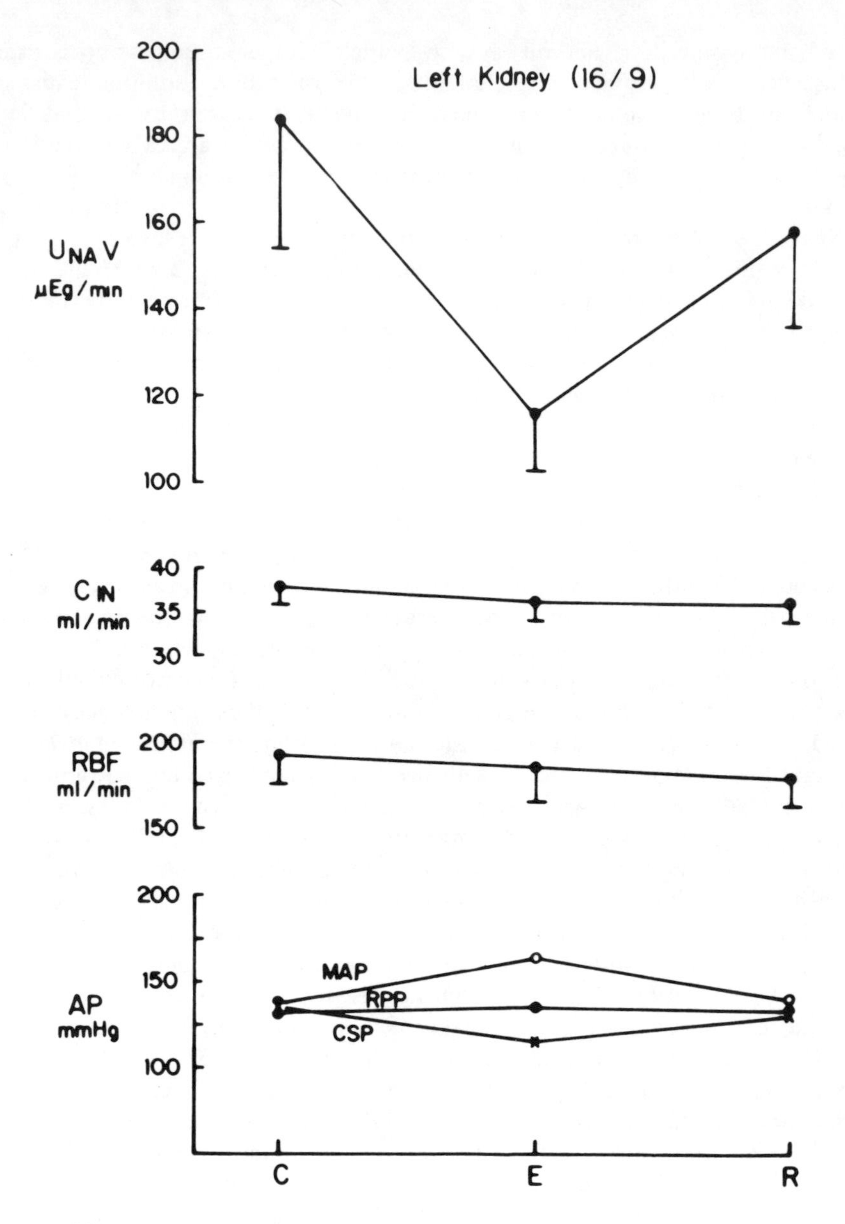

FIGURE 17. Summary of data for carotid sinus perfusion studies. $U_{Na}V$, urinary sodium excretion, µEq/min; C_{In}, inulin clearance, mℓ/min; RBF, renal blood flow, mℓ/min; AP, arterial pressure, mmHg; MAP, mean arterial pressure; RPP, renal perfusion pressure; CSP, carotic sinus pressure; C, control period; E, experimental period; R, recovery period. Data are means ± SE for left kidney representing 16 tests in 9 dogs. (From Zambraski, E. J., DiBona, G. F., and Kaloyanides, G. J., *J. Pharmacol. Exp. Ther.*, 198, 464, 1976. With permission of The Williams & Wilkins Co., Baltimore.

It appears that the site of the tubular effect of adrenergic activity is the proximal nephron. Bello-Reuss and colleagues,[131,132] utilizing clearance and micropuncture techniques in anesthetized rats, observed an ipsilateral increase in urinary sodium excretion without changes in GFR and RBF in denervated as compared to contralateral innervated kidneys. Using micropuncture techniques, proximal tubular salt and water reabsorption was decreased by 40% while SNGFR remain unchanged. Enhanced distal reabsorption of sodium only partially compensated for this decrement in proximal reabsorption, thus accounting for the resultant natriuresis.

This localization of the proximal nephron as the site of effect of changes in adrenergic tone was confirmed by the same group during renal nerve stimulation.[133] In studies on anesthetized, saline-loaded rats in which low-level splanchnic nerve stimulation was applied, the urinary flow rate and sodium excretion decreased, but GFR and RBF did not change. With micropuncture techniques the investigators found an increase in proximal tubular fractional and absolute water reabsorption in the absence of changes in SNGFR. Thus, these micropuncture studies in the rat support a direct effect of renal sympathetic activity on proximal tubular sodium and water reabsorption.

In spite of the emergence of the above results, questions still remain regarding the true physiologic significance of these findings. These doubts persist because of questions about the relevance of acute studies to the chronic state and of studies in anesthetized animals to the awake state. Suffice it to say that some studies on chronic denervation in conscious animals have found no effect on sodium excretion;[120,134] however, in these studies the adequacy or permanence of surgical denervation was not verified. In addition, Lifschitz[135] has found that in awake dogs, renal denervation did not blunt either the antinatriuretic response to hemorrhage or alter the natriuretic response to volume expansion. Smith,[120] in a discussion of experiments performed over three decades ago, then wrote "the phenomenon of 'denervation diuresis' is observable only under the abnormal conditions of his experiment (i.e., anesthesia), which are such as to excite vasoconstriction in the kidney."

It therefore appears that the definition of the true importance of the sympathetic nervous system in regulating sodium excretion awaits further investigation. The clinical implications of the question are vast, for if this neurogenic control mechanism contributes substantially to normal sodium homeostasis, perhaps derangements in the normal sympathetic physiology contribute to the positive sodium balance of the edema-forming states.

H. Hormonal Influences on Renal Sodium Excretion

With the discovery of the mineralocorticoid hormone, aldosterone, it was apparent that hormonal influences could effect the tubular sodium reabsorption independent of changes in GFR. Subsequently, the classic experiments of de Wardener et al.[47] focused attention on factor(s) other than GFR and aldosterone. Though initially this "third factor" was considered to be a humoral factor, it is now recognized that a large number of regulatory influences beyond GFR and aldosterone contribute to the category of "third factor". Thus, blood compositional changes and renal innervation might be considered as contributory. In addition, the entire question of hormonal influences on sodium excretion remains an area of active investigative ferment. This research has examined the effect of known, identified hormones on sodium transport, and, of course, includes the search for a natriuretic hormone. The latter has been an immense, as yet unrequited, effort which has captivated the attention of some of the foremost investigators in renal physiology. The following is a summary of the work with well-established hormones and their effects on sodium excretion and of evidence for the still elusive natriuretic hormone.

1. Nonmineralocorticoid Steroids

The reader is referred to the pertinent reviews of Knox and Diaz-Buxo[136] and Levinsky.[137] As discussed in these reviews, and as will be discussed more fully, it now appears that the importance of the proximal tubule in the final regulation of sodium excretion is limited. Thus, when considering hormones with effects on isolated nephron segments, it should be noted that their impact on total regulation of sodium excretion may be minimal or unclear. The mineralocorticoid hormones were discussed earlier and will not be discussed here.

2. Estrogens and Progesterones

The common clinical occurrence of sodium retention in such estrogenic states as pregnancy and premenstrual edema has suggested an association between estrogens and sodium homeostasis. Though the concomitant increase in progesterone secretion in these states precludes definitive interpretation of the role of estrogens per se, evidence in both animals and man suggests that estrogen produces a modest sodium-retaining effect. Whether this effect derives from alterations in GFR, extrarenal effects on the fluid content of other tissues, or a mineralocorticoid effect which could be either direct or mediated through the renin-angiotensin-aldosterone system is unclear. Though estrogens have been shown to increase hepatic angiotensinogen (substrate) production and thereby to increase plasma renin activity, the absence of an accompanying kaliuresis mitigates against a role of aldosteronism in estrogen-induced sodium retention.[136]

Results regarding the effect of progesterone on sodium excretion come from normal pregnant women[138] and normal males[139] in whom a correlation has been demonstrated between progesterone and aldosterone secretion. The physiologic importance of this relationship is suggested by the observation that pregnant women given mineralocorticoids and ACTH develop sodium retention and suppress endogenous aldosterone production. When the exogenous steroids are discontinued, a natriuresis may be observed.[140] At the renal level, progesterone has been shown to antagonize the antinatriuretic properties of aldosterone.[136]

3. Parathyroid Hormone

Parathyroid hormone (PTH) exerts a regulatory effect on proximal tubular anion transport, namely, by inhibiting phosphate and bicarbonate reabsorption. As a result, PTH secondarily inhibits proximal tubular sodium reabsorption through a mechanism not yet established. Some evidence suggests that PTH inhibits carbonic anhydrase in a manner analogous to acetazolamide.[136] The speculation that this proximal tubular inhibition is mediated through the adenylate cyclase system was strengthened by the demonstration that the infusion of dibutyl cyclic AMP results in inhibition of sodium and phosphate reabsorption by the proximal tubule.[141] Notwithstanding the demonstrable proximal inhibition by PTH, the importance of PTH in modulating final sodium excretion is minimal. This derives from the studies of Agus et al.[141] in which the PTH-induced decrease in proximal tubular reabsorption of sodium was virtually totally compensated for by enhanced reabsorption in more distal nephron segments.

4. Thyroid Hormone

In spite of the clinical observation that myxedematous patients excrete sodium normally, a number of animal experiments imply a possible effect of thyroid hormone on tubular sodium transport. Clearance and micropuncture techniques in hypothyroid rats have demonstrated an abnormally high fractional excretion of sodium under some circumstances, an effect likely due both to proximal and distal tubular dysfunction. Whether these observations relate to a direct effect of hormonal deficiency or to a generalized metabolic defect is unclear.[136]

5. Bradykinin

Bradykinin is but one of several vasodilators (including acetylcholine, dopamine, papaverine, and prostaglandins) which causes a natriuresis when infused into the renal artery. Recent reviews show the kallikrein-kinin system to be a hormonal cascade residing partly in the kidney, which results in a number of potent physiologic effects including vasodilatation.[142,143] Bradykinin is a nonapeptide released from the alpha-2-globulin kininogen by kallikrein, a proteolytic enzyme. It is well established that bradykinin infusion diminishes renal vascular resistance and increases RBF, without exerting a systematic effect on GFR. Though bradykinin has been demonstrated to influence transepithelial sodium and water transport in vitro,[144] it is currently felt that the natriuresis it induces in vivo derives from its vasoactive effect. Though its vasodilatory effect would be expected to increase intrarenal hydrostatic pressure and decrease filtration fraction, and therefore peritubular oncotic pressure, a decrease in proximal tubular reabsorption has not been observed with bradykinin during micropuncture experiments.[145,146] Thus, the increase in sodium excretion must be due to a decrease in distal sodium reabsorption, or to the inhibition of proximal reabsorption in nephrons not accessible to micropuncture. It appears, therefore, that a poorly understood direct distal effect, an effect related to nephron heterogeneity or a secondary effect relating to enhanced medullary blood flow, may account for the augmented sodium excretion observed with bradykinin infusion.[143]

6. Glucagon

A natriuretic effect of glucagon is well documented in animals and man. Whether this results from a direct tubular effect of glucagon, an indirect humoral effect promoted by the hormone, or a hemodynamic effect is not clear. Results exist both to implicate and to negate the observed increases in GFR as a mediator of the natriuresis. Some authors have suggested that impurities in the glucagon preparations were responsible for the discrepant results. While the question remains unresolved, the experiments of Pullman et al.[147] demonstrated a unilateral natriuresis after intrarenal glucagon injection in spite of comparable hemodynamic changes in the contralateral kidney, thus suggesting a direct tubular effect of glucagon to decrease tubular sodium reabsorption.

7. Insulin

Nearly half a century ago, Atchley et al.[148] linked insulin administration in diabetic patients with urinary conservation of sodium and cessation of insulin therapy with an augmented urinary sodium excretion. Subsequent observations confirmed and refined these original findings.[149,150]

A direct effect of insulin on sodium transport was later suggested by in vitro studies on amphibian epithelia and on the isolated perfused dog kidney.[151] More recently, in a series of elegant metabolic studies utilizing clearance techniques in man and micropuncture techniques in dogs, de Fronzo and his associates[151,152] have examined the in vivo role of insulin in regulating sodium excretion. They found that intravenous insulin administration was associated with a marked fall in urinary sodium excretion, in the absence of hemodynamic changes or alterations in filtered load of glucose or plasma aldosterone concentration. Their subjects also increased their free water excretion during the insulin infusion thus implicating enhanced sodium chloride reabsorption in the diluting segment of the distal nephron. The investigators then demonstrated that dogs maintained hyperinsulinemic but euglycemic also demonstrated this antinatriuretic response, and again that the sodium-conserving influence was a distal phenomenon. Though the daily physiologic relevance of these findings is uncertain, the authors note that antinatriuretic effects were sometimes noted at plasma insulin levels within the physiologic range, and at levels comparable to those observed during hyperglycemia. Therefore, that changes in plasma insulin levels may substantially influence renal sodium handling is an intriguing but unproven prospect. It must be remembered, however, that the effects of insulin are vast and complex, and that secondary changes in intermediary metabolism may exert a more immediate influence on sodium handling than insulin itself. The recent demonstration[153] that urinary ketoacid and sodium excretion correlate during nonnutritive maneuvers, and that plasma insulin and urinary sodium excretion did not correlate in that circumstance, reinforces the insecurity appropriate to the interpretation of studies in this area.

8. Angiotensin II

Angiotensin II, a known potent stimulator of aldosterone secretion, has been investigated for an additional, direct tubular effect of the peptide. The results reflect both some dose dependency in the biologic effect and some confusion as to whether the humoral influence is direct or mediated through hemodynamic changes. In any event, a number of in vivo studies, as well as correlative in vitro techniques, have yielded heterogenous results which suggest that angiotensin II may demonstrate either natriuretic or antinatriuretic properties under specific experimental conditions. However, at least one investigational model of pathologic sodium-retaining states is not altered with angiotensin blockade.[154] Thus, the direct pathophysiologic impact of the hormone appears to be minimal.[136]

9. Prostaglandins

Prostaglandins are ubiquitous, naturally occurring unsaturated fatty acids which are found, amongst other places, in the renal medulla and papilla of all species studied. Their protean physiologic effects include vasodilatation, influences on the renin-angiotensin-aldosterone system, and inhibition of vasopressin-mediated water transport in the mammalian collecting duct. The relationships of prostaglandin activity to normal renal function are extremely complex and have been reviewed recently.[155-158] The natriuresis accompanying the administration of some prostaglandins (PGE and PGI) appears to be mediated largely through a nonspecific vasodilatory effect, resulting in decreased proximal and probably distal sodium reabsorption. Whether there exists a direct natriuretic effect, independent of hemodynamic alterations, remains the focus of an ongoing controversy. Results on both sides of the issue are difficult to interpret in view of species differences, differences in methodology, anesthesia, and acute surgery. A recent study by Gagnon and Felipe[159] in which prostaglandin synthetase inhibition caused an antinatriuresis in anesthetized, nonhydrated animals, but not in anesthetized, water-loaded, or unanesthetized dogs, underscores the importance of the investigative setting as a determinant of the results obtained. At this time there is no way to reconcile the contradictory data, and a direct tubular effect of prostaglandins on sodium excretion remains unproven.

10. Vasopressin

Though the primary effect of vasopressin is antidiuretic, it has been shown to be natriuretic in a number of laboratory circumstances.[136] Though recent studies have considered an indirect, hemodynamic effect,[160] earlier work suggested an independence of the natriuretic effect from changes in renal hemodynamics and mineralocorticoid action. Though the precise location of this natriuretic effect is unclear, evidence exists supporting both proximal and distal sites of action.[136] The importance of this phenomenon in normal physiology remains to be proven, however.

11. Natriuretic Hormone

Ever since the classical experiments of de Wardener et al.,[47] extraordinary energies have been expended on the search for a natriuretic hormone. The data supporting the existence of such a hormone have been reviewed extensively,[9,52,137,161-166] so will be summarized briefly here. The evidence to date is mostly indirect, and derives from several experimental settings. These include cross circulation experiments, micropuncture experiments, studies utilizing a variety of in vivo and in vitro assays, and biochemical efforts at isolating the putative hormone.

a. Cross-circulation Studies

The cross-circulation experiments have generally involved volume expansion of a mineralocorticoid treated "donor" animal, with perfusion by the donor's bloodstream

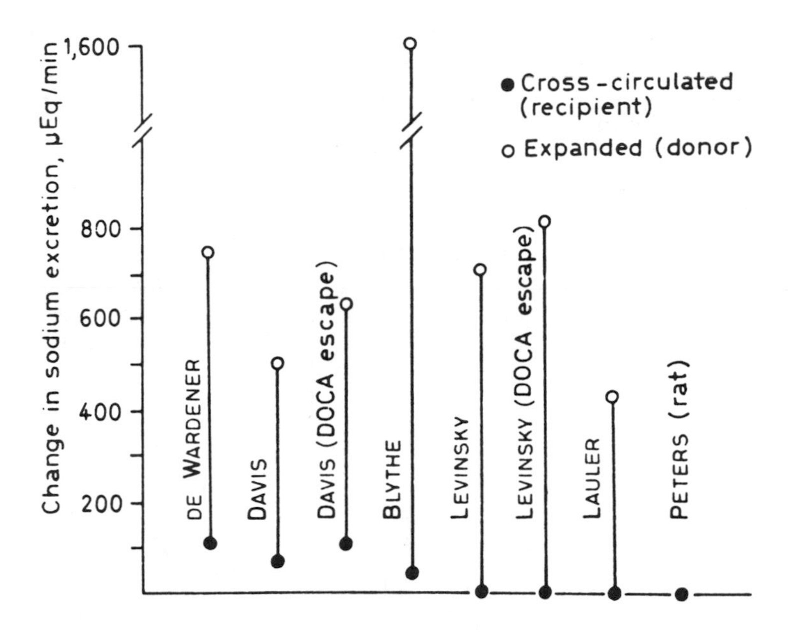

FIGURE 18. Excretion of sodium in cross-circulation studies; expansion of donor with saline. Data (recalculated in some cases) are from the following sources, left to right: *Clin. Sci.*, 21, 250, 196; *Proc. Soc. Exp. Biol. Med.*, 121, 1058, 1966; *Circ. Res.*, 20, 1, 1967; *Circ. Res.*, 28, (Suppl. 2), 21, 1971; unpubl.; *Nephron*, 4, 1, 1967; *Pflugers Arch. Ges. Physiol.*, 318, 21, 1970. (From Levinsky, N. G., *Proc. 5th Int. Cong. Nephrol. Mexico*, 2, 162, 1972. With permission of S. Karger AG, Basel.)

of either a denervated "recipient" animal or an isolated assay kidney. In either case the perfusion pressure could be controlled, renal nerves eliminated, and the "recipient's" blood volume held isovolemic. As the "donor" was mineralocorticoid treated, decrements in aldosterone activity could not explain a natriuresis in the "recipient" animal resulting from the donor's blood volume expansion. In addition, the studies utilizing the infusion of equilibrated blood likely averted artifacts from hemodilution. Thus, this experimental preparation theoretically allows the exclusion of all factors other than a circulating humoral substance in the mediation of the recipient's natriuresis. Levinsky[137] has summarized several of the early studies in this area in Figures 18 and 19. As is apparent, the data from these experiments is contradictory. A number of laboratories have observed no "recipient" natriuresis with donor expansion. However, other investigators have found such a natriuresis. Advocates of the hormone have contended that the "negative" studies are explained by either a short circulating half-life of the putative hormone, or the lack of sustained volume expansion in the donor animal because of its natriuresis. To ensure adequacy of volume expansion some investigators have utilized urine reinfusion or urinary replacement with isotonic fluid.

Perhaps the most convincing of the cross-circulation experiments are the series of studies performed by Kaloyanides and his colleagues. Initially, Kaloyanides and Azer[167] utilized an isolated dog kidney preparation, perfused with the blood of a donor animal which had been pretreated with DOCA and vasopressin. As pictured in Figure 20, their elegant methodology allowed the expansion of the donor animal with equilibrated blood, and the perfusion of the isolated kidney at constant perfusion pressures. The isolated kidney was obviously denervated, and in addition, its renal venous pressure was constant. Thus, during volume expansion of the donor animal, all known compositional, hormonal, physical and neural factors known to influence sodium ex-

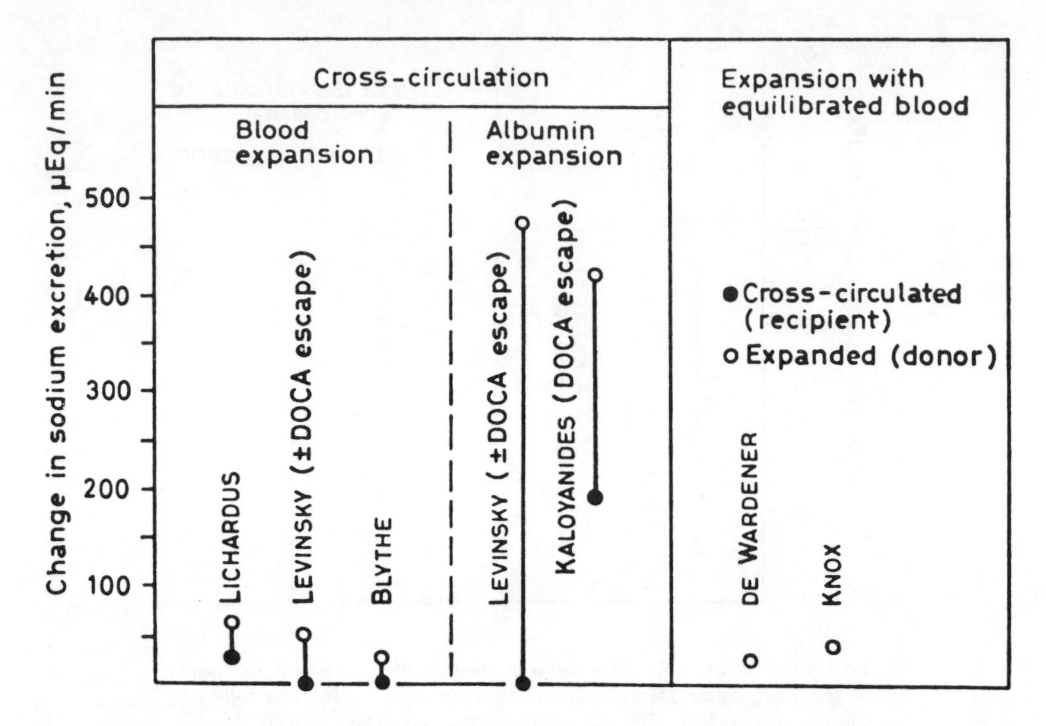

FIGURE 19. Excretion of sodium in cross-circulation or expansion studies; expansion of donor with albumin or blood. Data (recalculated in some cases) are from following sources, left to right: *Nature*, 209, 407, 1966; unpubl.; *Circ. Res.*, 28 (Suppl. 2), 21, 1971; unpubl.; *J. Clin. Invest.*, 50, 1603, 1971; *Clin. Sci.*, 32, 403, 1967; *Am. J. Physiol.*, 215, 1041, 1968. (From Levinsky, N. G., *Proc. 5th Int. Cong. Nephrol. Mexico*, 2, 162, 1972. With permission of S. Karger AG, Basel.)

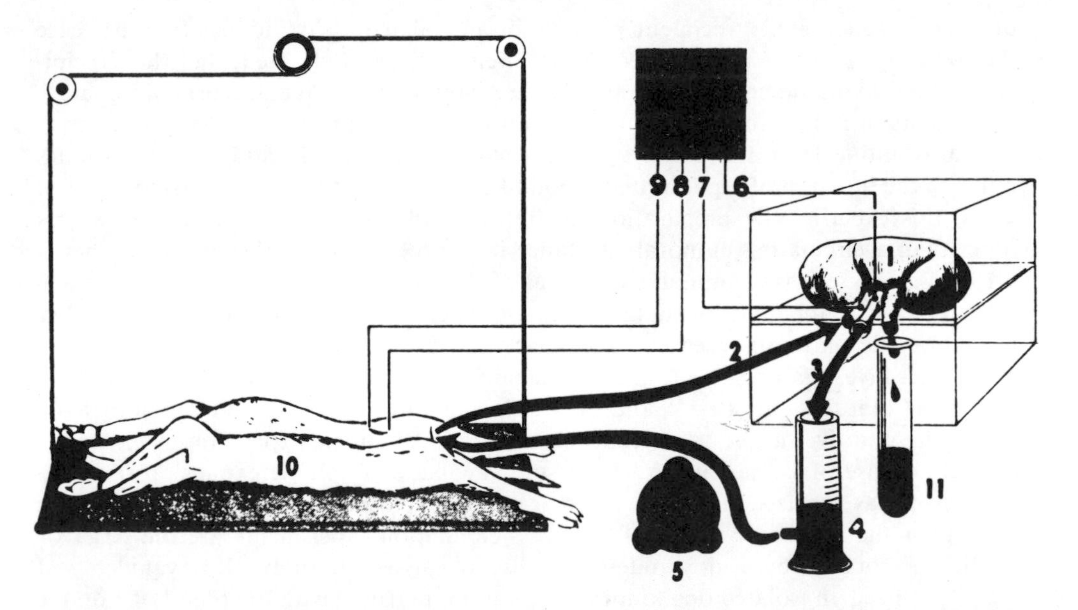

FIGURE 20. Diagram of the isolated kidney preparation. The isolated kidney (1) is placed in a constant temperature-humidity chamber where it is perfused with blood from the femoral artery (2) of a second dog (10). Renal venous blood (3) flows by gravity into a reservoir (4) from which it is pumped (5) to the femoral vein of the perfusion animal. The perfusion animal (10) rests on an adjustable platform and by raising or lowering the platform with respect to the isolated kidney, renal arterial pressure in the isolated kidney can be regulated. Pressure in the femoral artery (9) and vein (8) and renal artery (7) and vein (6) are monitored with pressure transducers. Urine (11) is collected from a catheter secured in the ureter. (From Kaloyanides, G. J. and Azer, M., *J. Clin. Invest.*, 50, 1603, 1971. With permission.)

cretion had been maintained constant by experimental design. Any natriuresis from the perfused assay kidney would therefore have to result from changes in an unidentified hormonal system.

The results of these experiments are summarized in Figure 21. Assay kidneys perfused by control (unexpanded) and expanded animals demonstrated a decline in inulin clearance and RBF with time; however, in only the latter group was there a fall in renal arterial pressure. In spite of these hemodynamic changes, a natriuresis was observed in the isolated perfused kidney after volume expansion of the donor animals, with urinary sodium excretion increasing from 153.6 ± 27.9 (SEM) to 345.5 ± 57.8 μEq/min ($p < 0.001$) and fractional sodium excretion increasing from 3.4 ± 0.6 to 8.1 ± 1.2% ($p < 0.01$). Such a natriuresis was not seen in the kidneys perfused by the time-control, unexpanded animals, thus leading to the conclusion that some function of volume expansion caused the natriuresis. By exclusion it was felt that the responsible variable must be an increase in a natriuretic factor or a decrease in an antinatriuretic factor. The experiments of Pearce and Veress[168] in which plasma extracts from volume-expanded animals were natriuretic in assay animals, provide support for the former postulate.

In subsequent studies, Kaloyanides et al.[169] attempted to localize the source of the natriuretic factor through a series of ablation experiments. They found that the natriuresis observed from an isolated kidney perfused by a donor, volume-expanded dog was unaffected by preparation of the donor with thyroparathyroidectomy, hypophysectomy, or adrenalectomy. These and other ablation experiments[46] argue against the kidney and the adrenal, parathyroid, thyroid, and pituitary glands as the source of the hormone. Recently, however, volume expansion of decapitated dogs failed to result in a natriuretic response in an isolated perfused assay kidney, suggesting that the brain is either the source of natriuretic hormone or a vital link in its physiology.[170] On the basis of theoretical concepts relating to comparative physiology and evolution, as well as extensive prior investigation summarized elsewhere,[161,162] a number of authors agree that the source of the postulated natriuretic hormone is probably the central nervous system.[161,162,171]

b. Micropuncture Studies

As summarized by Knox et al.,[172] early micropuncture studies confirmed the finding of influences on sodium excretion independent of filtration rate, mineralocorticoids, and peritubular physical factors. In these studies, hydropenic animals were volume expanded, and changes in the determinants of peritubular Starling forces and sodium excretion were recorded. Through the use of intrarenal protein infusions and an aortic clamp, the investigators then restored peritubular oncotic and hydrostatic pressure to control levels. With the return of peritubular oncotic force to above control values, the decrement in proximal tubular sodium reabsorption resulting from volume expansion was only partially reversed with intrarenal protein infusion. With the return of net Starling forces to levels equaling those in hydropenic animals, fractional excretion of sodium still exceeded that observed during hydropenia (Figure 22). These changes occurred in the absence of alterations in GFR, SNGFR, and renal plasma flow which could explain the natriuresis. Thus, with constancy of hemodynamics and peritubular physical forces, sodium excretion was augmented by volume expansion. It appeared that sodium reabsorption was inhibited at some distal site in the nephron, and that this site is not responsive to changes in peritubular Starling forces. Though a direct role of renal nerves was not excluded and the possible importance of nephron heterogeneity was not fully explored, these micropuncture studies did implicate some natriuretic mechanisms that could not then be identified. The authors speculated that these mechanisms may be humoral, thus encouraging efforts to better establish the hormone in assay systems, and ultimately, to purify the hormone.

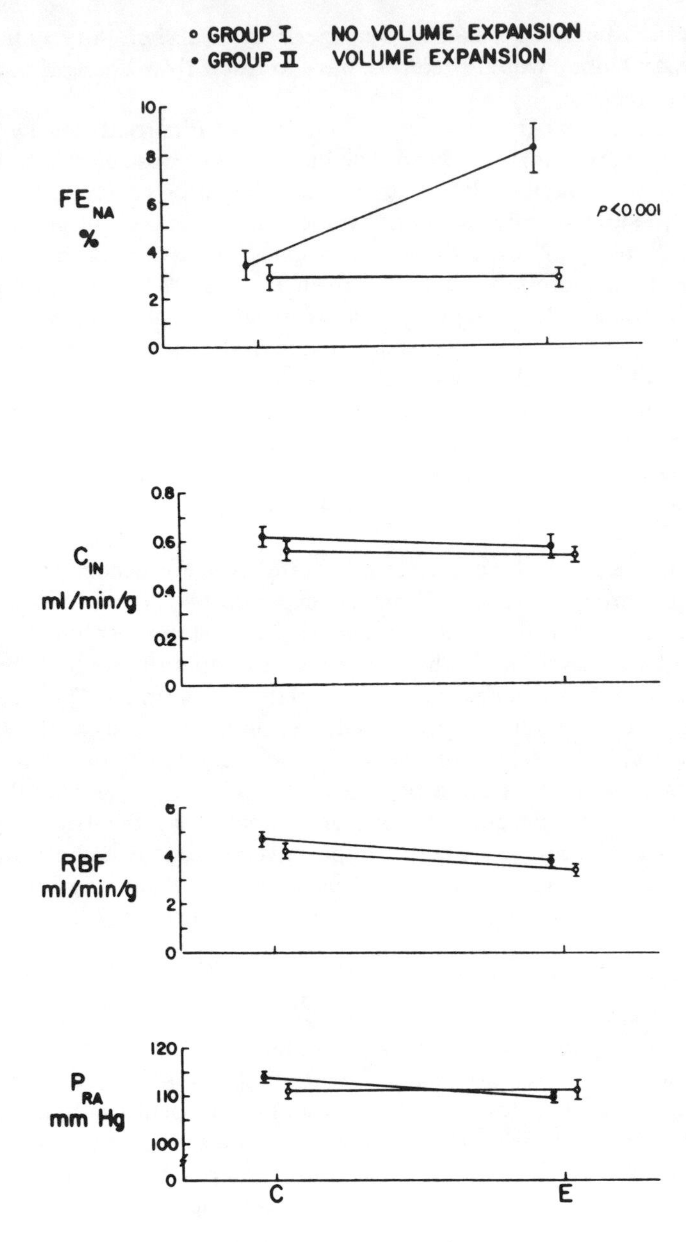

FIGURE 21. Comparison between groups I and II of function in the isolated kidney during the control (C) and experimenta (E) periods. FE_{Na} equals fractional sodium excretion, C_{In} equals insulin clearance factored by kidney weight, RBF equals renal blood flow factored by kidney weight and P_{RA} equals renal arterial pressure. (From Kaloyanides, G. J. and Azer, M., *J. Clin. Invest.*, 50, 1603, 1971. With permission.)

C. Assays for Natriuretic Hormone

In addition to the cross-circulation experiments discussed earlier, a number of other techniques have been employed to assay for the presence of a natriuretic hormone. Generally the substance in question is a plasma or urinary extract from volume-expanded animals, which is then applied to an in vivo or in vitro system. The former has included dogs or rats under widely varying experimental conditions, without stand-

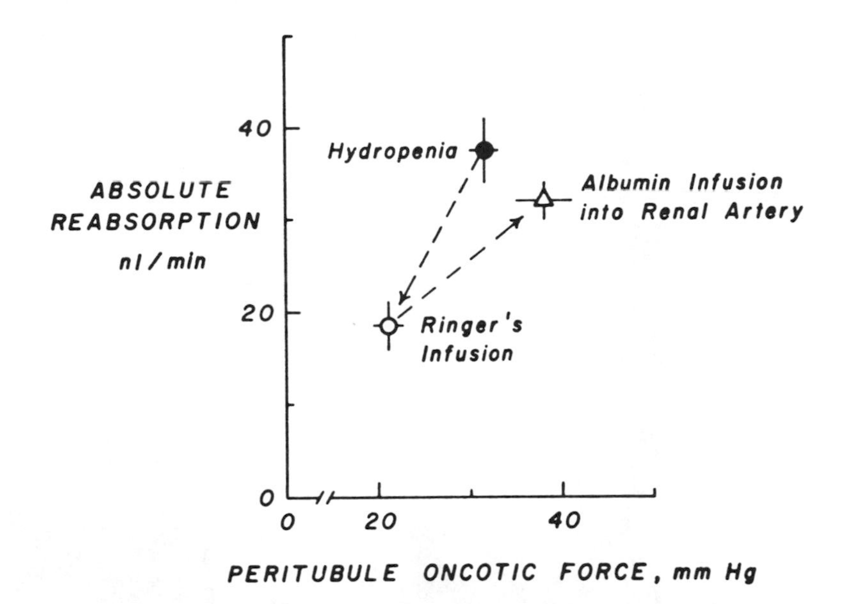

FIGURE 22. The relationship between absolute reabsorption by the proximal tubule and oncotic pressure in the peritubule microcirculation is shown for hydropenia, volume expansion, and protein infusion into the renal artery during continued volume expansion. Note that reabsorption is not restored to control values despite increases in oncotic pressure in the peritubule microcirculation above control values. Results from 6 dogs, 25 tubules. (From Knox, F. G., Schneider, E. G., Strandhoy, J. W., Willis, L. R., Warms, P. C., and Davis, B. B., *Proc. 5th Int. Cong. Nephrol., Mexico,* 2, 151, 1972. With permission of S. Karger AG, Basel.)

ardization of anesthesia, volume state, surgical preparation, or renal function. After injection into in vivo assay animals, increases in absolute urinary sodium excretion were detected; however, many of the studies lacked renal hemodynamic measurements. Hence, it is unclear whether the natriuretic response to the test material is a direct tubular effect or is mediated by changes in renal hemodynamics. The absence of this pertinent information and the heterogeneity of the experimental preparations unfortunately preclude conclusive interpretation of these data.

In an effort to avoid the many nonhormonal mediators of natriuresis in vivo, many investigators have resorted to in vitro assay systems to detect inhibitors of sodium transport in the test material. A number of techniques have been utilized, including the measurement of the short-circuit current (a measure of transepithelial sodium transport) across isolated frog skin[173-175] and toad bladder.[176,177] Other assays have involved measuring the sodium and potassium concentration of isolated kidney tubules or kidney slices, and measuring the renal uptake of PAH, a sodium-dependent process. In these systems, the plasma or urine extracts of volume-expanded animals have variably demonstrated an inhibition of sodium transport. In alluding to these and other in vitro assays, de Wardener has written, "The bewildering variety of these assays has caused a great deal of difficulty in interpretation. The best has still to emerge."[161] Recent results from de Wardener's group and others reported at a satellite symposium of the 26th International Congress of Physiology in Bratislava, Czechoslovakia, on Hormonal Regulation of Sodium Excretion suggest that inhibition of Na-K-ATPase in an in vitro assay system may provide the best assay system for the natriuretic hormone.[196] In spite of the inherent difficulties of all the assay systems, both the in vivo and in vitro methodologies have been instrumental in stimulating attempts to isolate the hormone biochemically. It is clear that in a number of experimental circumstances, plasma and urine extracts from volume expanded subjects or animals have promoted a natriuresis in test animals or have been shown to inhibit in vitro sodium transport.

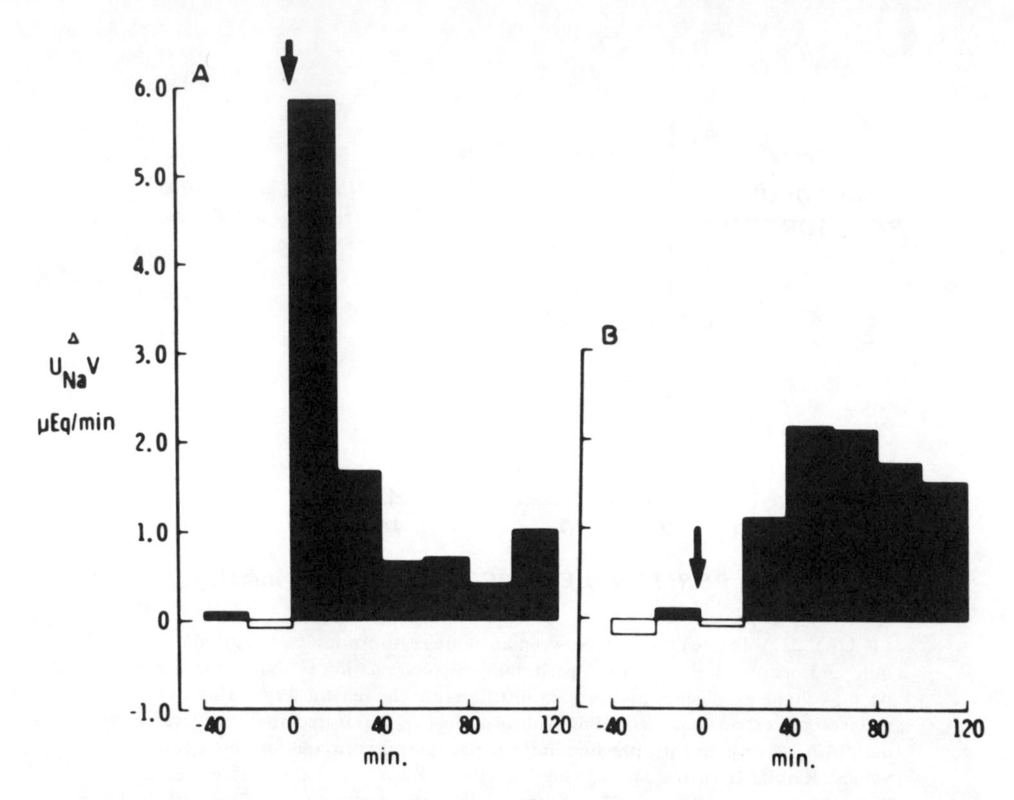

FIGURE 23. Mean change in urinary sodium excretion of A, 22 rats given a small natriuretic substance, and B, 39 rats given a larger natriuretic substance, both substances having been extracted from urine of 9 salt-loaded normal human subjects. (From Clarkson, E. M., Raw, S. M., and de Wardener, H. E., *Kidney. Int.*, 10, 381, 1976. With permission.)

d. Nature of the Hormone

The efforts to demonstrate natriuretic activity in extracts of thoracic duct lymph, plasma, urine, kidney, neurohypophysis, and hypothalamus, and liver have been summarized.[161] A detailed consideration of the extraction techniques is beyond the scope of this chapter, but the current state of knowledge can be summarized. Two natriuretic substances have been detected in the urine of man and dog. One has a molecular weight between 10,000 and 30,000 daltons, is obtained through ultrafiltration or Sephadex G-50® column chromatography, and produces a natriuresis 1 hr after bolus injection which persists for over 2 hr.[9] The other natriuretic substance may have a molecular weight of less than 500 daltons and produces a more prompt natriuresis of shorter duration. The different natriuretic patterns are illustrated in Figure 23. Recently, Clarkson et al.[178] demonstrated that the natriuretic response to the smaller substance may be sustained when the substance is given by continuous infusion, but is very short-lived as a bolus. After considering a number of physiochemical characteristics, the authors suggest that the natriuretic substance is a very small peptide, which may coincide with a material obtained by Dr. Bricker's group. The latter has been proposed as the agent mediating the increases in fractional excretion of sodium seen in chronic renal failure.[9,159]

Thus, while the hormone has not yet been isolated or characterized, progress has been made. Until such time as the hormone is purified and established, nagging doubts about its very existence will persist. Any data confirming the existence of natriuretic hormone will have to avoid rediscovering already well-established hormones or other natriuretic mechanisms. Though these efforts are painstaking, the rewards, should they be successful, will be of the greatest physiologic (and likely clinical) importance.

IV. QUALITATIVE AND QUANTITATIVE ANALYSIS OF RENAL SEGMENTAL SODIUM TRANSPORT

Having considered the large number of factors which regulate in vivo sodium excretion, it is now appropriate to discuss the transport characteristics of different nephron sites. A detailed examination of transmembrane and transepithelial transport per se is beyond the scope of this chapter, and the interested reader is referred to recent pertinent reviews.[96,179-183] The following summarizes briefly our current understanding of segmental sodium reabsorption along the nephron.

Sodium reabsorption occurs in four major sites: the proximal tubule, loop of Henle, distal tubule, and collecting duct. Broadly speaking, the total tubular reabsorption of sodium is composed of two serial components. First, reabsorption of large amounts of filtered sodium must occur to maintain the constancy of total body sodium. This bulk sodium reabsorption appears to occur primarily in the proximal tubule and the loop of Henle. Second, a fine regulation of sodium reabsorption is necessary to maintain precise sodium balance; this process requires adjustment in reabsorptive capacity of the relatively small amount of sodium delivered to the distal nephron. Substantial evidence now underscores the importance of the distal tubule and collecting duct in the final modulation of urinary sodium excretion.

A. Proximal Tubule

The proximal tubule normally reabsorbs between 50% and 75% of the filtered sodium load. It does so isosmotically, as the tubular fluid to plasma osmolar and sodium concentration ratios remain at unity.[184] Several observations establish that at least part of the reabsorptive process is active. First, in the early portion of the proximal tubule, where there exists a lumen-negative potential difference, sodium transport occurs against an electrical gradient.[185-187] Second, in situations where the intraluminal sodium concentration is lowered by the addition of such substances as mannitol, sodium transport occurs against a concentration gradient.[188] Finally, inhibitors of Na-K-ATPase inhibit proximal tubular sodium reabsorption by approximately 35%.[189]

Sodium reabsorption in the proximal tubule involves two steps:

1. Entry of sodium from tubular fluid into tubular epithelial cells
2. Transport of sodium from within the cell into peritubular capillaries.[185,187]

While the latter is an active step, at least partly mediated by Na-K-ATPase, the former is the rate-limiting step. This passive entry of sodium into the cells is facilitated by an electrochemical gradient, as the intracellular sodium concentration is lower and the intracellular voltage more negative than in the luminal fluid. Though passive, this entry process does not occur by simple diffusion. Instead, the sodium transport in the early proximal nephron is linked to the reabsorption of bicarbonate, glucose, amino acids, and other organic substrates, through an electrogenic co-transport of organic solute and sodium ions. In this process, movement of the positively charged sodium-solute complex across the luminal cell membrane leads to a 2-mV, lumen negative, transepithelial potential difference in the early proximal tubule.

The nature of sodium reabsorption then changes in the late proximal tubule. Glucose, amino acids, and organic substrates are predominantly reabsorbed in the early proximal tubule. In addition, bicarbonate is preferentially reabsorbed over chloride, probably secondary to hydrogen ion secretion. The resultant high luminal chloride concentration in the late proximal tubule then promotes diffusion of chloride down its concentration gradient and establishes a lumen positive potential difference. This intraluminal positivity enables the passive transport of sodium, which, according to

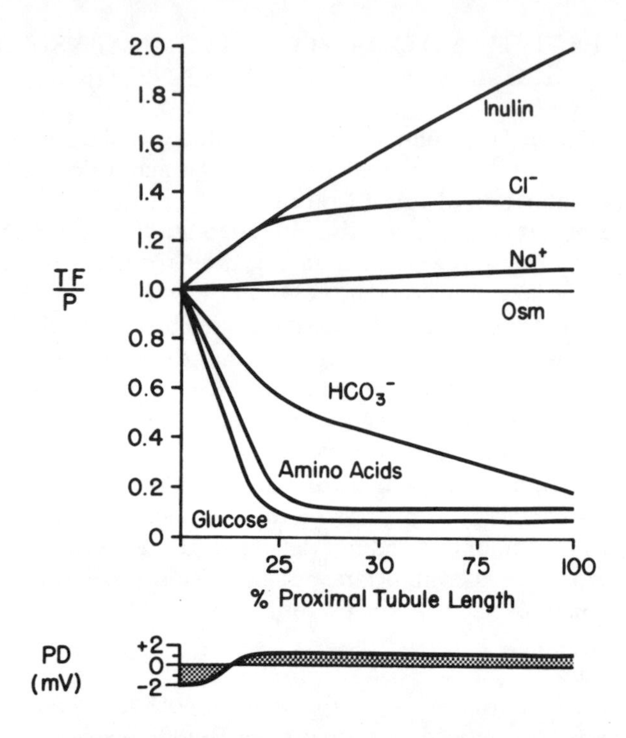

FIGURE 24. Reabsorption of various solutes in the proximal tubule in relationship to proximal tubular length. A change in proximal tubular potential difference (PD) from negative to positive occurs early in the proximal tubule. TF/P = ratio of tubular fluid to plasma concentration. (From Ganong, W. F., *Review of Medical Physiology,* 9th ed., Lange, Los Altos, Calif., 1979, 547. With permission of F. C. Rector.)

some metabolic and thermodynamic considerations, may account for up to two-thirds of the salt and water reabsorption in the proximal tubule.[32] In addition, active sodium transport has been demonstrated in the late proximal tubule as well as in the pars recta.[185,190] Proximal tubular solute transport is depicted in Figure 24 in relationship to length and potential difference along that nephron segment.

Electrophysiologic studies have demonstrated the proximal tubular epithelia to be "leaky", with a low resistance at intercellular shunt sites. These shunt pathways consist of a tight junction connecting juxtapositioned epithelial cells at the luminal side, a lateral intercellular space between the cells, and the basement membrane. As discussed earlier, it is currently thought that sodium is extruded from the cell into the intercellular space, establishing an osmotic gradient which attracts water. This results in an increased intercellular hydrostatic pressure, favoring egress from the intercellular space into the peritubular capillary. Alternatively, under some conditions, a "back-leak" reabsorbate into the tubular lumen may occur. Thus, net proximal tubular fluid reabsorption is likely determined by the rate of active reabsorption and a back-leak into the lumen. It is unclear to what extent this reabsorption is governed by changes in active transport, blood compositional factors, renal hemodynamics, adrenergic nerve activity, natriuretic hormones, or a combination thereof. It does appear, however, that aldosterone does not exert any physiologically significant effect on this segment.[46]

Though the relative importance of these regulatory factors in modulating proximal reabsorption remains to be established, it is known that the proximal tubule does respond to changes in ECFV, increasing wth ECFV depletion and decreasing with ECFV expansion. In addition, in some animal models of pathophysiologic sodium retention, enhanced proximal reabsorption has been demonstrated.[1] However, as substantial data now suggests that proximal alterations alone have little or no effect on sodium excretion,[136,137,191] more distal nephron sites have been implicated in both physiologic sodium homeostasis and abnormal states of sodium retention.

B. Loop of Henle

As the loop of Henle is inaccessible to micropuncture, our current understanding of its transport behavior derives largely from in vitro microperfusion studies on isolated rabbit tubules. The descending limb of Henle's loop has been found to be virtually impermeable to sodium. The thin ascending limb, however, is permeable to sodium and is a site where passive sodium transport may occur down a concentration gradient for sodium between tubular lumen and interstitium.[1]

The bulk of sodium reabsorption in Henle's loop, accounting for roughly 25% of the filtered load of sodium, occurs in the thick ascending limb. A unique transport process in the thick ascending limb actively pumps chloride out of the tubular lumen, generating a lumen-positive potential difference. Thus, sodium transport can proceed passively down an electrochemical gradient.[1,32] This active chloride pump is inhibitable with furosemide and ethacrynic acid, which explains the potent natriuretic effect of these agents.

Absolute sodium reabsorption in the thick ascending limb varies directly with delivered sodium load but is not affected by the ECFV per se.[1,192] Thus, at this time there is no convincing evidence that sodium reabsorption in the loop of Henle, at least of superficial nephrons, is inhibited. Recent studies by Stein and colleagues[193,194] also imply a potentially significant role for nephron heterogeneity, with changes in sodium reabsorption in the inner cortical nephrons contributing to the natriuresis of volume expansion.[193,194] These investigators demonstrated that under conditions of volume expansion in rats, sodium delivery was greater to the base of the papillary collecting duct than to the late distal tubule of superficial nephrons.[193] This addition of sodium between the late distal tubule and papillary base was attributed to a greater inhibition of sodium reabsorption in deep cortical nephrons during volume expansion.

In an elegant series of subsequent studies these investigators demonstrated that fractional sodium delivery from the proximal nephron to the loop of Henle was comparable between superficial and deep nephrons.[194] In addition, furosemide, but not chlorothiazide, given during Ringer loading, reversed the pattern of sodium addition between late distal tubule and papillary base. These data were interpreted as localizing the greater inhibition of sodium transport in deep nephrons to the loop of Henle. The authors propose a model, depicted in Figure 25, which incorporates the concept of passive sodium reabsorption from the thin ascending limb along a sodium concentration gradient, which exists between the thin ascending limb and ascending vasa recta during hydropenia.[194,195] Volume expansion with Ringers solution markedly decreases the hypertonicity of the medullary interstitium, and may therefore abolish this concentration gradient. Thus, during volume expansion, the passive reabsorption occurring along a concentration gradient in the thin ascending limb of deep nephrons during hydropenia is abolished. By contrast, superficial nephrons in the rat have only thick ascending limbs, where transport is active and not impaired by volume expansion.

C. Distal Convoluted Tubule

The distal convoluted tubule begins at the macula densa and ends at the cortical collecting duct. It is a segment with a high electrical resistance, a property which allows the maintenance of a high transepithelial sodium concentration gradient. Electrical measurements have demonstrated a lumen negative potential difference, the magnitude of which increases along the length of the distal convoluted tubule. The intraluminal

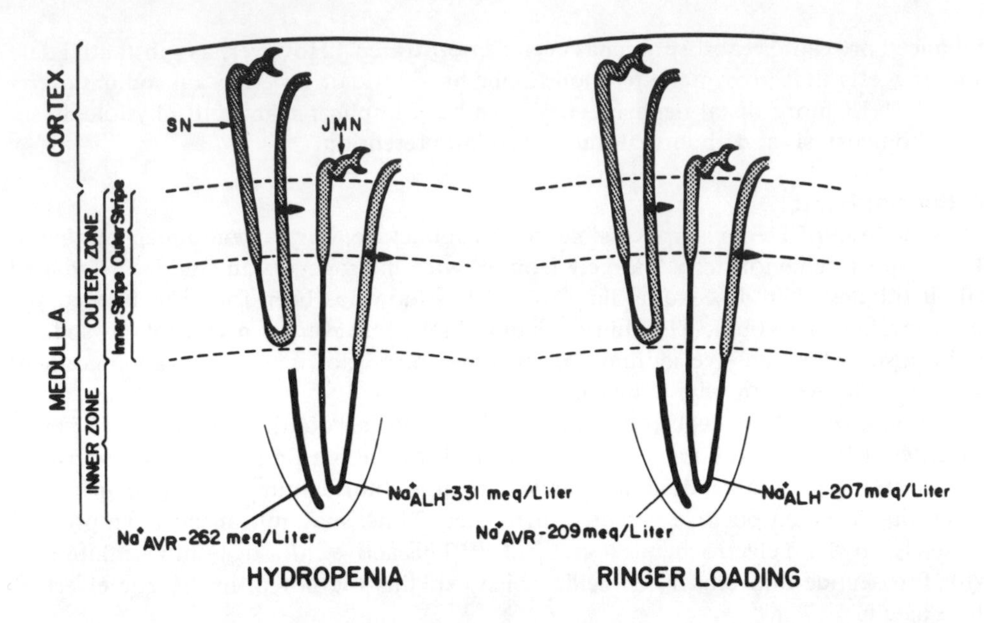

FIGURE 25. Proposed model to explain heterogeneity of Na^+ transport during Ringer loading. → Denotes active salt transport in thick ascending link of both SN and JMN. In hydropenia (left panel), a sodium concentration gradient exists between thin ascending limb (ALH) and ascending vasa recta (AVH), 331 vs. 262 mEq/ℓ. During Ringer loading, however, this gradient is abolished. Thus, sodium transport would be markedly decreased in the thin ascending limb of JMN, because transport in this segment presumably occurs totally by passive means. (From Osgood, R. W., Reineck, H. J., and Stein, J. H., *J. Clin. Invest.*, 62, 311, 1978. With permission.)

sodium concentration is less than that in the surrounding plasma, and falls progressively along the segment's length. Thus, sodium transport is active and occurs against an electrochemical gradient. Studies of distal convoluted tubular transport during volume expansion have demonstrated an increased absolute sodium reabsorption but a decrease in fractional sodium reabsorption. In vivo microperfusion studies have demonstrated a direct correlation between perfusion rate and sodium reabsorption in the distal convoluted tubule, a relationship not altered by volume expansion.[1]

D. Collecting Duct

Based on physiologic differences regarding sodium reabsorption, the collecting duct can be divided into two segments: the cortical collecting duct and the medullary-papillary collecting duct. The cortical collecting duct is a segment of high electrical resistance with little "back-leak" of sodium. Sodium reabsorption there appears to be an active, electrogenic transport process on the basis of several observations. First, a lumen-negative potential difference has been observed, and the segment has been demonstrated to establish and maintain a large sodium concentration gradient. The transport process is dependent upon the presence of luminal sodium and is inhibited by ouabain. Of note is the finding that the lumen-negative potential difference exists only in the presence of mineralocorticoids, suggesting that this hormonal influence is necessary for active sodium transport in the cortical collecting duct.[1]

The medullary-papillary collecting duct is also capable of active electrogenic sodium transport. The factors governing terminal collecting duct sodium transport are not clear, however, and the response of the segment to changes in volume status has been difficult to discern because of the probable existence of nephron heterogeneity. It has been suggested that the natriuretic hormone(s) may exert their primary effect by inhibiting sodium reabsorption in this nephron segment.[9] Although the regulatory influences of sodium reabsorption in the terminal collecting duct await further elucidation, it is appealing to postulate that, by virtue of its strategic location, this segment is a major determinant of final urinary sodium concentration and excretion.

V. SUMMARY

This discussion has surveyed the current understanding of renal sodium handling. The importance of sodium in the maintenance of ECFV has been considered. The numerous mechanisms governing sodium reabsorption and excretion have been discussed, and the nephron sites of the regulatory process have been reviewed briefly. As a result of rigorous and vigorous investigative scrutiny, a greater appreciation of the immense yet precise machinery of renal sodium regulation exists, but nevertheless is still incomplete. It is hoped that as the normal physiology is further clarified, our ability to intervene in pathophysiologic circumstances of sodium retention will be suitably enhanced.

REFERENCES

1. **Reineck, H. J. and Stein, J. H.,** Renal regulation of sodium balance, in *Clinical Disorders of Fluid and Electrolyte Metabolism,* 3rd ed., Maxwell, M. and Kleeman, C., Eds., McGraw-Hill, New York, 1980, 89.
2. **McCance, R. A.,** Experimental sodium chloride deficiency in man, *Proc. R. Soc. London, Ser. B,* 119, 245, 1936.
3. **Strauss, M. B., Lamdin, E., Smith, W. P., and Bleifer, D. J.,** Surfeit and deficit of sodium, *Arch. Intern. Med.,* 102, 527, 1958.
4. **Sadowski, J. and Wocial, B.,** Renin release and autoregulation of blood flow in a new model of nonfiltering nontransporting kidney, *J. Physiol. (London),* 266, 219, 1977.
5. **Morgan, T. and Gillies, A.,** Factors controlling the release of renin, *Pfluegers Arch.,* 368, 13, 1977.
6. **Barajas, L.,** The ultrastructure of the juxtaglomerular apparatus as disclosed by three-dimensional reconstruction from serial sections: the anatomical relationship between the tubular and vascular components, *J. Ultrastruct. Res.,* 33, 116, 1970.
7. **Barajas, L.,** Renin secretion: an anatomical basis for tubular control, *Science,* 172, 485, 1971.
8. **Blendstrup, K., Leyssac, P. P., Paulsen, K., and Skinner, S. L.,** Characteristics of renin release from isolated superfused glomeruli in vitro, *J. Physiol. (London),* 246, 653, 1975.
9. **de Wardener, H. E.,** The control of sodium excretion, *Am. J. Physiol.,* 235, F163, 1978.
10. **Vander, A. J. and Miller, R.,** Control of renin secretion in the unanesthetized dog, *Am. J. Physiol.,* 207, 537, 1964.
11. **Landwehr, D. M., Schnerman, J., Klose, R. M., and Giebisch, G.,** Effect of reduction in filtration rate on renal tubular sodium and water reabsorption, *Am. J. Physiol.,* 215, 687, 1968.
12. **Glabman, S., Aynedjian, H. S., and Bank, N.,** Micropuncture study of the effect of acute reductions in glomerular filtration rate on sodium and water reabsorption by the proximal tubules of the rat, *J. Clin. Invest.,* 44, 1410, 1965.
13. **Bartoli, E. and Earley, L. E.,** Effects of saline infusion on glomerulotubular balance, *Kidney Int.,* 1, 67, 1972.
14. **Daugharty, T. M., Zweig, S. M., and Earley, L. E.,** Assessment of renal hemodynamic factors in whole kidney glomerulotubular balance, *Am. J. Physiol.,* 220, 2021, 1971.
15. **Schrier, R. W. and Humphreys, M. H.,** Role of distal reabsorption and peritubular environment in glomerulotubular balance, *Am. J. Physiol.,* 222, 379, 1972.
16. **Buentig, W. E. and Earley, L. E.,** Demonstration of independent roles of proximal tubular reabsorption and intratubular load in the phenomenon of glomerulotubular balance during aortic constriction in the rat, *J. Clin. Invest.,* 50, 77, 1971.
17. **Burg, M. B. and Orloff, J.,** Control of fluid absorption in the renal proximal tubule, *J. Clin. Invest.,* 47, 2016, 1968.

18. **Morel, F. and Murayama, Y.**, Simultaneous measurement of unidirectional and net sodium fluxes in microperfused rat proximal tubules, *Pfluegers Arch.*, 320, 1, 1970.
19. **Morgan, T. and Berliner, R. W.**, In vivo perfusion of proximal tubules of the rat, Glomerulotubular balance, *Am. J. Physiol.*, 217, 992, 1969.
20. **Radtke, H. W., Rumrich, G., Kloss, S., and Ullrich, K. J.**, Influence of luminal diameter and flow velocity on the isotonic fluid absorption and ^{36}Cl permeability of the proximal convolution of the rat kidney, *Pfluegers Arch.*, 324, 288, 1971.
21. **Wiederholt, M., Hierholzer, K., Windhager, E. E., and Giebisch, G.**, Microperfusion study of fluid reabsorption in proximal tubules of rat kidneys, *Am. J. Physiol.*, 213, 809, 1967.
22. **Kelman, R. B.**, A theoretical note on exponential flow in the proximal part of the mammalian nephron, *Bull. Math. Biophys.*, 24, 303, 1962.
23. **Gertz, K. H., Mangos, J. A., Braun, G., and Pagel, H. D.**, On the glomerular tubular balance in the rat kidney, *Pfluegers Arch.*, 285, 360, 1965.
24. **Leyssac, P. P.**, The in vivo effect of angiotensin and noradrenaline of the proximal tubular reabsorption of salt in the mammalian kidneys, *Acta Physiol. Scand.*, 64, 167, 1965.
25. **Lowitz, H. D., Stumpe, K. O., and Ochwadt, B.**, Micropuncture study of the action of angiotensin II on tubular sodium and water reabsorption in the rat, *Nephron*, 6, 173, 1969.
26. **Johnson, M. D. and Malvin, R. L.**, Stimulation of renal sodium reabsorption by angiotensin II, *Am. J. Physiol.*, 232, F298, 1977.
27. **Steiner, R. W., Tucker, B. J., and Blantz, R. C.**, Glomerular hemodynamics in rats with chronic sodium depletion. Effect of saralasin, *J. Clin. Invest.*, 64, 503, 1979.
28. **Bartoli, E. and Earley, L. E.**, Importance of ultrafilterable plasma factors in maintaining tubular reabsorption, *Kidney Int.*, 3, 142, 1973.
29. **Bartoli, E., Conger, J. D., and Earley, L. E.**, Effect of intraluminal flow on proximal tubular reabsorption, *J. Clin. Invest.*, 52, 843, 1973.
30. **Frazier, H. S.**, Renal regulation of sodium balance, *N. Engl. J. Med.*, 279, 868, 1968.
31. **Earley, L. E. and Daugharty, T. M.**, Sodium metabolism, *N. Engl. J. Med.*, 281, 72, 1969.
32. **Klahr, S. and Slatopolsky, E.**, Renal regulation of sodium excretion. Function in health and in edema-forming states, *Arch. Intern. Med.*, 131, 780, 1973.
33. **Mills, I. H.**, Renal regulation of sodium excretion, *Ann. Rev. Med.*, 21, 75, 1970.
34. **Davis, B. B., Jr. and Knox, F. G.**, Current concepts of the regulation of urinary sodium excretion — a review, *Am. J. Med. Sci.*, 259, 373, 1970.
35. **Williams, G. H. and Dluhy, R. G.**, Aldosterone biosynthesis: interrelationships of regulatory factors, *Am. J. Med.*, 53, 595, 1972.
36. **Al-Awqati, Q.**, H^+ transport in urinary epithelia, *Am. J. Physiol.*, 235, F77, 1978.
37. **Feldman, D., Funder, J. W., and Edelman, I. S.**, Subcellular mechanisms in the action of adrenal steroids, *Am. J. Med.*, 53, 545, 1972.
38. **Ludens, J. H. and Fanestil, D. D.**, Aldosterone stimulation of acification of urine by isolated urinary bladder of the Columbian toad, *Am. J. Physiol.*, 226, 1321, 1974.
39. **Al-Awqati, Q., Norby, L. H., Mueller, A., and Stinmetz, P. R.**, Characteristics of stimulation of H^+ transport by aldosterone in turtle urinary bladder, *J. Clin. Invest.*, 58, 351, 1976.
40. **Al-Awqati, Q.**, Effect of aldosterone on the coupling between H^+ transport and glucose oxidation, *J. Clin. Invest.*, 60, 1240, 1977.
41. **Lifschitz, M. D., Schrier, R. W., and Edelman, I. S.**, Effect of actinomycin D on aldosterone-mediated changes in electrolyte excretion, *Am. J. Physiol.*, 224, 376, 1973.
42. **Wright, F. S., Knox, F. G., Howards, S. S., and Berliner, R. W.**, Reduced sodium reabsorption by the proximal tubule of DOCA-escaped dogs, *Am. J. Physiol.*, 216, 869, 1969.
43. **Haas, J. A., Berndt, T. J., Youngberg, S. P., and Knox, F. G.**, Collecting duct sodium reabsorption in deoxycorticosterone-treated rats, *J. Clin. Invest.*, 63, 211, 1979.
44. **Sonnenberg, H.**, Proximal and distal tubular function in salt-deprived and in salt-loaded deoxycorticosterone acetate-escaped rats, *J. Clin. Invest.*, 52, 263, 1973.
45. **Sonnenberg, H.**, Collecting duct function in deoxycorticosterone acetate-escaped, normal, and salt-deprived rats. Response to hypervolemia, *Circ. Res.*, 39, 282, 1976.
46. **Knox, F. G., Burnett, J. C., Kohan, D. E., Spielman, W. S., and Strand, J. C.**, Escape from the sodium-retaining effects of mineralocorticoids, *Kidney Int.*, 17, 263, 1980.
47. **de Wardener, H. E., Mills, I. H., Clapham, W. F., and Hayter, C. J.**, Studies on efferent mechanism of sodium diuresis which follows administration of intravenous saline in the dog, *Clin. Sci.*, 21, 249, 1961.
48. **Earley, L. E. and Friedler, R. M.**, Observations on the mechanism of decreased tubular reabsorption of sodium and water during saline loading, *J. Clin. Invest.*, 43, 1928, 1964.
49. **Goldberg, M., Staum, B. B., and Cristol, J. L.**, Separation of humoral from hemodynamic factors affecting proximal sodium reabsorption during blood volume expansion, *J. Clin. Invest.*, 47, 41a, 1968.

50. **Levinsky, N. G. and Lalone, R. C.,** The mechanism of sodium diuresis after saline infusion in the dog, *J. Clin. Invest.,* 42, 1261, 1963.
51. **Rector, F. C., Giesen, G. V., Kiil, F., and Seldin, D. W.,** Influence of expansion of extracellular volume on tubular reabsorption of sodium independent of changes in glomerular filtration rate and aldosterone activity, *J. Clin. Invest.,* 43, 341, 1964.
52. **Schrier, R. W. and de Wardener, H. E.,** Tubular reabsorption of sodium ion: influence of factors other than aldosterone and glomerular filtration rate, *N. Engl. J. Med.,* 285, 1231, 1971.
53. **Earley, L. E. and Schrier, R. W.,** Intrarenal control of sodium excretion by hemodynamic and physical factors, in *Handbook of Physiology,* Orloff, J., Berliner, R. W., and Geiger, S. R., Eds., Williams & Wilkins, Baltimore, 1973, 721.
54. **Bresler, E. H.,** Ludwig's theory of tubular reabsorption: the role of physical factors in tubular reabsorption, *Kidney Int.,* 9, 313, 1976.
55. **Starling, E. H. and Verney, E. B.,** The secretion of urine as studied on the isolated kidney, *Proc. R. Soc. London, Ser. B,* 97, 321, 1925.
56. **Craig, G. M., Mills, I. H., Osbaldiston, G. W., and Wise, B. L.,** The effect of change of perfusion pressure and hematocrit in the perfused isolated dog kidney, *J. Physiol. (London),* 186, 113P, 1966.
57. **Nizet, A.,** Influence of serum albumin and dextran on sodium and water excretion by the isolated dog kidney, *Pfluegers Arch.,* 301, 7, 1968.
58. **Nizet, A., Godon, J. P., and Mahieu, P.,** Quantitative excretion of water and sodium load by isolated dog kidney: autonomous renal response to blood dilution factors, *Pfluegers Arch.,* 304, 30, 1968.
59. **Martino, J. A. and Earley, L. E.,** Demonstration of a role of physical factors as determinants of the natriuretic response to volume expansion, *J. Clin. Invest.,* 46, 1963, 1967.
60. **Daugharty, T. M., Belleau, L. J., Martino, J. A., and Earley, L. E.,** Interrelationship of physical factors affecting sodium reabsorption in the dog, *Am. J. Physiol.,* 215, 1442, 1968.
61. **Earley, L. E., Martino, J. A., and Friedler, R. M.,** Factors affecting sodium reabsorption as determined during blockade of distal sodium reabsorption, *J. Clin. Invest.,* 45, 1668, 1966.
62. **Lewy, J. E. and Windhager, E. E.,** Peritubular control of proximal tubular fluid reabsorption in the rat kidney, *Am. J. Physiol.,* 214, 943, 1968.
63. **Spitzer, A. and Windhager, E. E.,** Effect of peritubular oncotic pressure changes on proximal tubular fluid reabsorption, *Am. J. Physiol.,* 218, 1188, 1970.
64. **Brenner, B. M., Falchuk, K. H., Keimowitz, R. I., and Berliner, R. W.,** The relationship between peritubular capillary protein concentration and fluid reabsorption by the renal proximal tubule, *J. Clin. Invest.,* 48, 1519, 1969.
65. **Windhager, E. E., Lewy, J. E., and Spitzer, A.,** Intrarenal control of proximal tubular reabsorption of sodium and water, *Nephron,* 6, 247, 1969.
66. **Brenner, B. M. and Troy, J. L.,** Postglomerular vascular protein concentration: evidence for a causal role in governing fluid reabsorption and glomerulotubular balance by the renal proximal tubule, *J. Clin. Invest.,* 50, 336, 1971.
67. **Falchuk, K. H., Brenner, B. M., Tadokoro, M., and Berliner, R. W.,** Oncotic and hydrostatic pressures in peritubular capillaries and fluid reabsorption by the proximal tubule, *Am. J. Physiol.,* 220, 1427, 1971.
68. **Brenner, B. M., Troy, J. L., and Daugharty, T. M.,** On the mechanism of inhibition in fluid reabsorption by the renal proximal tubule of the volume expanded rat, *J. Clin. Invest.,* 50, 1596, 1971.
69. **Brenner, B. M., Troy, J. L., and Daugharty, T. M.,** Quantitative importance of changes in postglomerular colloid osmotic pressure in mediating glomerulotubular balance in the rat, *J. Clin. Invest.,* 52, 190, 1973.
70. **Brenner, B. M. and Galla, J. H.,** Influence of postglomerular hematocrit and protein concentration on rat nephron fluid transfer, *Am. J. Physiol.,* 220, 148, 1971.
71. **Rumrich, G. and Ullrich, K. J.,** The minimum requirements for the maintenance of sodium chloride reabsorption in the proximal convolution of the mammalian kidney, *J. Physiol., London,* 197, 69, 1968.
72. **Holzgreve, H. and Schrier, R. W.,** Evaluation of peritubular capillary microperfusion method by morphological and functional studies, *Europ. J. Physiol.,* 356, 59, 1975.
73. **Holzgreve, H. and Schrier, R. W.,** Variation of proximal tubular reabsorptive capacity by volume expansion and aortic constriction during constancy of peritubular capillary protein concentration in rat kidney, *Europ. J. Physiol.,* 356, 73, 1975.
74. **Conger, J. D., Bartoli, E., and Earley, L. E.,** A study 'in vivo' of peritubular oncotic pressure and proximal tubular reabsorption in the rat, *Clin. Sci. Mol. Med.,* 51, 379, 1976.
75. **Ott, C. E., Haas, J. A., Cuche, J. L., and Knox, F. G.,** Effect of increased peritubule protein concentration on proximal tubule reabsorption in the presence and absence of extracellular volume expansion, *J. Clin. Invest.,* 55, 612, 1975.
76. **Blake, W. D., Wegria, H. P., Ward, H. P., and Frank, C. W.,** Effect of renal arterial concentration on excretion of sodium and water, *Am. J. Physiol.,* 163, 422, 1950.

77. **Handley, C. A. and Moyer, J. H.**, Unilateral renal adrenergic blockade and the renal response to vasopressor agents and to hemorrhage, *J. Pharmacol. Exp. Ther.*, 112, 1, 1954.
78. **Earley, L. E. and Friedler, R. M.**, The effects of combined renal vasodilatation and pressor agents on renal hemodynamics and the tubular reabsorption of sodium, *J. Clin. Invest.*, 45, 542, 1966.
79. **Martino, J. A. and Earley, L. E.**, Relationship between intrarenal hydrostatic pressure and hemodynamically induced changes in sodium excretion, *Circ. Res.*, 23, 371, 1968.
80. **Hall, P. W. and Selkurt, E. E.**, Effects of partial graded venous obstruction on electrolyte clearance by the dog's kidney, *Am. J. Physiol.*, 164, 143, 1951.
81. **Rodicio, J., Herrera-Acosta, J., Sellman, J. C., Rector, F. C., Jr., and Seldin, D. W.**, Studies on glomerulotubular balance during aortic constriction, ureteral obstruction, and venous occlusion in hydropenic and saline-loaded rats, *Nephron*, 6, 437, 1969.
82. **Wathen, R. L. and Selkurt, E. E.**, Intrarenal regulatory factors of salt excretion during renal venous pressure elevation, *Am. J. Physiol.*, 216, 1517, 1969.
83. **Blake, W. D., Wegria, R., Keating, R. P., and Ward, H. P.**, Effect of increased renal venous pressure on renal function, *Am. J. Physiol.*, 157, 1, 1949.
84. **Schrier, R. W. and Humphreys, M. H.**, Factors involved in the antinatriuretic effects of acute constriction of the thoracic inferior and abdominal vena cava, *Circ. Res.*, 29, 479, 1971.
85. **Bahlmann, J., MacDonald, S. J., Dunningham, J. G., and de Wardener, H. E.**, The effect on urinary sodium excretion of altering the packed cell volume with albumen solutions without changing the blood volume in the dog, *Clin. Sci.*, 32, 395, 1967.
86. **Nashat, F. S. and Portal, R. W.**, The effects of changes in hematocrit on renal function, *J. Physiol. (London)*, 193, 513, 1967.
87. **Nashat, F. S., Scholifield, F. R., Tappin, J. W., and Wilcox, C. S.**, The effect of acute changes in haematocrit in the anesthetized dog on the volume and character of the urine, *J. Physiol. (London)*, 205, 305, 1969.
88. **Schrier, R. W. and Earley, L. E.**, Effects of hematocrit on renal hemodynamics and sodium excretion in hydropenic and volume-expanded dogs, *J. Clin. Invest.*, 49, 1656, 1970.
89. **Knox, F. G., Howards, S. S., Wright, F. S., Davis, B. B., and Berliner, R. W.**, Effect of dilution and expansion of blood volume on proximal sodium reabsorption, *Am. J. Physiol.*, 215, 1041, 1968.
90. **Clapp, J. R.**, Factors effecting proximal sodium reabsorption in response to acute volume contraction, *Clin. Res.*, 18(Abstr.), 61A, 1970.
91. **Schrier, R. W., McDonald, K. M., Wells, R. E., and Lauler, D. P.**, Influence of hematocrit and colloid on whole blood viscosity during volume expansion., *Am. J. Physiol.*, 218, 346, 1970.
92. **Kamm, D. E. and Levinsky, N. G.**, Inhibition of renal tubular sodium reabsorption by hypernatremia, *J. Clin. Invest.*, 44, 1144, 1965.
93. **Goldsmith, C., Rector, F. C., Jr., and Seldin, D. W.**, Evidence for a direct effect of serum sodium concentration on sodium reabsorption, *J. Clin. Invest.*, 41, 850, 1962.
94. **Blythe, W. B. and Welt, L. G.**, Plasma sodium concentrations and urinary sodium excretion, *Trans. Assoc. Am. Physicians*, 78, 90, 1965.
95. **Davis, B. B., Knox, F. G., Wright, F. S., and Howards, S. S.**, Effect of expansion of extracellular fluid volume on proximal sodium reabsorption in hyponatremic dogs, *Metabolism*, 19, 291, 1970.
96. **Windhager, E. E. and Giebisch, G.**, Proximal sodium and fluid transport, *Kidney Int.*, 9, 121, 1976.
97. **Earley, L. E., Humphreys, M. H., and Bartoli, E.**, Capillary circulation as a regulator of sodium reabsorption and excretion, *Circ. Res.*, 31, II-1, 1972.
98. **Green, R., Windhager, E. E. and Giebisch, G.**, Protein oncotic pressure effects on proximal tubular fluid movement in the rat, *Am. J. Physiol.*, 226, 265, 1974.
99. **Willis, L. R., Ludens, J. H., and Williamson, H. E.**, Dependence of the natriuretic action of acetylcholine upon an increase in renal blood flow, *Proc. Soc. Exp. Biol. Med.*, 128, 1069, 1968.
100. **Willis, L. R., Ludens, J. H., Hook, J. B., and Williamson, H. E.**, Mechanism of natriuretic action of bradykinin, *Am. J. Physiol.*, 217, 1, 1969.
101. **Pomeranz, B. H., Britch, A. G., and Barger, A. C.**, Neural control of intrarenal blood flow, *Am. J. Physiol.*, 215, 1067, 1968.
102. **Selkurt, E. E., Womack, I., and Dailey, W. N.**, Mechanism of natriuresis and diuresis during elevated renal arterial pressure, *Am. J. Physiol.*, 209, 95, 1965.
103. **Earley, L. E. and Friedler, R. M.**, Changes in renal blood flow and possibly the intrarenal distribution of blood during natriuresis accompanying saline loading in the dog, *J. Clin. Invest.*, 44, 929, 1965.
104. **Earley, L. E. and Friedler, R. M.**, Studies on the mechanism of natriuresis accompanying increased renal blood flow and its role in the renal response to extracellular volume expansion, *J. Clin. Invest.*, 44, 1857, 1965.
105. **Lamiere, N. H., Lifschitz, M. D., and Stein, J. H.**, Heterogeneity of nephron function, *Ann. Rev. Physiol.*, 39, 159, 1977.
106. **Barger, A. C.**, Renal hemodynamic factors in congestive heart failure, *Ann. N.Y. Acad. Sci.*, 139, 276, 1961.

107. **Handler, J. S., Bensinger, R., and Orloff, J.,** Effect of adrenergic agents on toad bladder response to ADH, 3′,5′-AMP, and theophylline, *Am. J. Physiol.,* 215, 1024, 1968.
108. **Field, M. and McCall, I.,** Ion transport in rabbit ileomucosa. III. Effects of catecholamines, *Am. J. Physiol.,* 225, 852, 1975.
109. **Schrier, R. W.,** Effects of adrenergic nervous system and catecholamines on systemic and renal hemodynamics, sodium and water excretion and renin secretion, *Kidney Int.,* 6, 291, 1974.
110. **McDonald, K. M., Rosenthal, A., Schrier, R. W., Galicich, J., Lauler, D. P.,** Effect of interruption of neural pathways on renal response to volume expansion, *Am. J. Physiol.,* 218, 510, 1970.
111. **Pearce, J. W. and Sonnenberg, H.,** Effects of spinal section and renal denervation and the renal response to blood volume expansion, *Canad. J. Physiol. Pharmacol.,* 43, 211, 1965.
112. **Gilmore, J. P. and Daggett, W. M.,** Response of chronic cardiac denervated dog to acute volume expansion, *Am. J. Physiol.,* 210, 509, 1966.
113. **Knox, F. G., Davis, B. B., and Berliner, R. W.,** Effect of chronic cardiac denervation on renal response to saline infusion, *Am. J. Physiol.,* 213, 174, 1967.
114. **Schrier, R. W., McDonald, K. M., Jagger, P. I., Lauler, D. P.,** The role of the adrenergic nervous system in the renal response to acute extracellular fluid volume expansion, *Proc. Soc. Exp. Biol. Med.,* 125, 1157, 1967.
115. **Barger, A. C., Muldowney, F. P., and Liebowitz, M. R.,** Role of the kidney in the pathogenesis of congestive heart failure, *Circulation,* 20, 273, 1959.
116. **Gill, J. R. and Casper, A. G. T.,** Role of the sympathetic nervous system in the renal response to hemorrhage, *J. Clin. Invest.,* 48, 915, 1969.
117. **Kamm, D. E. and Levinsky, N. G.,** The mechanism of denervation natriuresis, *J. Clin. Invest.,* 44, 93, 1965.
118. **Kaplan, S. A. and Rapoport, S.,** Urinary excretion of sodium and chloride after splanchnicotomy: effect on the proximal tubule, *Am. J. Physiol.,* 164, 175, 1951.
119. **Bonjour, J. P., Churchill, P. C., and Malvin, R. L.,** Change of tubular reabsorption of sodium and water after renal denervation in the dog, *J. Physiol, (London),* 204, 571, 1969.
120. **DiBona, G. F.,** Neurogenic regulation of renal tubular sodium reabsorption, *Am. J. Physiol.,* 233, F73, 1977.
121. **Barajas, L. and Muller, J.,** The innervation of the juxtaglomerular apparatus and surrounding tubules: a quantitative analysis by serial section electron microscopy, *J. Ultrastruct. Res.,* 43, 107, 1973.
122. **Muller, J. and Barajas, L.,** Electron microscopic and histochemical evidence for a tubular innervation in the renal cortex of the monkey, *J. Ultrastruct. Res.,* 41, 533, 1972.
123. **LaGrange, R. G., Sloop, C. H., and Schmid, H. E.,** Selective stimulation of renal nerves in the anesthetized dog, *Circ. Res.,* 33, 704, 1973.
124. **Slick, G. L., Aguilera, A. J., Zambraski, E. J., DiBona, G. F., and Kaloyanides, G. J.,** Renal neuroadrenergic transmission, *Am. J. Physiol.,* 229, 60, 1975.
125. **Slick, G. L., DiBona, G. F., and Kaloyanides, G. J.,** Renal sympathetic nerve activity in sodium retention of acute caval constriction, *Am. J. Physiol.,* 226, 925, 1974.
126. **Zambraski, E. J., DiBona, G. F., and Kaloyanides, G. J.,** Specificity of neural effect on renal tubular sodium reabsorption, *Proc. Soc. Exp. Biol. Med.,* 151, 543, 1976.
127. **Zambraski, E. J. and DiBona, G. F.,** Angiotensin II in antinatriuresis of low level renal nerve stimulation, *Am. J. Physiol.,* 231, 1105, 1976.
128. **DiBona, G. F., Zambraski, E. J., Aguilera, A. J., and Kaloyanides, G. J.,** Neurogenic control of renal tubular sodium reabsorption in the dog, *Circ. Res.,* 40, Suppl. I, I-127, 1977.
129. **Zambraski, E. J., DiBona, G. F., and Kaloyanides, G. J.,** Effect of sympathetic blocking agents on the antinatriuresis of reflex renal nerve stimulation, *J. Pharmacol. Exp. Ther.,* 198, 464, 1976.
130. **Prosnitz, E. H. and DiBona, G. F.,** Effect of decreased renal sympathetic nerve activity on renal tubular sodium reabsorption, *Am. J. Physiol.,* 235, F557, 1978.
131. **Bello-Reuss, E., Colindres, R. E., Pastoriza-Munoz, E., Mueller, R. A., and Gottschalk, C. W.,** Effects of acute unilateral denervation in the rat, *J. Clin. Invest.,* 56, 208, 1975.
132. **Bello-Reuss, E., Pastoriza-Munoz, E., and Colindres, R. E.,** Acute unilateral renal denervation in rats with extracellular volume expansion, *Am. J. Physiol.,* 232, F26, 1977.
133. **Bello-Reuss, E., Trevino, D. L., and Gottschalk, C. W.,** Effect of renal sympathetic nerve stimulation on proximal water and sodium reabsorption, *J. Clin. Invest.,* 57, 1104, 1976.
134. **Berne, R. M.,** Hemodynamics and sodium excretion of denervated kidney in anesthetized and unanesthetized dog, *Am. J. Physiol.,* 171, 148, 1952.
135. **Lifschitz, M. D.,** Lack of a role for the renal nerves in renal sodium reabsorption in conscious dogs, *Clin. Sci. Mol. Med.,* 54, 567, 1978.
136. **Knox, F. G. and Diaz-Buxo, J. A.,** The hormonal control of sodium excretion, in *International Review of Physiology. Endocrine Phys. II,* Vol. 16, McCann, S. M., Ed., University Park Press, Baltimore, 1977, 173.

137. **Levinsky, N. G.,** Natriuretic hormones, in *Advances in Metabolic Disorders,* Vol. 7, Levine, R. and Luft, R., Eds., Academic Press, New York, 1974, 37.
138. **Jones, K. M., Lloyd-Jones, R., Riondel, A., Tait, J. F., Tait, S. A. S., Bulbrook, R. D., and Greenwood, F. C.,** Aldosterone secretion and metabolism in normal men and women and in pregnancy, *Acta Endocrinol.,* 30, 321, 1959.
139. **Laidlaw, J. C., Ruse, J. L., and Gornall, A. G.,** The influence of estrogen and progesterone on aldosterone excretion, *J. Clin. Endocrinol. Metab.,* 22, 161, 1962.
140. **Ehrlich, E. N. and Lindheimer, M.,** Effect of administered mineralocorticoids or ACTH in pregnant women: attenuation of kaliuretic influence of mineralocorticoids during pregnancy, *J. Clin. Invest.,* 51, 1301, 1972.
141. **Agus, Z. S., Puschett, J. B., Senesky, D., and Goldberg, M.,** Mode of action of parathyroid hormone and cyclic 3′,5′-monophosphate on renal tubular phosphate reabsorption in the dog, *J. Clin. Invest.,* 50, 617, 1971.
142. Editorial, Kinins and blood pressure, *Lancet,* 1, 663, 1978.
143. **Levinsky, N. G.,** The renal kallikrein-kinin system, *Circ. Res.,* 44, 441, 1979.
144. **Crocker, A. D. and Willavays, P.,** Effect of bradykinin on transepithelial transfer of sodium and water in vitro, *J. Physiol.,* 253, 401, 1975.
145. **Stein, J. H., Ferris, T. F., Huprich, J. E., Smith, T. C., and Osgood, R. W.,** Effect of renal vasodilatation on the distribution of cortical blood flow in the kidney of the dog, *J. Clin. Invest.,* 50, 1429, 1971.
146. **Schneider, E. G., Strandhoy, J. W., Willis, L. R., and Knox, F. G.,** Relationship between proximal sodium reabsorption and excretion of calcium, magnesium, and phosphate, *Kidney Int.,* 4, 369, 1973.
147. **Pullman, T. N., Lavender, A. R., and Aho, I.,** Direct effects of glucagon on renal hemodynamics and excretion of inorganic ions, *Metabolism,* 16, 358, 1967.
148. **Atchley, D. W., Loeb, R. F., Richards, D. W., Benedict, E. M., and Driscoll, M. E.,** On diabetic acidosis. A detailed study of electrolyte balances following the withdrawal and reestablishment of insulin therapy, *J. Clin. Invest.,* 12, 297, 1933.
149. **Miller, J. H. and Bagdonoff, M. D.,** Antidiuresis associated with administration of insulin, *J. Appl. Physiol.,* 6, 509, 1954.
150. **Sandek, C. D., Boulter, P. R., Knopp, R. H., and Acky, R. A.,** Sodium retention accompanying insulin treatment of diabetes mellitus, *Diabetes,* 23, 240, 1974.
151. **De Fronzo, R. A., Cooke, C. R., Andres, R., Faloona, G. R., and Savis, P. J.,** The effect of insulin on renal handling of sodium, potassium, calcium, and phosphate in man, *J. Clin. Invest.,* 55, 845, 1975.
152. **De Fronzo, R. A., Goldberg, M., and Agus, Z. S.,** The effects of glucose and insulin on renal electrolyte transport, *J. Clin. Invest.,* 58, 83, 1976.
153. **Davidson, M. B. and Erlbaum, A. I.,** Role of ketogenesis in urinary sodium excretion: elucidation by nicotinic acid administration during fasting, *J. Clin. Endocrinol. Metab.,* 49, 818, 1979.
154. **Spielman, W. S., Davis, J. O., Freeman, R. H., and Lobmeier, T. R. E.,** Systemic and intrarenal arterioral action of angiotension II in dogs with experimental high output heart failure, *Physiologist,* 17, 335, 1974.
155. **Anderson, R. J., Berl, T., McDonald, K. M., and Schrier, R. W.,** Prostaglandins: effects on blood pressure, renal blood flow, sodium, and water excretion, *Kidney Int.,* 10, 205, 1976.
156. **Dunn, M. J. and Hood, V. L.,** Prostaglandins and the kidney, *Am. J. Physiol.,* 233, 169, 1977.
157. **Zins, G. R.,** Renal prostaglandins, *Am. J. Med.,* 58, 14, 1975.
158. **Horton, R. and Zipser, R.,** Prostaglandins: renin release and renal function, *Contr. Nephrol.,* 14, 87, 1978.
159. **Gagnon, J. A. and Felipe, I.,** Effect of prostaglandin synthetase inhibition on renal sodium excretion in the water-loaded conscious and anesthetized dog, *Mineral Electrolyte Metab.,* 2, 293, 1979.
160. **Smith, M. J., Cowley, A. W., Guyton, A. C., and Manning, R. D.,** Acute and chronic effects of vasopressin on blood pressure, electrolytes, and fluid volumes, *Am. J. Physiol.,* 237, F232, 1979.
161. **de Wardener, H. E.,** Natriuretic hormone, *Clin. Sci. Mol. Med.,* 53, 1, 1977.
162. **Klahr, S. and Rodriguez, H. J.,** Natriuretic hormone, *Nephron,* 15, 387, 1975.
163. Editorial, *Lancet,* 1, 537, 1977.
164. **Clarkson, E. M.,** General evidence for the natriuretic hormone, *Proc. 5th Int. Cong. Nephrol.,* 2, 146, 1972.
165. **Levinsky, N. G.,** A critical appraisal of some of the evidence for a natriuretic hormone, *Proc. 5th Int. Cong. Nephrol.,* 2, 162, 1972.
166. **Bricker, N. S., Schmidt, R. W., Favre, H., Fine, L., and Bourgoignie, J. J.,** On the biology of sodium excretion: the search for a natriuretic hormone, *Yale J. Biol. Med.,* 48, 293, 1975.
167. **Kaloyanides, G. J. and Azer, M.,** Evidence for a humoral mechanism in volume expansion natriuresis, *J. Clin. Invest.,* 50, 1603, 1971.

168. **Pearce, J. W. and Veress, A. T.**, Concentration and bioassay of a natriuretic factor in plasma of volume expanded rats, *Canad. J. Physiol. Pharmacol.*, 53, 742, 1975.
169. **Kaloyanides, G. J., Cohen, L., and DiBona, G. F.**, Failure of selected endocrine organ ablation to modify the natriuresis of blood volume expansion in the dog, *Clin. Sci. Mol. Med.*, 52, 351, 1977.
170. **Kaloyanides, G. J., Balabanian, M. B., and Bowman, R. L.**, Evidence that the brain participates in the humoral natriuretic mechanism of blood volume expansion in the dog, *J. Clin. Invest.*, 62, 1288, 1978.
171. **Lee, J. and de Wardener, H. E.**, Neurosecretion and sodium excretion, *Kidney Int.*, 6, 323, 1974.
172. **Knox, F. G., Schneider, E. G., Strandhoy, J. W., Willis, L. R., Warms, P. C., and Davis, B. B.**, Micropuncture evidence for natriuretic hormone, *Proc. 5th Int. Cong. Nephrol.*, 2, 151, 1972.
173. **Cort, J. H., Dousa, T., Pliska, V., Lichardus, B., Safarova, J., Vranesic, M., and Rudinger, J.**, Saluretic activity of blood during carotid occlusion in the cat, *Am. J. Physiol.*, 215, 921, 1968.
174. **Kramer, H. J., Gostodinov, B., and Kruck, F.**, Humorale Hemmund des epithelialen Natriurectransports nach akutr Expansion des Extracellularvolemens, *Klin. Wochenschr.*, 52, 801, 1974.
175. **Nutbourne, D. M., Howse, J. D., Schrier, R. W., Talner, L. B., Ventom, M. G., Verroust, P. J., and de Wardener, H. E.**, The effect of expanding the blood volume of a dog on the short circuit current across an isolated frog skin incorporated in the dog's circulation, *Clin. Sci.*, 38, 629, 1970.
176. **Buckalew, V. M., Martinez, F. J., and Green, W. E.**, The effect of dialysates and ultrafiltrates of plasma of saline-loaded dogs on toad bladder sodium transport, *J. Clin. Invest.*, 49, 926, 1970.
177. **Buckalew, V. M. and Nelson, D. B.**, Natriuretic and sodium transport inhibitory activity in plasma of volume expanded dogs, *Kidney Int.*, 5, 12, 1974.
178. **Clarkson, E. M., Raw, S. M., and de Wardener, H. E.**, Further observations on a low-molecular-weight natriuretic substance in the urine of normal man, *Kidney Int.*, 16, 710, 1979.
179. **Sweadner, K. J. and Goldin, S. M.**, Active transport of sodium and potassium ions: mechanism, function, and regulation, *N. Engl. J. Med.*, 302, 777, 1980.
180. **Hays, R. M.**, Principles of ion and water transport in the kidney, *Hosp. Prac.*, Sept., 79, 1978.
181. **Sachs, G.**, Ion pumps in the renal tubule, *Am. J. Physiol.*, 233, F359, 1977.
182. **Kinne, R.**, Membrane molecular aspects of tubular transport. Kidney and urinary tract physiology, *Int. Rev. Physiol.*, II, 169, 1976.
183. **Whittembury, G. and Grantham, J. J.**, Cellular aspects of renal sodium transport and cell volume regulation, *Kidney Int.*, 9, 103, 1976.
184. **Malnic, G., Klose, R., and Giebisch, G.**, Micropuncture study of distal tubular potassium, and sodium transport in the rat nephron, *Am. J. Physiol.*, 211, 529, 1966.
185. **Burg, M. B.**, The renal handling of sodium chloride, in *The Kidney*, Brenner, B. M. and Rector, F. C., Jr., Eds., W. B. Saunders, Philadelphia, 1976, 272.
186. **Burg, M. B. and Grantham, J. J.**, *Ion Movement in Renal Tubules: Membranes in Ion Transport*, Vol. 3, Wiley Interscience, New York, 1970, 49.
187. **Kinne, R.**, Current concepts of renal proximal tubular function, *Contr. Nephrol.*, 14, 14, 1978.
188. **Windhager, E. E. and Giebisch, G.**, Micropuncture study of renal tubular transfer of sodium cloride in the rat., *Am. J. Physiol.*, 200, 581, 1961.
189. **Gyory, A. Z., Bundel, V., and Kinne, R.**, Effect of cardiac glycosides and sodium ethacrynate on transepithelial sodium transport in in vivo micropuncture experiments and on isolated plasma membranes sodium-potassium-ATPase in vitro of the rat, *Pfluegers Arch.*, 335, 287, 1972.
190. **Schaffer, J. A., Patlak, C. S., and Andreoli, T. E.**, A component of fluid absorption linked to passive ion flow in the superficial pars recta, *J. Gen. Physiol.*, 66, 445, 1975.
191. **Stein, J. H. and Reineck, H. J.**, Effect of alterations in extracellular fluid volume on segmental sodium transport, *Physiol. Rev.*, 55, 128, 1975.
192. **Stein, J. H., Kirschenbaum, M. A., Bay, W. H., Osgood, R. W., and Ferris, T. F.**, Role of the collecting duct in the regulation of sodium balance, *Circ. Res.*, Suppl. I, 36, 37, I-119, 1975.
193. **Stein, J. H., Osgood, R. W., and Kunau, R. T.**, Direct measurement of papillary collecting duct sodium transport in the rat., *J. Clin. Invest.*, 58, 767, 1976.
194. **Osgood, R. W., Reineck, H. J., and Stein, J. H.**, Further studies on segmental sodium transport in the rat kidney during expansion of the extracellular fluid volume, *J. Clin. Invest.*, 62, 311, 1978.
195. **Johnston, P. A., Battilana, C. A., Lacy, F., and Jamison, R.**, Evidence for a concentration gradient favoring outward movement of sodium from the thin loop of Henle, *J. Clin. Invest.*, 59, 234, 1977.
196. **Alaghband-Zadeh, J., Fenton, S., Clarkson, E., MacGregor, G. A., and de Wardener, H. E.**, Preliminary studies for the possible detection in plasma of the small molecular weight urinary natriuretic material, in *Hormonal Regulation of Sodium Excretion*, Lichardus, B., Schrier, R. W., and Ponec, J., Eds., Elsevier North-Holland, 1980, 341.

Sodium Excess

Edema

Chapter 3

EDEMA — AN INTRODUCTORY OVERVIEW

Solomon Papper

TABLE OF CONTENTS

I. DEFINITION

Edema is an increase in the volume of interstitial fluid. The latter is the extravascular portion of the extracellular compartment, formed as an ultrafiltrate of plasma. Plasma volume may or may not be increased in patients with edema.

II. PATHOPHYSIOLOGY

Normally there is a balance of the Starling forces to maintain the distribution of fluid between the capillary (i.e., vascular) and the interstitial compartments (see Figure 1). These forces include hydrostatic pressure, colloid osmotic pressure (i.e., oncotic pressure), tissue turgor pressure, and lymphatic flow. The net forces favor ultrafiltration at the arteriole end of the capillary and reabsorption at the venous side.

Interstitial fluid volume may increase (i.e., edema may form) if there is increased formation of interstitial fluid or its removal is impaired. Increased formation of interstitial fluid implies one or more of the following events: increased capillary hydrostatic pressure, increased capillary permeability, or a longer time for ultrafiltration. Decreased removal of interstitial fluid is favored when one or more of the following occurs: diminished oncotic pressure, reduced tissue turgor pressure, and impaired lymphatic flow.

Whatever the etiology of edema, it is ultimately manifested by alterations in the Starling forces at the capillary level such that there is increased formation of interstitial fluid, or its removal is diminished, or both.

If the physiological stage is set for an increase in the volume of interstitial fluid, then the amount of sodium available is a major determinant of the magnitude of edema that is formed.

III. ETIOLOGY

"Localized" edema refers to edema produced by regional obstruction of venous or lymphatic flow, or both. The former causes increased formation of interstitial fluid and the latter results in its impaired removal. Examples include lower extremity edema due to deep venous thrombosis, and unilateral lymphedema due to pelvic neoplasm. Hydrothorax and ascites may be localized phenomena or occur in any generalized edematous state. "Generalized" edema is a symptom or sign of a primary clinical disorder which has set in motion a chain of events resulting in decreased renal excretion of sodium. The most common primary disorders associated with generalized edema are heart disease in congestive failure, cirrhosis of the liver, and certain renal diseases.

IV. CLINICAL MANIFESTATIONS

There may be a considerable increase in the interstitial fluid volume before it is clinically appreciated. The symptoms and signs of edema are unexplained weight gain, tightness of a ring or shoe, puffiness of the face, swollen extremities, enlarged abdominal girth, and persistence of indentation of the skin following pressure.

The distribution of generalized edema is influenced by gravity and local tissue turgor. For example, in the ambulatory patient edema is generally most prominent in the lower extremities and may worsen as the day goes by. On the other hand, postsacral edema may be more evident if the patient is bedridden. Periorbital tissue has reduced turgor and is prone to become edematous, but usually not in patients with heart failure who cannot lie flat in bed. The scrotum also has low tissue turgor and may become edematous; in adults this is generally seen in patients with ascites.

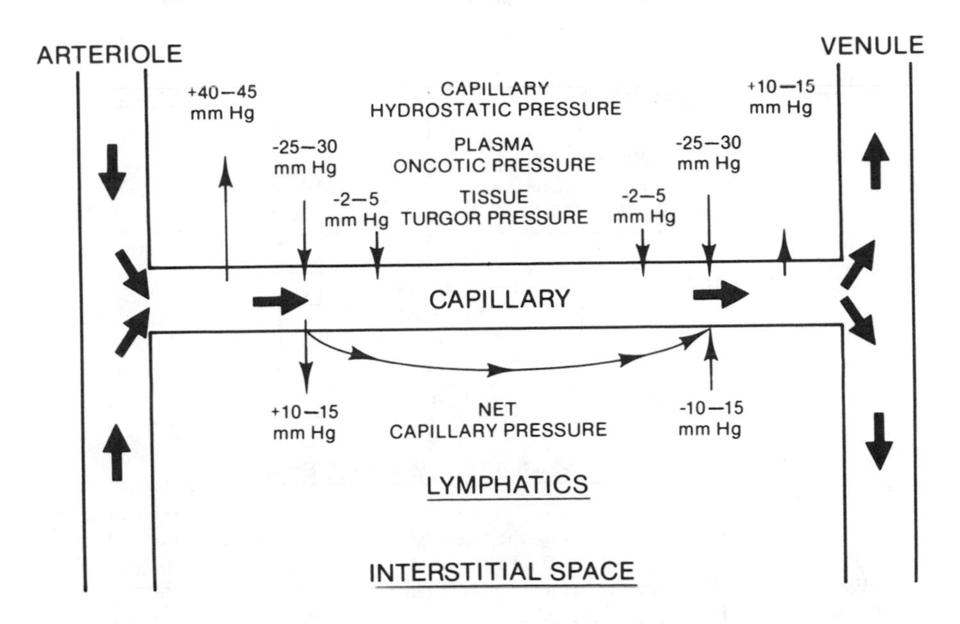

FIGURE 1. Capillary Starling forces that influence fluid distribution between plasma and interstitial compartment. (Adapted from Schrier, R. W., *Renal and Electrolyte Disorders,* Little Brown, Boston, 1976. With permission.)

V. PATHOGENESIS OF GENERALIZED EDEMA

Some years ago the explanations of generalized edema focused entirely on mechanisms that seemed quite directly to alter Starling forces. Thus, attempts were made to explain the edema of heart failure as due exclusively to increased capillary hydrostatic pressure secondary to elevated venous pressure. Similarly, diminished removal of interstitial fluid secondary to reduced oncotic pressure was seen as the sole mechanism of nephrotic edema. In cirrhosis where ascites predominated, attention tended to be limited to decreased oncotic pressure, increased portal vein hydrostatic pressure, and impaired flow of liver lymphatics. Although these mechanisms are still relevant in edema formation, the situation in each instance is more complex. Perhaps the major complexity was perceived when it was recognized that all conditions characterized by generalized edema have in common decreased renal excretion of sodium and water. Much effort has been directed towards understanding the renal mechanisms of sodium retention and their relationship to the capillary Starling forces.

Heart failure, cirrhosis of the liver, and the nephrotic syndrome are all suspect of having a contracted "effective" ECFV despite the obvious expansion of total ECFV. In all three disease states, a decreased "effective" ECFV has been postulated to set in motion the same physiologic events as a reduction in dietary sodium in the normal individual with resultant increase in sodium reabsorption (see Figure 2). The evidence in heart failure, cirrhosis, and the nephrotic syndrome supporting this suspicion is reviewed for each condition in the subsequent three chapters in this section. However, in each of these instances, there are data that question whether reduced "effective" ECFV is the sole mechanism of renal sodium reabsorption. For example, it has been conceptualized that heart failure may result in a lower volume in that portion of the circulation (perhaps the fullness of the arterial tree) which contains baroreceptors that

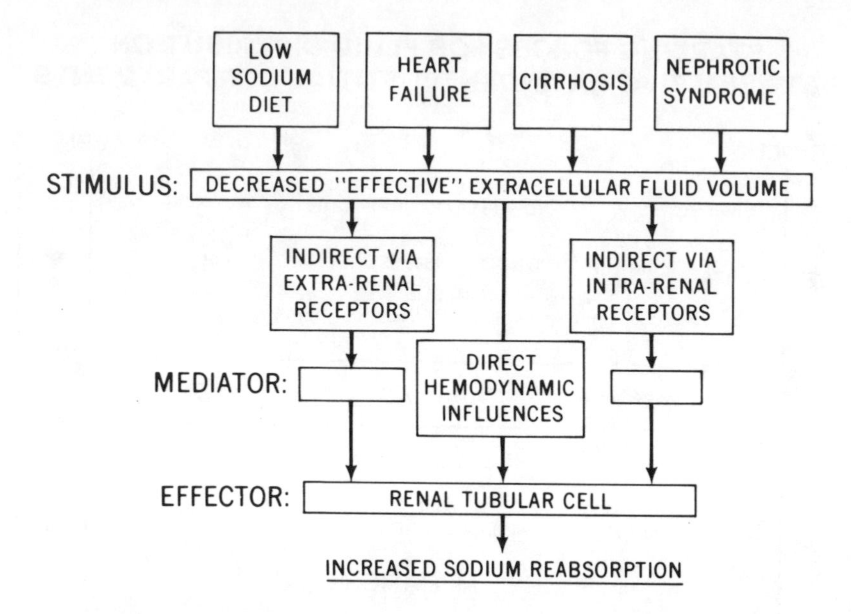

FIGURE 2. A framework depicting the "general" ways in which the presumed reduction in effective extracellular fluid volume of edematous states results in increased tubular reabsorption of sodium and reduced excretion of the cation. (From Papper, S., *Clinical Nephrology,* 2nd ed., Little, Brown, Boston, 1978. With permission.)

perceive the volume reduction as reduced pressure. It is, however, difficult to invoke this explanation in instances of heart failure with high cardiac output. Most data in heart failure derive from animal models and the analogy to man is incomplete. For example, in ligation of the thoracic inferior vena cava model, the role of cardiac output itself is not established. In the case of cirrhosis, fluid may be viewed as physiologically "sequestered" in the peritoneal space and hence not perceived by the organism, which on the contrary, appreciates the "effective" ECFV as reduced. On the other hand, there is also evidence for an inadequately explained primary renal retention of sodium. In this context, there is thought to be secondary "overflow" into the peritoneal cavity because of altered hepatic hemodynamics. In the nephrotic syndrome, the reduced oncotic pressure secondary to urinary protein loss causes leakage of fluid out of the intravascular compartment, which might therefore be sensed as a reduction in effective volume. However, in the instance of the nephrotic syndrome, this mechanism should result in reduced plasma volume, which is not always the case. Furthermore, there are other states with hypoalbuminemia not accompanied by generalized edema.

How the normal organism appreciates changes in ECFV and signals the renal tubular reabsorption of sodium is considered in detail in Chapter 2. A host of "afferent" pathways have been considered to perceive changes in ECFV. These include "volume receptors" — probably really strategically located baroreceptors—in the carotid circulation, heart chambers, aorta, and great veins of the thorax. The liver has been postulated to have its own receptor, a reasonable supposition when one considers that sodium is normally ingested, and the liver is the organ that joins the intestine and portal circulation to the systemic circulation. "Efferent" pathways proposed to signal the renal tubular cell to reabsorb more sodium are many and include: increased secretion of aldosterone, decreased natriuretic hormone, and increased adrenergic activity. Finally, it is also likely that reduced ECFV has direct circulatory effects that might increase reabsorption of sodium, such as cortico-medullary redistribution of renal blood flow and altered Starling forces in the peritubular capillaries. Although much is known about afferent and efferent mechanisms, the chapters in this section make it clear that there is even more that is not known.

There is some repetition in the four chapters on edema, especially as they consider the matter of ECFV and the various afferent and efferent mechanisms involved in sodium retention. It is our belief that repetition in these instances is necessary because the evidence differs among the pathophysiological states characterized by the formation of edema, and because it allows presentation of a more complete picture in each instance.

There are other states with generalized edema that are even more difficult to fit neatly into the concept of reduced "effective" ECFV. These include especially chronic renal failure, acute nephritic syndrome, and idiopathic edema.

VI. TREATMENT

Because the initiating event in the development of generalized edema is usually heart failure, cirrhosis, or renal disease, optimal treatment of the underlying condition whenever possible is important. Some degree of dietary sodium restriction is generally advisable. In some patients with impaired ability to excrete water, fluid intake needs to be restricted as well. Diuretic drugs are useful agents but are not always indicated in patients with edema.

Chapter 4

HEART FAILURE

David B. Bernard and Edward A. Alexander

TABLE OF CONTENTS

I. INTRODUCTION

It has been known for many years that the kidneys play a major role in the pathogenesis of congestive heart failure (CHF) and that most of the symptoms and signs that occur in this disease state result from renal sodium and water retention.[1] While the congestion of the circulation which characterizes CHF was initially thought to be merely a mechanical phenomenon, later workers realized that a more complex series of events is involved and that the kidneys play a key role in this process. Indeed, since failure of the heart to perform adequately its pumping action has potentially disastrous consequences for the whole organism, it is not surprising that a large number of compensatory hemodynamic and neurohumoral mechanisms are brought into play to minimize the impact of pump failure. The kidney is significantly involved in this compensatory process and acts to restore the adequacy of the circulation by retaining salt and water which expands the extracellular fluid volume (ECFV). Continued retention eventually leads to overcompensation and what is recognized clinically as edema. It is clear that regardless of the initial changes in cardiac or systemic hemodynamics which occur in CHF, expansion of the ECFV requires the participation of the kidney in reducing sodium excretion and that an inability to excrete ingested sodium is a fundamental component of this disease process. Studies have shown that the ability of the kidney to excrete sodium and water diminishes as the pumping function of the heart declines. While patients with mild heart failure may not be edematous, a reduced ability to excrete a sodium load can still be demonstrated. In advanced heart failure, marked renal sodium retention is present and only a small fraction of an infused or ingested salt load can be excreted.[2]

This chapter will review the mechanisms responsible for renal sodium retention in heart failure. Four aspects, as they relate specifically to heart failure, will be considered: the afferent stimulus that initiates the process; the afferent receptors that signal the kidney to retain sodium; the efferent limb involved in this process; and the nephron sites where tubular function is altered. Much of the available information has been derived from studies of experimental heart failure in animals. Primarily, two models have been used: ligation of the inferior vena cava in the thorax (TIVC) in dogs and rabbits as an example of a low-output state; and creation of arteriovenous fistulae, predominantly in rats and dogs, as a model of high-output failure. Although the TIVC model has been widely studied and is generally accepted as an example of low-output cardiac failure, it differs from human heart failure in that the myocardium and its vascular supply are intact and cardiac hemodynamics are different from those found in CHF. Nevertheless, since renal hemodynamic features are similar, it seems at least possible that mechanisms of sodium retention are sufficiently comparable to make the TIVC model useful for the study of renal function in CHF. Further advances in the understanding of the pathogenesis of renal sodium retention in CHF are hindered, however, by the lack of an animal model which is more analogous to human heart failure.

II. THE AFFERENT STIMULUS

Two main theories have been offered to explain the initial events leading to sodium retention in CHF. Neither adequately covers all types of failure. The "backward failure theory" postulates that enhanced peripheral venous pressure promotes net transudation of fluid from the intravascular to the interstitial space. This results in hypovolemia, which signals the kidney to conserve salt and water. However, while it is true that the kidney behaves in CHF as if extracellular volume (i.e., "effective" ECFV) is reduced, and renal blood flow changes are similar to those found in animals placed on severely salt restricted diets, ECFV is increased rather than decreased in most cases of CHF.[3]

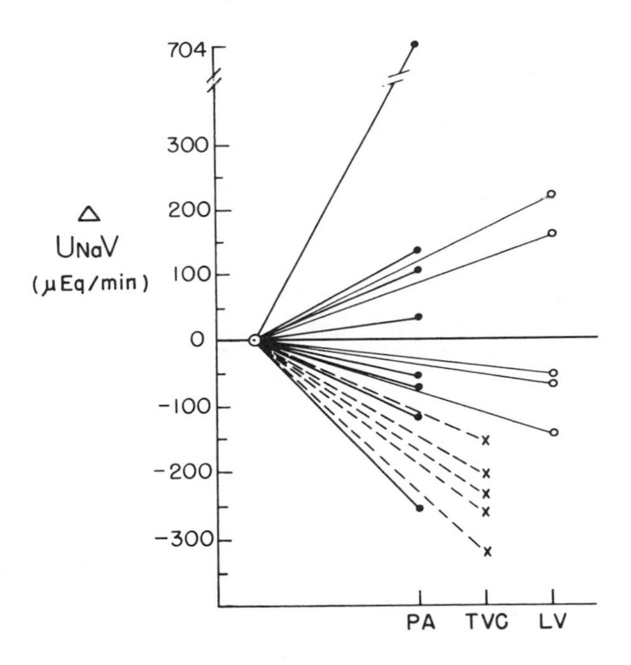

FIGURE 1. The change in sodium excretion in dogs after similar reduction in cardiac output by pulmonary artery constriction (PA), thoracic inferior vena cava constriction (TIVC), or left ventricular infarction (LV). Only the TIVC group showed a consistent antinatriuresis. (From Migdal, S., Alexander, E. A., and Levinsky, N. G., Evidence that decreased cardiac output is not the stimulus to sodium retention during acute constriction of the vena cava, *J. Lab. Clin. Med.*, 89, 809, 1977. With permission.)

The "forward failure theory" postulates that a falling cardiac output (CO) initiates the process. This results in a reduction in renal perfusion such that the kidney senses the need to restore circulatory adequacy and attempts to do so by retaining salt and water. The possibility that a reduction in CO is the afferent stimulus has been studied in the TIVC model in the dog. Sodium retention occurs in this model in association with a low CO. Ligation of the inferior vena cava in the abdomen produces similar changes in IVC pressures, but no change in CO and a significantly smaller reduction in sodium excretion.[4] That differences in hepatic venous pressure do not account for the differences between these models was confirmed when it was shown that ligation of the superior vena cava was associated with a fall in CO and sodium excretion without change in hepatic vein pressure.[5] A study in which cardiac hemodynamics were measured during sodium retention and again after release of the ligation seemed to establish the importance of a low CO.[6] During the development of ascites, CO fell and total peripheral resistance increased. With release of the ligation, the ascites dissipated and CO and peripheral resistance returned to normal. However, a different conclusion was reached in a study in which CO was reduced in dogs by three separate maneuvers.[7] When CO was lowered by 25 to 30% with pulmonary artery banding, TIVC ligation, or left ventricular infarction, sodium excretion dropped significantly only in the TIVC group (Figure 1). Measurements of central pressure changes showed the only difference to be in the right ventricular end-diastolic pressure which was reduced in the TIVC and increased in the pulmonary artery banded dogs. The authors concluded that changes in the intracardiac hemodynamics may be more important in the antinatriuresis of the TIVC dogs than arterial receptors sensitive to CO. Similar findings in studies on conscious nondiuretic dogs have supported these conclusions.[8]

Table 1
AFFERENT RECEPTORS SIGNALING SODIUM RETENTION IN HEART FAILURE

Receptors / maneuvers	Effect
Low pressure volume receptors in central veins	
Reduce venous return	
Prolonged standing	Decrease sodium excretion
Tourniquets to legs	
Positive pressure breathing	
Antihypertensive drugs	
? Increase systemic venous pressure	
Increase venous return	
Recumbency	Increase sodium excretion
Water immersion	
Negative pressure breathing	
High pressure volume receptors in arterial tree	
Baroreceptors	
Juxtaglomerular apparatus	
Others ?	

Although the ECFV is known to be an important determinant of renal sodium excretion, the kidney in CHF avidly retains sodium despite an expanded total ECFV. The usual explanation for this discrepancy is to invoke the concept of the "effectiveness" of the arterial plasma volume. This concept suggests that the afferent receptors sense the "relative" rather than the absolute filling of the arterial tree. Relative filling may be a representation of the relationships among CO, arterial plasma volume, peripheral run-off and resistance, compliance of arteries and arterioles, arterial filling pressure, the capacity of the vascular tree, and other unknown factors. A disturbance of this relationship could arise in several ways; for example, by a decline in CO, by a decrease in total plasma volume, by an increased capacity of the arterial tree, or by an increased rate of peripheral run-off of blood from arterial to venous systems. In each case, the afferent receptors recognize a deficiency in the "effective" ECFV, and signal the kidney to retain sodium and water in an attempt to restore the effectiveness of the circulation.

III. AFFERENT RECEPTORS

The pathways by which this reduction in "effective" ECFV is signaled to the kidney probably involve an afferent receptor system. Although receptors influencing renal sodium excretion have not been definitively identified, considerable evidence exists for their presence in both the low pressure central venous or pulmonary circulation, as well as in the high pressure arterial side of circulation (Table 1).

A. Low Pressure Receptors

Maneuvers which alter venous return to the heart change renal sodium excretion. Standing upright for prolonged periods, applying tourniquets to the legs, inducing positive pressure breathing, and using antihypertensive drugs all reduce venous return and decrease salt excretion by the kidneys. Clearly, a reduction in "effective" ECFV or perhaps an increase in systemic venous pressure itself, may do likewise. Recumbency, negative pressure breathing, or water immersion enhance venous return and augment urinary sodium excretion.

Considerable uncertainty exists as to the specific sites of these sensors in the low pressure circulation, but the atria are a likely location.[9,10] Numerous studies have shown that distention of either the left or right atrium resulted in enhanced excretion of salt and water. Distention of the left and right atrial appendages also promotes sodium excretion, while atrial tamponade or a decreased right atrial volume produces sodium retention. Certainly, both atria have a dense population of nerve fibers which could respond to changes in central volume or pressure.

The mechanisms of natriuresis following atrial distention are thought to involve a reduction in renal sympathetic nerve discharge (with left atrial distention[11]) and in plasma renin activity (with right atrial distention[12]). The exact way in which the anti-natriuresis of CHF, when right atrial pressure is increased, fits into this schema is not immediately clear, but one study suggested that a dampened response of atrial receptors to elevated venous pressure was present in CHF.[13]

B. High Pressure Receptors

Some evidence exists as well for the presence of volume or pressure receptors in the high pressure arterial circulation which may affect urinary sodium excretion. Experiments in which arteriovenous fistulae were opened and closed[14] suggested that renal sodium excretion responds as well to changes in the effective filling of the arterial tree. Once again, the responsible sensors have not yet been adequately defined. However, two receptors in the arterial tree are known to be activated by changes in volume or pressure and may be important in modifying tubular sodium transport (see following section). These are the baroreceptors of the aortic arch and carotid sinus which modify activity of the sympathetic nervous system and the juxtaglomerular apparatus in the kidney which modulates the rate of renin release into the circulation. The exact location of these receptors, the factors which stimulate them, and their interrelationships, remain important areas for future research.

IV. EFFERENT MECHANISMS

Irrespective of the initiating afferent stimulus in CHF and where it is perceived, stimulation of the efferent limb produces two major alterations in renal function; a reduction in renal blood flow (RBF), and a reduction in sodium excretion. The mechanisms thought to be responsible for each effect and their interrelationships will be considered next.

A. Renal Blood Flow

Several mechanisms may reduce RBF in CHF. Clearly, a drop in systemic blood pressure results in a decrease in renal perfusion pressure. More importantly, RBF decreases because of renal vasoconstriction and an increase in renal vascular resistance which occurs because of the activation of the renin-angiotensin and adrenergic nervous systems.

Several lines of evidence suggest that angiotensin plays an important role in the reduced RBF in CHF. Angiotensin II is known to be a powerful renal vasoconstrictor, and infusion of the hormone in sub-pressor doses into the renal artery produces alterations in renal hemodynamics which are similar to those seen in heart failure.[15-17] GFR is maintained in the face of a reduced RBF suggesting predominantly efferent arteriolar vasoconstriction by angiotensin II. In addition, interruption of the effects of antiotensin II with the use of the competitive antagonist saralasin (sar-l-ala-8 angiotensin II) in experimental models of low- and high-output failure has produced significant renal vasodilation and enhanced RBF in some,[18,19] but not all studies.[20] Similar results have been obtained following the use of the orally active converting enzyme inhibitor, captopril, in human heart failure.[21,22]

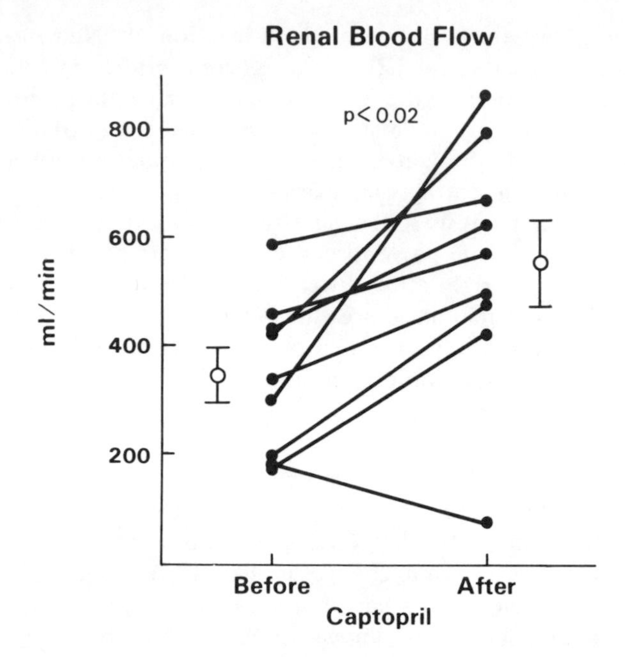

FIGURE 2. The effect of 100 mg of captopril on renal blood flow in 9 patients with congestive heart failure.

In Figure 2 the effect of 100 mg of captopril on RBF in nine patients with CHF unresponsive to conventional therapy (class 3 or 4) is depicted. One to 2 hr after treatment, RBF increased from 344 to 522 mℓ/min.[21] Similar results have been obtained with more chronic captopril therapy.[22] While the most likely explanation for these results is the inhibition of the vasoconstrictive action of angiotensin, other factors may have contributed, such as stimulation of kinins, known to be potent renal vasodilators[23] whose activity is increased by captopril.[24] While measurements of plasma renin activity and angiotensin in experimental or clinical heart failure have been variable, one study examined this system during different stages of CHF in the dog and may have reconciled these discrepancies. It was shown that angiotensin was active in the early stages of CHF of all severities but only in the later stages in the more severe forms.[25] This finding correlates as well with the RBF changes that have been observed at different stages of experimental heart failure, where blood flow is reduced early but has returned to normal later in the course of the disease.

In addition to promoting the release of angiotensin, the afferent stimulus also leads to increased activity of the sympathetic nervous system via the baroreceptors of the great vessels. These enhanced adrenergic discharges result in increased sympathetic nerve activity, as well as increased circulating catecholamines and may play an important role in the renal vasoconstriction and reduction in RBF in CHF. It is of interest that in the aforementioned acute captopril study, plasma norepinephrine concentration decreased from 818 ± 105 to 686 ± 103 pg/mℓ ($P < 0.05$), perhaps also contributing to the increment in RBF.[21]

B. Sodium Retention

The specific factors which signal the kidney to retain sodium in heart failure have not been clearly defined. Under normal conditions sodium excretion is regulated to equal dietary intake and thus the ECF volume is maintained within narrow limits. Clearly, in heart failure the pathways which link changes in the ECFV to changes in sodium reabsorption have become disrupted, since there is continued sodium reabsorp-

Table 2
FACTORS REGULATING SODIUM EXCRETION

Glomerular filtration rate
Peritubular physical factors (Starling's Forces)
Aldosterone
Autonomic nervous system
Circulating and intrarenal hormones

tion in the face of an expanded ECFV. A convenient way of analyzing the factors which may be responsible for this ''inappropriate'' sodium reabsorption is to consider them in the light of the factors known to be important in the normal regulation of sodium excretion as shown in Table 2. These have been discussed in Chapter 2. Although an extensive system of interdependent variable and overlapping regulatory responses clearly exists, a total understanding of the events and interrelationships that regulate sodium reabsorption by the kidney is not yet available.

1. Glomerular Filtration Rate (GFR)

Although changes in the GFR and hence in filtered sodium could theoretically have a considerable influence on urinary sodium excretion, there is substantial evidence indicating that changes in the filtered load do not necessarily result in concomitant changes in sodium excretion. On the other hand, significant decreases in GFR would obviously favor sodium retention or at least produce a diminished ability to excrete a sodium load. However, in the majority of patients with CHF, at least in the early stages, GFR is normal or only slightly reduced.[26-28] Experimentally as well, several studies have shown marked sodium avidity in the presence of a normal or near normal GFR, thus tending to exclude changes in GRF as a major factor in the sodium retention in this state.[29,30] Clearly, adjustments in tubular sodium transport are more important in the excess accumulation of sodium and water in heart failure.

2. Peritubular Physical Factors

There is considerable evidence that sodium reabsorption along the proximal tubule varies with the peritubular physical factors (Starling's forces), particularly the peritubular oncotic pressure.[31] Thus, an increase in pertubular capillary albumin concentration, or a decrease in hydrostatic pressure, favors the reabsorption of salt and water, while a decrease in protein or increase in pressure has the opposite effect. The concentration of albumin in the peritubular capillaries depends on its concentration in the afferent arterioles and in the filtration fraction (the portion of renal plasma flow [RPF] that undergoes filtration). As noted above, RPF is decreased frequently in CHF while GFR remains normal or is only slightly reduced. Thus, the filtration fraction (GFR/RPF) is increased in CHF from 0.02 to 0.30—0.40 or higher.[26-30,32] The increase in filtration fraction (FF) produces an increase in peritubular protein concentration and oncotic pressure, thus enhancing sodium and water reabsorption. Other physical factors including a decrease in hydrostatic pressure in the peritubular capillaries, possibly induced by efferent arteriolar constriction by angiotensin II, may also contribute to increased proximal tubular reabsorption. Thus, one logical and attractive schema to explain the mechanism of renal sodium avidity in CHF is to invoke hemodynamically induced changes in filtration fraction and peritubular forces. Indeed, several studies in man and animals have shown urinary sodium retention to be associated with an increased FF,[27-30,32] but it cannot be invoked in all cases. First, several studies of experimental heart failure have shown marked sodium avidity and edema formation in the absence of changes in FF.[6,7,33-36] Second, other studies have shown a rise in FF to be associated with no change in sodium excretion[8] or with an increase in sodium excretion.[37] Third, although some studies have shown that a fall in FF in CHF induced by pharmacologic means may be associated with an increase in urinary sodium excretion,[21,38] other studies have shown avidity to be unaltered by a drop in FF.[18,19,27,28,39] Finally, from several studies of experimental heart failure in which the tubular segments responsible for sodium avidity were studied, it was concluded that the altered function was occurring beyond the proximal tubule, where physical factors are not established as being important in sodium transport.[29,30,34,40]

3. Aldosterone

Although the importance of aldosterone in the regulation of sodium balance has been the subject of some debate, recent work suggests that its major effect is exerted along the collecting duct where a small but significant fraction of filtered sodium remains.[41] In the initial phase of experimental heart failure, aldosterone levels are elevated, but they later fall to normal when renin and angiotensin levels drop.[25] In more severe failure, aldosterone levels remain elevated. In human heart failure, aldosterone levels have been found to be elevated in the presence of right-sided failure in association with sodium retention.[42] While it is attractive to postulate an important role for aldosterone in the renal avidity of CHF, it is unlikely to be a critical factor in edema formation. Aldosterone levels in patients with CHF are not consistently increased.[43] Aldosterone inhibitors rarely produce significant natriuresis in these patients,[44] and the administration of aldosterone to normal people does not produce edema. In a recent study, aldosterone levels and the response to sodium loading were studied in eight patients with CHF.[45] Four of these patients continued to have elevated renin and aldosterone levels after retaining approximately 500 mEq of sodium. The other four, however, had similar weight gains and sodium retention without elevated renin and aldosterone levels, suggesting that aldosterone played little or no role in the reduced sodium excretion. We believe that while aldosterone may contribute to the sodium retention in heart failure, it is not an essential factor.

4. Miscellaneous Factors

A number of other factors have been suggested as being important in the sodium retention of heart failure. There is some evidence suggesting that *angiotensin II* may have a direct renal tubular action to enhance sodium transport. These data have been derived from experiments in which angiotensin II was infused directly[16,46,47] or angiotensin activity was inhibited pharmacologically[48-51] in normal and salt depleted animals. In two recent studies of experimental heart failure, inhibition of angiotensin by converting the enzyme inhibitor produced significant natriuresis,[38,39] although other studies in similar models, where angiotensin blockade was achieved with the competitive antagonist saralasin, showed no change in urinary sodium excretion.[18-20] In the recent study from this hospital, a significant natriuresis was obtained with captopril in nine patients with severe CHF.[21] As Figure 2 shows, urinary sodium doubled from 34 μEq/min before to 68 μEq/min after 100 mg of captopril. Whether this natriuresis was secondary to the alterations in hemodynamics or to a direct tubular action is not known.

A number of observations suggest that the *autonomic nervous system* plays an important role in the sodium retention of CHF. Stimulation of sympathetic activity could result in a reduction in RBF and produce changes in FF and physical factors in the peritubular environment as discussed above. Indeed, sympathetic blocking agents (phenoxybenzamine) have been shown to induce renal vasodilation and mild natriuresis in animals and patients with cardiac edema. Additionally, sympathetic nerves are known to promote the release of renin from the kidney with resultant elevations of angiotensin and aldosterone. Finally, renal tubular sodium handling has been shown to be influenced directly by the state of discharge of sympathetic nerves.[52] In this regard, stimulation of sympathetic efferents produce sodium retention which can be blocked by guanethidine in the absence of any alterations in intrarenal hemodynamics.[53] In the TIVC dog, sodium retention can be overcome by the α-adrenergic blocker phenoxybenzamine, guanethidine, by surgical renal denervation[54,55] and by ganglion blockade with pentolinium,[56] supporting an important role for sympathetic activity in this antinatriuretic state. One study showed that the antinatriuretic effect of acute TIVC constriction could not be abolished by surgical renal denervation, but could be overcome by autonomic blockade with hexamethonium or alpha-adrenergic receptor blockade with phenoxybenzamine,[5] suggesting a role as well for circulating catecholamines in the sodium avidity of this model. While the above evidence supports a role for enhanced sympathetic nervous activity and increased circulating catecholamines in the sodium retention of low output states, conflicting data is available which suggests a facilitative rather than a primary role for this system. In one study, the denervated kidney continued to retain sodium following chronic TIVC ligation,[6] and in another, renal tubular sodium absorption was clearly increased in the denervated kidney following acute TIVC ligation.[4] We believe that renal sympathetic nerve stimulation of circulating catecholamines contribute to the sodium retention of CHF, but that other factors are clearly involved.

It has been suggested that one or more *natriuretic hormones* are important in the renal excretion of sodium after volume expansion.[57] Evidence for the existence of such substances is largely circumstantial and no conclusive proof of their existence yet exists. In one study, however, plasma and urine fractions from healthy persons during extracellular volume expansion, as well as from patients with CHF, were tested for natriuretic activity by a rat bioassay.[58] Whereas the fractions from the normal persons with ECFV expansion had a definite natriuretic effect, no such effect was observed with the urine and plasma from patients with sodium retention and edema. Thus, edema formation in those subjects may have been mediated by the absence of a natriuretic factor. This possibility will need to be reevaluated if and when this factor has been isolated in patients with CHF.

Several other *circulating or intrarenal hormone systems* are known to influence sodium excretion. These include oxytocin, vasopressin, and the kallikrein-kinin and prostaglandin systems. No data is yet available implicating any of these hormones in the sodium avidity of heart failure. Vasopressin secretion, presumably in response to the reduced "effective" ECFV, is certainly important in the water retention which occurs in CHF however.

Since there is no evidence that *intrarenal redistribution of RBF* plays any role in the control of sodium excretion under normal or pathologic conditions, it seems unlikely to be a factor in the sodium retention of CHF.

Thus it is likely that several mechanisms are involved in the antinatriuresis of CHF and that different factors may be important at different stages in the disease. In the TIVC dog, for example, early acute caval constriction is associated with sodium retention and marked changes in renal hemodynamics, while in the more chronic state, RBF and FF frequently return to normal as sodium avidity continues.[18] The difficulty of defining precisely the factors involved in sodium retention in human CHF is highlighted by the studies performed recently to assess the role of angiotensin in this regard. In response to captopril, urinary sodium excretion doubled and there was a significant fall in circulating levels of aldosterone and norepinephrine. Yet, no statistically significant correlation was found between the increase in urine sodium and the changes in FF, GFR, RBF, aldosterone, or norephinephrine.

5. Tubular Site of Sodium Retention in Heart Failure

Two studies have examined the nephron sites of enhanced sodium reabsorption in humans with heart failure. In one it was shown that a mannitol infusion resulted in a greater increase in free-water excretion in patients with CHF than in normal subjects.[59] This observation suggests that mannitol exerted a greater effect in the nephron segments proximal to the diluting site, and favored the notion that enhanced sodium avidity was occurring in these segments as well. In a second study in which a distal tubular transport blockade technique was used to assess proximal tubular function, the authors concluded that proximal tubular reabsorption was set higher than normal at all levels of extracellular fluid volume.[60]

Limited micropuncture studies are available which allow a direct assessment of the nephron segments responsible for enhanced sodium reabsorption in experimental models of heart failure. Several studies have offered evidence that the proximal tubule is involved. An early study in the acute TVC dog showed no reduction in fractional reabsorption in the proximal tubule in response to a saline load as occurs in normal animals, and the authors concluded that this implicated the proximal tubule as the site of sodium avidity.[61] This was confirmed by a later study in the same model when proximal fractional reabsorption was found to be significantly increased in association with significant changes in intrarenal hemodynamics.[29] However, additional studies were done where the vena caval constriction had been performed a week or more before the experiment.[29] In these studies, GFR and RBF were only mildly and not significantly reduced. Nevertheless, the animals were avidly retaining sodium, both during hydropenia and following volume expansion. Proximal reabsorptive function was entirely normal during hydropenia and a normal depression in proximal fractional reabsorption occurred after saline loading. This suggested that the site of sodium avidity was distal to the proximal tubule. These investigators hypothesized that enhanced proximal reabsorption had occurred early, probably in response to the decreased effective circulation and reduced renal perfusion, but that once the circulation had returned to normal and blood pressure and GFR were restored, proximal tubular reabsorption had also returned to normal and maintenance of sodium retention was dependent on enhanced reabsorption in distal segments. Further support for altered reabsorption beyond the proximal tubule was offered in a later study of the same model where enhanced reabsorption in the loop of Henle was strongly suggested.[40] That some hemodynamic factor or factors were responsible was suggested by the observation that increased urinary sodium excretion and decreased loop reabsorption could be achieved by a combination of renal vasodilatation and increased systemic blood pressure, but not by either alone.[40] In the model of aorta to vena cava fistula in dogs, fractional reabsorption in the proximal tubule was found to be normal both during hydropenia or following saline expansion.[34] Interestingly, FF was decreased in this model. Since the animals were obviously retaining sodium, these authors also concluded that the responsible nephron segments were situated distal to the proximal tubule. The same model has been studied in the rat both early after the fistula was created (2 to 6 days, "acute") and later (15 to 20 days, "chronic").[30,35] In the acute model, GFR was the same in experimental and control, but RBF was reduced in the fistula rats so that FF was increased in this group.[30] Micropuncture studies revealed enhanced sodium reabsorption in the proximal tubule in these animals. However, in the more chronic model, GFR, RPF, and FF were not different between experimental and control. Proximal tubular function was entirely normal in these chronic animals, both during hydropenia and following volume expansion, yet the animals were clearly excreting less sodium than control.[30,35] Early distal micropuncture studies showed that the fraction of filtered sodium remaining by the early distal tubule was significantly lower in the experimental animals, again suggesting that the loop of Henle was the major site of sodium avidity in this model. The authors similarly concluded that during the acute stages of this model, proximal reabsorption is enhanced due to changes in systemic and renal hemodynamics, but that by the time several more days have elapsed, these factors are no longer present and proximal tubular function returns to normal. Sustained sodium retention seemed to be occurring in the distal nephron. In a different experimental model in dogs, namely chronic constrictive pericarditis, enhanced sodium reabsorption appeared to be occurring in the distal nephron.[62] Finally, the most distal part of the nephron, the collecting tubule, a site known to be important in modifying urinary sodium excretion, has not yet been studied directly in heart failure, but has been implicated in other antinatriuretic states.[63,64] Clarification of this point remains an important subject for future research.

REFERENCES

1. **Starling, E. H.**, Physiologic factors involved in the causation of dropsy, *Lancet,* 2, 1407, 1896.
2. **Braunwald, E., Plauth, W. H., Jr., and Morrow, A. G.**, A method for the detection and quantification of impaired sodium excretion: results of an oral sodium tolerance test in normal subjects and in patients with heart disease, *Circulation,* 32, 223, 1965.
3. **Samet, P., Fritts, H. W., Jr., Fishman, A. P., and Cournand, A.**, The blood volume in heart disease, *Medicine,* 36, 211, 1957.
4. **Schrier, R. W. and Humphreys, M. H.**, Factors involved in the antinatriuretic effects of acute constriction of the thoracic and abdominal inferior vena cava, *Circ. Res.,* 29, 479, 1971.
5. **Schrier, R. W., Humphreys, M. H., and Ufferman, R. C.**, Role of cardiac output and the autonomic nervous system in the antinatriuretic response to acute constriction of the thoracic superior vena cava, *Circ. Res.,* 29, 490, 1971.
6. **Lifschitz, M. D. and Schrier, R. W.**, Alterations in cardiac output with chronic constriction of thoracic inferior vena cava, *Am. J. Physiol.,* 225, 1364, 1973.
7. **Migdal, S., Alexander, E. A., and Levinsky, N. G.**, Evidence that decreased cardiac output is not the stimulus to sodium retention during acute constriction of the vena cava, *J. Lab. Clinc. Med.,* 89, 809, 1977.
8. **Yaron, M. and Bennett, C. M.**, Renal sodium handling in acute right-sided heart failure in dogs, *Miner. Electrolyte Metab.,* 1, 303, 1978.
9. **Gauer, O. H., Henry, J. P., and Behn, C.**, The regulation of extracellular fluid volume, *Ann. Rev. Physiol.,* 32, 547, 1970.
10. **Reinhardt, H. W., Kaczmarczyk, G., Eiselle, R., Arnold, B., Eigenheer, F., and Kuhl, U.**, Left atrial pressure and sodium balance in conscious dogs on a low sodium intake, *Pflügers Archiv.,* 370, 59, 1977.
11. **Prosnitz, E. H. and DiBona, G. F.**, Effect of decreased renal sympathetic nerve activity on renal tubular sodium reabsorption, *Am. J. Physiol.,* 235, F557, 1978.
12. **Brennan, L. A., Jr., Malvin, R. L., Jochim, K. E., and Roberts, D. E.**, Influence of right and left atrial receptors on plasma concentration of ADH and renin, *Am. J. Physiol.,* 221, 273, 1971.
13. **Greenberg, T. T., Richmond, W. H., Stocking, R. A., Gupta, P. D., Meehan, J. P., and Henry, J. P.**, Impaired atrial receptor responses in dogs with heart failure due to tricuspid insufficiency and pulmonary artery stenosis, *Circ. Res.,* 32, 424, 1973.
14. **Epstein, F. H., Post, R. S., and McDowell, M.**, Effects of an arteriovenous fistula on renal hemodynamics and electrolyte excretion, *J. Clin. Invest.,* 32, 233, 1953.
15. **Waugh, W. H.**, Angiotensin II: local renal effects of physiological increments in concentration, *Can. J. Physiol. Pharm.,* 50, 711, 1972.
16. **Hall, J. E., Guyton, A. C., Salgado, H. C., McCaa, R. E., and Balfe, J. W.**, Renal hemodynamics in acute and chronic angiotensin II hypertension, *Am. J. Physiol.,* 235(3), F174, 1978.
17. **Freeman, R. H. and Davis, J. O.**, Physiological actions of angiotensin II on the kidney, *Fed. Proc.,* 38, 2276, 1979.
18. **Freeman, R. H., Davis, J. O., Vitale, S. J., and Johnson, J. A.**, Intrarenal role of angiotensin II: homeostatic regulation of renal blood flow in the dog, *Circ. Res.,* 32, 692, 1973.
19. **Freeman, R. H., Davis, J. O., Spielman, W. S., and Lohmeier, T. E.**, High output heart failure in the dog: systemic and intra-renal role of angiotensin II, *Am. J. Physiol.,* 229(2), 474, 1975.

20. **Slick, G. L., DiBona, G. F., and Kaloyanides, G. J.**, Renal blockade to angiotensin II in acute and chronic sodium-retaining states, *J. Pharmacol. Exp. Ther.*, 195, 185, 1975.
21. **Creager, M. A., Halperin, J. L., Bernard, D. B., Faxon, D. P., and Ryan, T.**, Acute regional circulatory and renal hemodynamic effects of converting enzyme inhibitor in patients with congestive heart failure, *Circulation*, 64, October 1981 (in press).
22. **Dzau, V. J., Colucci, W. S., Williams, G. H., Curfman, G., Meggs, L., and Hollenberg, N. K.**, Sustained effectiveness of converting-enzyme inhibition in patients with severe congestive heart failure, *N. Engl. J. Med.*, 302, 1373, 1980.
23. **Barraclough, M. A. and Mills, I. H.**, Effects of bradykinin on renal function, *Clin. Sci.*, 28, 69, 1965.
24. **Murphey, U. S., Walden, T. L., and Goldberg, M. E.**, The mechanism of bradykinin potentiation after inhibition of angiotensin-converting enzyme by SQ14225 in conscious rabbits, *Circ. Res.*, 43, 140, 1978.
25. **Watkins, L., Jr., Burton, J. A., Haber, E., Cant, J. R., Smith, F. W., Barger, A. C., McNeil, S. E., and Sherrill, S. M.**, The renin-angiotensin-aldosterone system in congestive failure in conscious dogs, *J. Clin. Invest.*, 57, 1606, 1976.
26. **Merril, A. J.**, Mechanisms of salt and water retention in heart failure, *Am. J. Med.*, 6, 357, 1949.
27. **Sandler, H., Dodge, H. T., and Murdaugh, H. V.**, Effects of isoproterenol on cardiac output and renal function in congestive heart failure, *Am. Heart J.*, 62, 643, 1961.
28. **Cogan, J. J., Humphreys, M. H., Carlson, C. J., and Rapaport, E.**, Renal effects of nitroprusside and hydralazine in patients with congestive heart failure, *Circulation*, 61, 316, 1980.
29. **Auld, R. B., Alexander, E. A., and Levinsky, N. G.**, Proximal tubular function in dogs with thoracic caval constriction, *J. Clin. Invest.*, 50, 2150, 1971.
30. **Stumpe, K. O., Reinelt, B., Ressel, C., Klein, H., and Kruck, F.**, Urinary sodium excretion and proximal tubule reabsorption in rats with high-output heart failure, *Nephron*, 12, 261, 1974.
31. **Brenner, B. M., Falchuk, K. H., Keimowitz, R. I., and Berliner, R. W.**, The relationship between peritubular capillary protein concentration and fluid reabsorption in the renal proximal tubule, *J. Clin. Invest.*, 48, 1519, 1969.
32. **Barger, A. C., Rudolph, A. M., and Yates, E. F.**, Sodium excretion and hemodynamics in normal dogs, dogs with mild valvular lesions of the heart and dogs in congestive heart failure, *Am. J. Physiol.*, 183, 595, 1965.
33. **Hostetter, T. H., Pfeffer, J. M., Pfeffer, M. A., Fletcher, P. J., Braunwald, E., and Brenner, B. M.**, Impaired sodium excretion in rats with myocardial infarction-relation to cardiac function, *Clin. Res.*, 27, 176A, 1979.
34. **Schneider, E. G., Dresser, T. P., Lynch, R. E., and Knox, F. G.**, Sodium reabsorption by proximal tubule of dogs with experimental heart failure, *Am. J. Physiol.*, 220, 952, 1971.
35. **Stumpe, K. O., Sölle, H., Klein, H., and Krück, F.**, Mechanism of sodium and water retention in rats with experimental heart failure., *Kidney Int.*, 4, 309, 1973.
36. **Crumb, C. K., Minuth, A. N., Flasterstein, A. H., Hebert, C. S., Eknoyan, G., Martinez-Maldonado, M., and Suki, W. N.**, Sodium excretion and intrarenal hemodynamics in thoracic inferior vena cava constriction, *Am. J. Physiol.*, 232, F507, 1977.
37. **Mandin, H.**, Cardiac edema in dogs: proximal tubular and renal function, *Can. J. Physiol. Pharmacol.*, 57, 185, 1979.
38. **Freeman, R. H., Davis, J. O., Williams, G. M., DeForrest, J. M., Seymour, A. A., and Rowe, B. P.**, Effects of the oral converting enzyme inhibitor, SQ14225, in a model of low cardiac output in dogs, *Circ. Res.*, 45, 540, 1979.
39. **Williams, G. M., Davis, J. O., Freeman, F. H., DeForrest, J. M., Seymour, A. A., and Rowe, B. P.**, Effects of the oral converting enzyme inhibitor SQ14225, in experimental high output failure, *Am. J. Physiol.*, 236, F541, 1979.
40. **Levy, M.**, Effects of acute volume expansion and altered hemodynamics on renal tubular function in chronic caval dogs, *J. Clin. Invest.*, 51, 922, 1972.
41. **Gross, J. B. and Kokko, J. P.**, Effects of aldosterone and potassium sparing diuretics on electrical potential differences across the distal nephron, *J. Clin. Invest.*, 59, 82, 1977.
42. **Laragh, J. H. and Sealey, J. E.**, The renin-angiotensin aldosterone hormonal system and regulation of sodium potassium, and blood pressure homeostasis, in *Handbook of Physiology: Renal Physiology*, Orloff, J. and Berliner, R. W., Eds., American Physiological Society, Washington, D.C., 1973, 831.
43. **Laragh, J. H.**, Hormones and the pathogenesis of congestive heart failure: vasopressin, aldosterone and angtiotensin II. *Circulation*, 25, 1015, 1962.
44. **Gill, J. R., Jr.**, Edema, *Annual Review of Medicine*, 21, 269, 1970.
45. **Chonko, A. M., Bay, W. H., Stein, J. H., and Ferris, T. F.**, The role of renin and aldosterone in the salt retention of edema, *Am. J. Med.*, 63, 881, 1977.

46. **DeClue, J. W., Cowley, A. W., Jr., Coleman, T. G., McCaa, R. E., and Guyton, A. C.,** Influence of long-term low dosage infusion of angiotensin II on arterial pressure, plasma aldosterone concentration, and renal sodium excretion, *Physiologist,* 19, 165, 1976.
47. **Johnson, M. D. and Malvin, R. L.,** Stimulation of renal sodium reabsorption by angiotensin II, *Am. J. Physiol.,* 232, F298, 1977.
48. **Kimbrough, H. M., Jr., Vaughn, E. D., Jr., Carey, R. M., and Ayers, C. R.,** Effect of intrarenal angiotensin II blockade on renal function in conscious dogs, *Circ. Res.,* 40, 174, 1977.
49. **Lohmeier, T. E., Cowley, A. W., Trippodo, N. C., Hall, J. E., and Guyton, A. C.,** Effects of endogenous angiotensin II on renal sodium excretion and renal hemodynamics, *Am. J. Physiol.,* 233, F388, 1977.
50. **Hall, J. E., Guyton, A. C., Smith, M. J., Jr., and Coleman, T. C.,** Chronic blockage of angiotensin II formation during sodium deprivation, *Am. J. Physiol.,* 237, F424, 1979.
51. **McCaa, R. E., Hall, J. E., and McCaa, C. S.,** The effects of angiotensin I-converting enzyme inhibitors on arterial blood pressure and urinary sodium excretion. Role of the renal renin-angiotensin and kallikrein-kinin systems, *Circ. Res.,* 43(Suppl. I), 132, 1978.
52. **DiBona, G. F.,** Neurogenic regulation of renal tubular sodium reabsorption, *Am. J. Physiol.,* 233, F73, 1977.
53. **Slick, G. L., Aguilera, A. J., Zambraski, E. J., DiBona, G. F., and Kaloyanides, G. J.,** Renal neuroadrenergic transmission, *Am. J. Physiol.,* 229, 60, 1975.
54. **Slick, G. L., DiBona, G. F., and Kaloyanides, G. J.,** Renal sympathetic nerve activity in sodium retention of acute caval construction, *Am. J. Physiol.,* 226, 925, 1974.
55. **Azer, M., Gannon, R., and Kaloyanides, G. J.,** Effect of renal denervation on the antinatriuresis of caval construction, *Am. J. Physiol.,* 222, 611, 1972.
56. **Gill, J. R., Carr, A. A., Fleischmann, L. E., Casper, A. G. T., and Bartter, F. C.,** Effects of pentolinium on sodium excretion in dogs with construction of the vena cava, *Am. J. Physiol.,* 212, 191, 1967.
57. **Levinsky, N. G.,** Natriuretic Hormones, *Adv. Metab. Disord.,* 7, 37, 1974.
58. **Kruck, F. and Kramer, H. J.,** Third factor and edema formation, *Contrib. Nephrol.,* 13, 12, 1978.
59. **Bell, N. H., Schedl, H. P., and Bartter, F. C.,** An explanation for abnormal water retention and hypoosmolality in congestive heart failure, *Am. J. Med.,* 36, 351, 1964.
60. **Bennett, W. M., Bagby, G. C., Antonovic, J. N., and Porter, G. A.,** Influence of volume expansion on proximal tubular sodium reabsorption in congestive heart failure, *Am. Heart. J.,* 85, 55, 1973.
61. **Cirksena, W. J., Dirks, J. H., and Berliner, R. W.,** Effect of thoracic caval obstruction on response of proximal tubule sodium reabsorption to saline infusion, *J. Clin. Invest.,* 45, 179, 1966.
62. **Mandin, H.,** Cardiac edema in dogs. I. Proximal tubular and renal function, *Kidney Int.,* 10, 591, 1976.
63. **Alexander, E. A., Bengele, H. H., and McNamara, E. R.,** The effect of adrenal enucleation on sodium excretion in the rat, *Am. J. Physiol.,* 232, F566, 1977.
64. **Bernard, D. B., Alexander, E. A., Couser, W. G., and Levinsky, N. G.,** Renal sodium retention during volume expansion in experimental nephrotic syndrome, *Kidney Int.,* 14, 478, 1978.

Chapter 5

LIVER DISEASE*

Murray Epstein

TABLE OF CONTENTS

* Portions of this paper are taken from Epstein, M., Deranged sodium homeostasis in cirrhosis, *Gastroenterology,* 76, 622, 1979. Reproduced with permission.

I. INTRODUCTION

One of the most intriguing problems in clinical medicine is the relationship between the diseased liver and the anatomically normal kidney which exhibits a wide spectrum of deranged renal function.[1] Prominent among such derangements is the progressive impairment of renal sodium handling, leading to the formation of ascites and edema, which frequently complicates the course of patients with decompensated Laennec's cirrhosis. Such involvement of the kidney in cirrhosis has been known for many years. Despite interest in and familiarity with this clinical entity, the management of patients with liver disease and fluid retention has been largely empirical, consisting primarily of the administration of diuretics. The past several years have witnessed a resurgence of interest in the investigation of deranged renal function in liver disease in general, and renal sodium handling in particular. A number of these studies have succeeded in delineating further the pathophysiology of the abnormal renal sodium handling. The available data have prompted a reappraisal of our thinking with regard to the pathophysiology of this abnormality and its management.

This chapter reviews the clinical features and pathogenesis of the sodium retention of liver disease. The thrust of this review is to demonstrate that the attribution of deranged sodium handling to abnormalities of aldosterone is an oversimplification and that it is more likely that the disordered physiology is due to the combined effects of derangements of several regulatory systems.

Finally, although the intent of this review was to survey the aberrations of sodium homeostasis in liver disease of diverse etiology, it became readily apparent that this review must perforce consider deranged sodium homeostasis in cirrhosis. There is a paucity of information in man on this subject in hepatic conditions other than cirrhosis. Thus, discussion of the pathogenesis of renal sodium retention is restricted to the cirrhotic population.

II. CLINICAL FEATURES

One of the earliest published observations of the association of ascites and edema with liver disease is attributed to John Brown in 1685.[2] During the subsequent three hundred years, the clinical features of this derangement in sodium homeostasis have been delineated.[3-11] It is well established that patients with Laennec's cirrhosis manifest a remarkable capacity for sodium chloride retention; indeed, such patients frequently excrete urine which is virtually free of sodium.[3-5,11] Extracellular fluid accumulates excessively and eventually becomes evident as clinically detectable ascites and edema. It should be emphasized that cirrhotic patients who are unable to excrete sodium will continue to gain weight and accumulate ascites and edema as long as dietary sodium content exceeds maximal urinary sodium excretion. If access to sodium is not curtailed, the relentless retention of sodium may lead to the accumulation of vast amounts of ascites (on occasion up to 20 ℓ). Weight gain and ascites formation promptly cease when sodium intake is limited to approximate maximal sodium excretion.

The abnormality of renal sodium handling in cirrhosis should not be perceived as a static condition. Rather, cirrhotic patients frequently manifest an undulating clinical course; they may undergo a spontaneous diuresis followed by a return to avid salt retention.[12] Although a large number of patients who are maintained on a sodium restricted dietary program may demonstrate a spontaneous diuresis, there is inadequate information about the incidence with which this occurs.

In this regard, the report of Bosch et al.[13] is of interest. These observers reported that 35 of 124 patients with cirrhosis and ascites who were hospitalized, underwent a spontaneous diuresis in response to bedrest and dietary sodium restriction. Since all patients with liver disease and ascites in the area served by the hospital were hospitalized during the period of the report, the figure of 28% reported by these authors may be representative of the frequency with which spontaneous diuresis occurs.

It has been suggested that patients presenting with ascites and edema for the first time, who have retained salt and water isotonically, excrete more than 10 mEq of sodium daily in the urine, and have normal GFR, are more likely to undergo a spontaneous diuresis. Patients who have reversible liver disease, such as fatty liver of the alcohol abuser, also tend to respond favorably when abstinent, rested, and fed.

Finally, it cannot be stressed often enough that the primary renal excretory abnormality causing fluid retention is a disturbance of sodium rather than of water excretion. Thus, many sodium-retaining patients with ascites and edema can excrete urine of low osmolality when given excessive amounts of water without sodium.[14,15] However, when sodium is administered it is not excreted.

A. Etiology of Hepatic Disease as a Determinant of Renal Sodium Retention

Although this review deals with sodium retention in *liver* disease, it should be stressed that this derangement does not complicate all diseases of the liver equally. In contrast to the findings in patients with Laennec's cirrhosis in whom renal sodium retention is common, patients with primary biliary cirrhosis (PBC) do not appear to manifest this abnormality. Chaimovitz et al.[16] have recently assessed the natriuretic and diuretic response to expansion of extracellular fluid volume (ECFV) in five patients with PBC. Despite conspicuous evidence of portal hypertension, ascites and edema were absent. These investigators demonstrated that the natriuretic and diuretic response exceeded that observed in both healthy normal volunteers and in edema-free patients with Laennec's cirrhosis. The authors suggested that a common mechanism may underlie both the augmented natriuretic response to volume in PBC patients and the rarity of fluid retention observed in this type of cirrhosis.[17]

The reasons why patients with PBC and Laennec's cirrhosis, despite comparable portal hypertension, manifest markedly different responses to volume expansion require explanation. A careful examination of the differences between these two disease processes, including the relative preservation of hepatic parenchyma found in PBC, may have important implications for a further understanding of the mechanisms which underlie deranged sodium homeostasis in liver disease.

B. Fulminant Hepatic Failure

Another abnormality of the liver in which limited data are available is fulminant hepatic failure. While there are relatively limited data available on renal function in fulminant hepatic failure, in a preliminary communication, Wilkinson et al.[18] suggested that the latter condition is also associated with sodium retention. Only additional detailed observation in this patient population will permit substantiation of this suggestion.

III. PATHOGENESIS

A consideration of the pathogenetic events leading to the deranged sodium homeostasis of cirrhosis is simplified by a consideration of the "afferent" and the "efferent" events which eventuate in this derangement. The discussion of afferent events will include consideration of the detector element responsible for the recognition of the degree of volume alteration, as well as a consideration of the extracellular fluid translocations or sequestration into serous spaces or interstitial fluid compartments which characterize advanced liver disease. The section considering efferent events will encompass a survey of the hormonal mediators implicated in the pathogenesis of sodium retention.

A. Afferent Events

Before examining the "afferent" events, it would be worthwhile to consider two concepts which have been frequently cited in any synthesis of the pathogenesis of the abnormal sodium retention of liver disease: (1) the role of a diminished "effective" volume; and (2) the "overflow" theory of ascites formation.

1. Role of Diminished "Effective" Extracellular Volume

Traditionally, it has been proposed that ascites formation in cirrhotic patients begins when a critical imbalance of Starling forces in the hepatic sinusoids and splanchnic capillaries causes an excessive amount of lymph formation, exceeding the capacity of the thoracic duct to return this excessive lymph to the circulation.[19,20] Consequently, excess lymph accumulates in the peritoneal space as ascites, with a subsequent contraction of circulating plasma volume. Thus, as ascites develops, there is a progressive redistribution of ECFV. While total plasma volume may be increased in this setting, the physiologic circumstance may mimic a reduction in volume (a reduced "effective" ECFV). The diminished "effective" volume is thought to constitute an afferent signal to the renal tubule to augment salt and water reabsorption. Thus, the traditional formulation suggests that the renal retention of sodium is a secondary rather than a primary event.

In this context, it is important to emphasize that the term "effective" ECFV refers to that part of the total circulating volume which is effective in stimulating volume receptors. The concept is elusive because the actual volume receptors remain incompletely defined. A diminished "effective" volume may reflect subtle alterations in systemic hemodynamic factors such as decreased filling of the arterial tree, a diminished central blood volume,[21] or both. Because the stimulus is unknown and the afferent receptors are incompletely elucidated, alterations in "effective" ECFV must be defined in a functional manner such as the kinetic response to volume manipulation.[21]

2. "Overflow" Theory

Over the past decade, an alternative hypothesis has been proposed. Lieberman and associates[22-24] proposed the "overflow" theory for ascites formation. In contrast to the traditional formulation, the overflow theory suggests that the primary event is the inappropriate retention of excessive sodium by the kidneys with a resultant expansion of plasma volume. In the setting of abnormal Starling forces (both portal venous hypertension and a reduction in plasma colloid osmotic pressure) in the portal venous bed and hepatic sinusoids, the expanded plasma volume is sequestered preferentially in the peritoneal space with ascites formation. Thus, according to this formulation, renal sodium retention and expansion of ECFV precede rather than follow the formation of ascites.

Since the promulgation of the overflow theory of ascites formation, controversy has centered on which of the two hypotheses is correct. The demonstration that plasma volume is increased in cirrhosis with ascites and the finding that increases in measured plasma volume have not been observed in ascitic cirrhotic patients undergoing a spontaneous diuresis have been cited as evidence in support of the overflow hypothesis. This hypothesis has recently received additional support from a series of elegant investigations carried out by Levy[25,26] on dogs with experimental portal cirrhosis. Levy has succeeded in developing a canine model for cirrhosis by the feeding of dimethylnitrosamine and has demonstrated in sequential studies that renal sodium retention is the initial event that precedes ascites formation.

Recently, these investigators attempted to test this hypothesis further by studying dogs with experimental cirrhosis of the liver before and after insertion of a peritoneovenous shunt, thereby preventing the reformation of ascites. In essence, the authors sought to study a model where the liver remains damaged, but where vascular underfilling because of fluid leaving the circulation cannot occur. It was demonstrated that increasing oral sodium intake in such dogs resulted in the development of anasarca and weight gain, but not ascites. Since the cirrhotic animals retained sodium in the absence of peritoneal sequestration of fluid, the authors interpreted their data as supporting the overflow theory of ascites formation.

While these observations collectively support the overflow theory of ascites formation, a number of clinical observations in *man* are inconsistent with such a formulation. Thus, rapid volume expansion with exogenous solutions including saline, mannitol, and albumin frequently result in a transient improvement in renal sodium and water handling.[27-29] However, the lack of specificity of these maneuvers that increase the volume of all fluid compartments and the presumed concomitant alterations in plasma composition have precluded definitive statements regarding the etiologic role of a diminished effective plasma volume.

Studies from our laboratory over the past 10 years have succeeded in circumventing many of the aforementioned methodologic problems by applying a newly developed investigative tool, the water immersion model, to the assessment of renal sodium and water homeostasis in diverse edematous states. Such an approach has recently been undertaken by our laboratory utilizing the model of head-out water immersion in cirrhotic man. Before proceeding to enumerate these findings, it may be worthwhile to briefly describe the salient features of the model (Table 1) and to underscore the differences between water immersion and more traditional attempts to achieve ECVE such as saline administration.

Head-out water immersion in the seated posture results in a prompt redistribution of blood volume with a sustained increase in central blood volume.[30,31] Furthermore, studies of the "efferent" limb of immersion have demonstrated a marked natriuretic and kaliuretic response during immersion that is indistinguishable from that of saline administration (2 ℓ/120 min) in normal subjects, assuming an identical seated posture.[32] While water immersion constitutes a potent means of inducing central volume expansion comparable to that induced by the infusion of isotonic saline,[31] it is capable of circumventing many of the problems associated with the infusion of exogenous volume expanders.

In contrast to saline administration:

1. Water immersion is associated with a decrease in body weight, rather than the increase which attends saline infusion.
2. The "volume stimulus" of immersion is promptly reversible after cessation of immersion, in contrast to the relatively sustained hypervolemia which follows saline administration; thus, this reversibility constitutes an important attribute in minimizing any risk to the patients.
3. In contrast to saline administration, the "volume stimulus" of immersion occurs in the absence of changes in plasma composition.

The delineation of the immersion model and the demonstration that it constitutes a potent "central volume stimulus" without the necessity of infusing exogenous volume expanders, commended its use for assessing the role of alterations of "effective" ECFV in the derangements of renal sodium homeostasis in cirrhosis. Studies in 34 patients with decompensated cirrhosis demonstrated a striking "normalization" of renal sodium handling. As shown in Figure 1, immersion resulted in marked natriuresis

Table 1
SALIENT FEATURES OF THE MODEL OF HEAD-OUT WATER IMMERSION

Immersion produces a prompt redistribution of circulating blood volume with a relative central hypervolemia

Cardiac output is increased by 25 to 33% and central blood volume by approximately 700 mℓ

The alterations in central hemodynamics are sustained throughout a 4-hr immersion period and are promptly reversible following cessation of immersion

Immersion-induced central hypervolemia is associated with a profound and progressive natriuresis, kaliuresis, and diuresis; these alterations are promptly reversible following cessation of immersion

The central hemodynamic and renal effects of immersion are equal in magnitude to those induced by acute saline administration (2ℓ of saline/2 hr)

In the sodium-replete state, the alterations in renal sodium, potassium, and water handling occur in the absence of changes in renal plasma flow and GFR

Immersion is associated with a prompt and profound suppression of PRA (62%) and plasma aldosterone (66%); cessation of immersion is associated with a prompt return of both PRA and PA to prestudy levels

The aforementioned alterations in renal function and renin-aldosterone responsiveness occur in the absence of changes in plasma composition

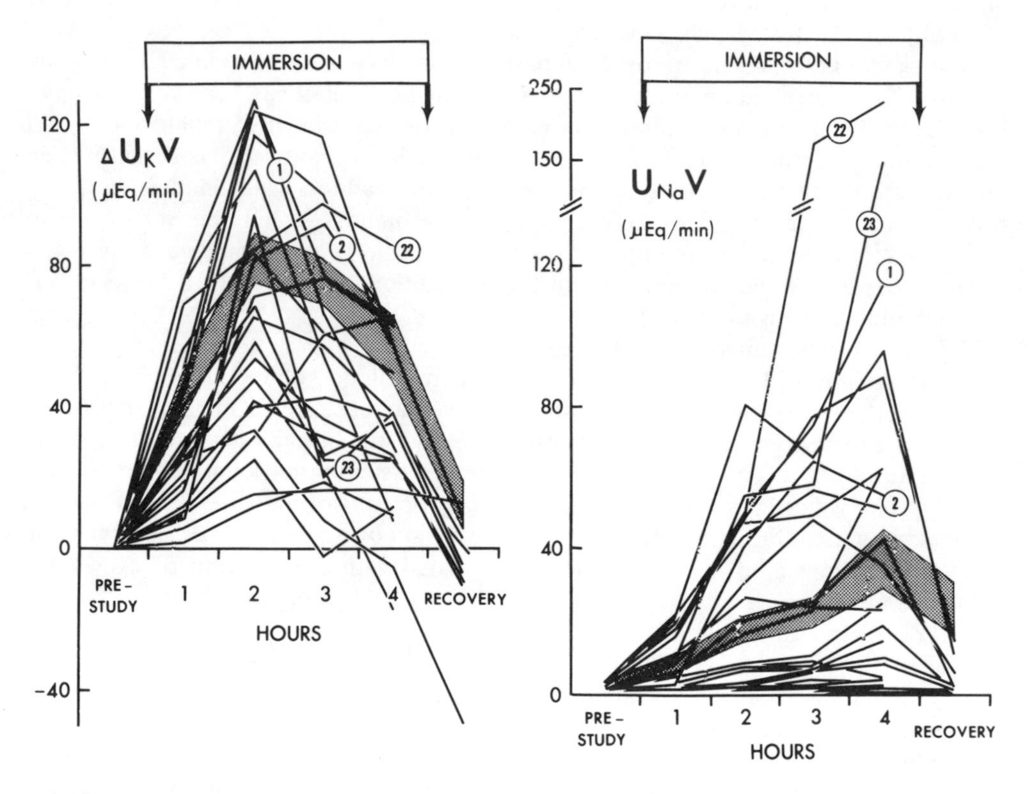

FIGURE 1. Effects of water immersion following 1 hr of quiet sitting (prestudy) on rate of sodium excretion ($U_{Na}V$) and potassium excretion (U_KV) in 25 patients with alcoholic liver disease. The circled numbers represent individual patients. Data for U_KV are expressed in terms of the absolute changes from prestudy hour (ΔU_KV). The shaded area represents the mean ± SE for 14 normal control subjects undergoing an identical immersion study while ingesting an identical 10 mEq sodium / 100 mEq potassium diet. Fourteen of the cirrhotic patients manifested an appropriate or "exaggerated" natriuretic response. In general, the increase in $U_{Na}V$ was associated with a concomitant increase in ΔU_KV.

and kaliuresis in the majority of these patients. During the final hour of immersion, $U_{Na}V$ was 20-fold greater than it was during the prestudy hour. Thus, the marked antinatriuresis of cirrhosis was promptly reversed by a manipulation which merely altered the distribution of ECFV without increasing (and often decreasing) total volume.

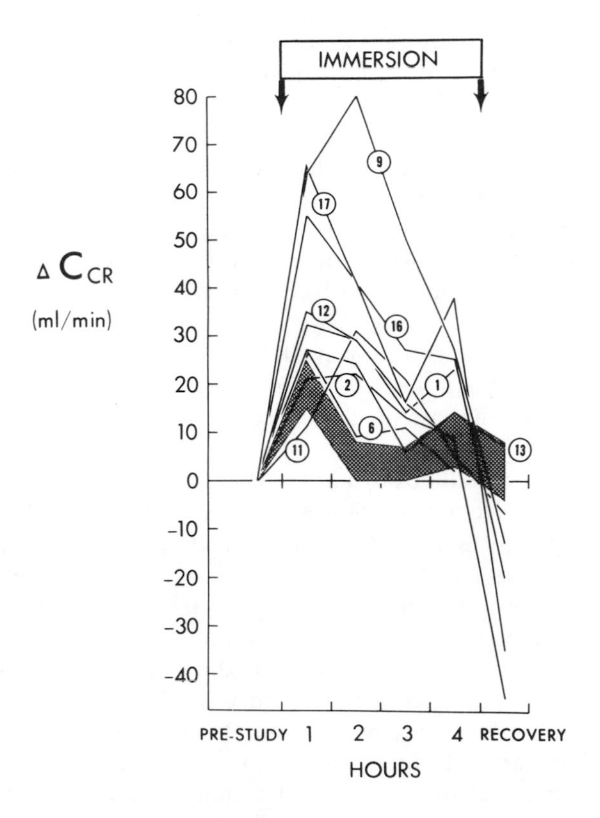

FIGURE 2. Effect of water immersion to the neck on creatinine clearance (C_{Cr}) in 13 cirrhotic patients. Shaded area represents the mean ±SE for 14 normal subjects undergoing an identical study while receiving an identical 10 mEq Na/100 mEq K diet. Data are expressed as the absolute changes from the pre-immersion hour (ΔC_{Cr}). The immersion-induced increments in C_{Cr} exceeded those seen in normal subjects. Cessation of immersion (recovery hour) was associated with prompt decrements in C_{Cr}. (From Epstein, M. et al., Characterization of the renin-aldosterone system in decompensated cirrhosis, *Circ. Res.*, 41, 818, 1977. By permission of the American Heart Association, Inc.)

These studies lend strong support to the concept that a diminished "effective" intravascular volume is a major determinant of the enhanced tubular reabsorption of sodium in cirrhosis.

That the immersion-induced natriuretic response is indeed supportive of a role for diminished "effective" volume and not merely an appropriate response comparable to that of normal subjects can be adduced from the concomitant changes in C_{Cr} during immersion.[33] In marked contrast to the demonstration that inulin and PAH clearances are unchanged during immersion in normal subjects,[34] immersion was associated with significant increments in C_{Cr} in a majority of cirrhotic patients.[33] A comparison of the individual changes in C_{Cr} of cirrhotic patients during immersion with the C_{Cr} changes manifest by sodium-depleted normal subjects undergoing an identical protocol disclosed that two thirds of the cirrhotic patients manifested increments in C_{Cr} which exceeded markedly (three- to fivefold) the increments observed in the normal subjects (Figure 2). These observations suggest that immersion tends to "normalize" the diminished effective volume of cirrhotic man with a resultant normalization of renal vascular tone.

Additional support for the concept that the diminished "effective" volume participates in the sodium retention of cirrhosis is derived from the studies of Schroeder et al. utilizing the LeVeen shunt.[35] In essence, this procedure constitutes a means of redistributing body fluid between compartments without increasing total plasma volume. Thus, the continuous peritoneovenous shunt resulted in marked increments in renal blood flow and GFR and profound suppression of PRA and PA.[35] Since, in essence, this procedure withdraws ascitic fluid from the peritoneal cavity and redistributes it to the circulating volume, the striking findings with regard to renal hemodynamics and renin-aldosterone are consistent with the interpretation that the initially low GFRs were attributable to a diminished "effective" blood volume.

Although I believe that the presently available evidence favors a prominent role for diminished effective volume in mediating the avid sodium retention of many cirrhotic patients, it is worthwhile emphasizing that these two formulations may not be mutually exclusive.

Virtually all the available clinical studies of deranged sodium homeostasis were carried out at a time when decompensation was well established, with little information available during the incipient stage of sodium retention. In contrast, the studies with the canine cirrhosis model deal with the relatively early stage of sodium retention. It seems to me that any formulation which suggests that the same antinatriuretic forces are operative throughout the evolution of sodium retention in cirrhotic man, is a marked oversimplification. Rather, one should adopt a more global view of the pathogenesis of abnormal sodium retention in cirrhosis in which differing forces participate in varying degrees as the derangement in sodium homeostasis evolves. Thus, it is quite conceivable that a primary defect in renal sodium handling may assume a more prominent role in the early stages of cirrhosis and a diminished effective volume may constitute the major determinant of sodium retention in many patients, once the derangement is established. The demonstration that the sodium retention in the canine cirrhosis model occurs in the absence of a concomitant decrement in GFR, in contrast to the situation in cirrhotic man, lends credence to such a formulation.

B. Efferent Events

The efferent factors which mediate the tubular retention of sodium in cirrhosis and their relative participation in the avid sodium retention have not been elucidated completely. Several mechanism(s) have been implicated or suggested, including (1) hyperaldosteronism; (2) alterations in intrarenal blood flow distribution; (3) the possible role of a humoral natriuretic factor; (4) an increase in sympathetic nervous system activity; (5) alterations in the endogenous release of renal prostaglandins; and (6) changes in the kallikrein-kinin system.

While the impairment to renal sodium handling in cirrhosis generally correlates with a moderate diminution in glomerular filtration rate (GFR), the impairment of sodium excretion can frequently be dissociated from changes in GFR.[8,36] Such observations suggest that while a decrease in filtered sodium load may contribute to the sodium retention of cirrhosis, it is unlikely that it constitutes the predominant determinant of this marked abnormality of renal function. Rather, the impaired sodium excretion is attributable primarily to an increased tubular reabsorption of sodium.[10] The mechanism(s) for enhanced sodium reabsorption remain controversial.

1. Hyperaldosteronism

With the isolation and characterization of aldosterone in the early 1950s, the task of explaining the abnormal sodium retention of cirrhosis was ostensibly rendered simple. Initially, several investigators documented increased levels of aldosterone in the urine of cirrhotic patients and increased adrenal secretion of aldosterone.[37-39] With the development of radioimmunoassay techniques, aldosterone has been demonstrated to be markedly elevated in peripheral plasma as well.[33,40] Many observers seized on this observation and proposed that aldosterone is the *major* determinant of sodium retention.

A number of recent studies have questioned the validity of such a formulation, suggesting that the elevated PA levels and the sodium retention of cirrhosis may be pari passu events and not necessarily etiologically related. Thus, Rosoff et al.[40] recently demonstrated dissociations between sodium excretion and PA in a group of cirrhotic patients. They noted that two of their patients underwent spontaneous diuresis without a concomitant change in PA concentration. Furthermore, these authors administered aminoglutethimide to three cirrhotic patients and noted that despite profound and rapid decreases in PA levels, neither a natriuresis nor a decrease in body weight ensued.[40]

Chonko et al.[41] examined the response of 11 cirrhotic patients to chronic dietary sodium loading (300 mEq of Na/day) for 5 days and demonstrated a striking dissociation between plasma aldosterone levels and renal sodium handling. While five patients manifested a normal suppression of PRA and aldosterone in response to sodium loading, the remaining six patients (designated Group A) manifested persistent hypersecretion of PRA and PA despite a 6.1 kg weight gain and a sodium retention of 881 mEq.

Wernze et al.[42] simultaneously determined PRA, PRC, angiotensinogen and plasma aldosterone levels in a large series of patients with cirrhosis and demonstrated a further dissociation between hyperaldosteronism and renal sodium handling. They noted that mean plasma aldosterone in 23 patients with moderately tense ascites (their Group I) was 12.8 ± 6.8 ng/100 mℓ, a value which did not differ from controls. These examples are representative of a number of studies demonstrating a dissociation between PA and renal sodium handling.

Although the above studies demonstrated a striking dissociation between elevated PA levels and increasing sodium excretion, none attempted to assess in a kinetic manner the responses of PA and renal Na excretion to *acute* volume manipulation. While an assessment of the relationship of PA to renal sodium handling during acute volume expansion was clearly important, methodologic considerations precluded investigations of the renal and hormonal response to acute volume expansion in cirrhotic man.[42] The delineation of the water immersion model has recently facilitated such kinetic studies. During the past several years, our laboratory has systematically applied the immersion model to a kinetic assessment of the relationship of PA responsiveness to renal sodium handling (Figure 3). It was demonstrated that immersion resulted in a prompt and profound (66%) suppression of PA in 15 of 16 patients. Despite suppression of PA to similar nadir levels in all patients, eight of the patients manifested an absent or markedly blunted natriuretic response during immersion.[33] This demonstration of a dissociation between the suppression of circulating aldosterone and the absence of a natriuresis lends strong support to the interpretation that aldosterone is not the primary determinant of the impaired sodium excretion of cirrhosis.

Additional studies utilizing the immersion model complemented this conclusion. Immersion studies were carried out during chronic aldosterone blockade.[9] We proposed that the administration of spironolactone to patients undergoing both control and immersion studies would permit an assessment of the relative contribution of aldosterone to the changes in renal sodium, potassium, and water handling. It was demonstrated that spironolactone administration without immersion resulted in only a modest increase in sodium excretion. If the encountered sodium retention was attributable to elevated aldosterone levels per se, one would have anticipated a more marked natriuresis with spironolactone administration alone. In contrast, there was a dramatic increase in sodium excretion when immersion was carried out during chronic spironolac-

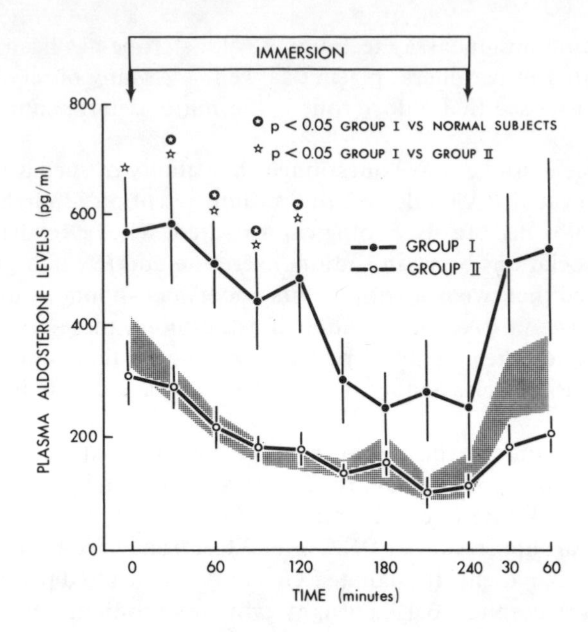

FIGURE 3. Effect of water immersion on plasma aldosterone (PA) in 16 cirrhotic subjects in balance on a 10 mEq sodium diet. Shaded area represents the mean ± SE for nine normal control subjects undergoing immersion while receiving an identical 10 mEq sodium and 100 mEq potassium diet. The immersion-induced suppression of PA in Group II subjects paralleled that of normal controls undergoing an identical study. Although prestudy PA levels in Group I subjects were twofold greater than both Group II subjects and normals, PA suppressed markedly during immersion, attaining levels not different from Group II or normal subjects. (From Epstein, M. et al., Characterization of the renin-aldosterone system in decompensated cirrhosis, *Circ. Res.*, 41, 818, 1977. By permission of the American Heart Association, Inc.)

tone administration, thereby indicating that the major contribution to the natriuresis was an enhanced distal delivery of filtrate. This explanation was supported by the documentation of a concomitant kaliuresis and an increase in free-water clearance.[9]

Taken together, the evidence presently available, therefore, favors the postulate that the hyperaldosteronism of cirrhosis is a permissive factor only, and that predominant component of the abnormal renal sodium handling is a diminished distal delivery of filtrate. Only when distal filtrate delivery is enhanced by an experimental or pharmacological maneuver does aldosterone exert an active role in the renal sodium handling in cirrhosis.

The demonstration that hyperaldosteronism cannot account completely for sodium retention in cirrhosis has prompted a search for other hormonal mediators which might participate in this derangement.

2. Renal Prostaglandins

That renal prostaglandins may be altered in chronic liver disease has been suggested by several reports indicating that PGE is elevated in patients with cirrhosis and ascites. Zipser et al.[43] reported that urinary immunoassayable PGE (iPGE) was markedly elevated in subjects with cirrhosis and ascites as compared to normal control subjects. Zusman et al.[44] determined plasma prostaglandin E concentration in three patients with cirrhosis and renal failure and observed that the levels were markedly elevated in all three subjects. However, not all observers have documented an elevation in renal PGE.[45,46]

The mechanism(s) responsible for the elevated PGE levels have not been elucidated, but possibilities include (1) increased renal synthesis, and (2) deranged metabolism. While the available evidence favors increased renal synthesis as the major cause of the elevated PGE levels, preliminary studies in a small group of cirrhotic patients have at least raised the possibility that alterations in the metabolism and/or handling of PGE in cirrhosis may also contribute to the encountered PGE elevation. In the course of attempting to investigate the effects of intrarenal PGE_1 on the renal vasoconstriction of the hepatorenal syndrome, Zusman et al.[44] noted striking and unanticipated changes in plasma PGE. Following the infusions, plasma PGE rose significantly in all three patients and achieved extremely high levels in two of the patients. Since animal studies have revealed that in excess of 95% of PGE is metabolized during a single passage through the pulmonary circulation, and since studies by Robertson reveal that normal humans demonstrated an insignificant rise in plasma levels after intravenous infusion of doses of PGE_2 up to 20 μg/min,[47] it is unlikely that the observed increments in plasma PGE were attributable to exceeding the metabolic capacity of the lung. Rather, the encountered results are consistent with (1) shunting from the systemic venous circulation to the arterial circulation (i.e. a right-to-left shunt), thereby bypassing the pulmonary metabolizing system; or (2) the possibility that the rate of metabolism of prostaglandin E is markedly diminished in patients with hepatic and/or renal failure. Clearly, additional studies are necessary to confirm and extend these preliminary observations.

Attempts to investigate the role of renal prostaglandins in mediating the sodium retention in cirrhosis have encompassed two manipulations: (1) administration of exogenous prostaglandins; and (2) alteration of the endogenous production of prostaglandins by inhibition of prostaglandin synthesis. Initially, the problem was approached by examining the renal hemodynamic response to the administration of exogenous prostaglandins. Thus, Arieff and Chidsey reported that the administration of prostaglandin A_1 (PG-A_1) when given intravenously in incremental doses of from 0.01 to 1.0 μg/kg/min resulted in increases in both GFR and effective renal plasma flow (ERPF) in many but not all of their patients.[48] Thus, the largest increase in ERPF occurred in patients with baseline ERPF values greater than 450 mℓ/min, with lesser increments in patients with baseline ERPF's between 150 and 450 mℓ/min. In contrast, patients with ERPF values less than 150 mℓ/min did not respond to PG-A_1 infusions. Similarly, sodium excretion was enhanced in patients with ERPF greater than 150 mℓ/min, but was unaltered in patients with lesser ERPF.

Unfortunately, the relevance of such studies to cirrhotic man is tenuous since the physiological significance of PGA in both rabbit and man has been challenged.[49] Furthermore, the interpretation of studies in which the effects of prostaglandin administration on renal sodium and water handling are examined, is difficult. The demonstration that PGE is almost completely inactivated in each passage across the lung[50] suggests that any action of this lipid on the kidney must be as a local tissue hormone. Thus, any evaluation of the physiological role of protaglandins on urinary sodium excretion necessitates an experimental design in which the endogenous production of the lipids is altered. Indeed, recent investigations of the role of prostaglandins in renal sodium and water handling have focused on comparisons before and after the administration of inhibitors of prostaglandin synthetase.

Boyer and Reynolds administered indomethacin (50 mg every 6 hr), a potent inhibitor of prostaglandin synthetase, to patients with alcoholic liver disease.[51] In two patients without sodium retention, indomethacin administration did not alter either creatinine clearance or ERPF. In contrast, indomethacin resulted in a decline in C_{Cr} and ERPF in cirrhotic patients with sodium retention. Since indomethacin is known to inhibit prostaglandin synthetase with a resultant decrease in renal blood flow in a number of experimental animals, these interesting observations raise the possibility that endogenous prostaglandin production may be a major determinant of ERPF and renal sodium handling in decompensated cirrhosis.

Recently, Zipser et al.[43] have documented markedly elevated prostaglandin E_2 levels in 11 cirrhotic patients with avid sodium retention. The administration of inhibitors of prostaglandin synthetase (both indomethacin and ibuprofen) resulted in a lowering of PRA and PA and a 57% decrement in creatinine clearance. It is apparent that additional studies including attempts to enhance endogenous renal prostaglandins are necessary to characterize further the role of prostaglandins as determinants of the abnormal sodium retention of cirrhosis.

3. Kallikrein-Kinin System

Several lines of evidence have suggested that bradykinin and other kinins synthesized in the kidney may participate in the modulation of intrarenal blood flow and renal sodium handling. Experimental studies[52,53] have demonstrated that i.v. infusion of kinins into humans or into the renal artery of dogs produces an increase in renal blood flow, urine flow, and urinary sodium excretion. Furthermore, alterations of plasma bradykinin and of urinary kallikrein excretion have been shown to correlate with changes in sodium balance and mineralocorticoid activity in laboratory animals and in normal subjects.[54]

It should be appreciated that the kallikrein-kinin system is highly complex and that there are a number of controversial areas concerning the control and importance of its activity.[55] There are structural and perhaps physiologic differences between *plasma* and *urinary* kallikreins and their kinin end products. There is considerable controversy as to whether kinins act on blood vessels, causing vasodilation, or directly on the tubular epithelial handling of salt and water.

Renal kallikrein releases the kinin, kallidin, from plasma kininogens. At present, the activity of the renal kallikrein-kinin system usually is inferred from measurements of urinary kallikrein. It should be borne in mind that the implicit assumption that kallikrein is the rate-limiting step in the system, while reasonable, is unproven. Moreover, urinary excretion may not reflect renal synthesis accurately, because some kallikrein may leave the kidney in blood or lymph or be metabolized in the kidney or urine. Regardless, most investigators believe that urinary kallikrein excretion constitutes an appropriate index of the activity of the renal kallikrein-kinin system.

Despite these questions, it seems appropriate to inquire whether the kallikrein-kinin system may be implicated in the pathogenesis of the abnormal sodium retention of liver disease. That the kallikrein-kinin system may indeed be involved in mediating sodium retention of liver disease is suggested by several preliminary reports of abnormalities of the *plasma* kallikrein system. Bagdasarian et al.[56] observed that in mild cirrhosis, plasma prekallikrein activity (by specific immunoassay) and plasma esterase activity were reduced proportionately by about one half. In severe cirrhosis, immunological kallikrein was not further depressed, but functional kallikrein by the esterase method decreased to 8% of normal. Although the data lend themselves to several interpretations, the investigators suggested that the liver in advanced cirrhosis secreted an inactive molecule. It is possible that renal salt retention in cirrhosis is due partly to decreased activity of the kallikrein-kinin system.

Wong et al.[57] extended these studies to patients with the hepatorenal syndrome. It was noted that "functional plasma prekallikrein" (esterase) was reduced still further to about 30% of that in patients with comparably severe cirrhosis, and plasma bradykinin was also decreased. As would be expected, plasma renin activity was greatly increased in both groups. The authors speculated that the vasodilator (kinin) deficiency (perhaps in conjunction with vasoconstrictor excess) may contribute to the pathogenetic mechanisms mediating both sodium retention as well as renal failure.

Although these observations are provocative, additional studies are necessary to characterize further the role of the kallikrein-kinin system as a determinant of the abnormal sodium retention of cirrhosis. It should be emphasized that the aforementioned preliminary results have examined the plasma kallikrein system, and an evaluation of the renal kallikrein system would prove more informative. It is to be hoped that additional studies aimed at evaluating the renal kallikrein-kinin system, in concert with a further characterization of the role of the system in normal sodium homeostasis, will permit further delineation of the role of the kallikrein-kinin system in the deranged sodium homeostasis of liver disease.

4. Increase in Sympathetic Nervous System Activity

It is possible that an increase in sympathetic nervous system activity may contribute to the sodium retention of cirrhosis. Alterations in adrenergic nervous activity can modify urinary sodium excretion by several mechanisms:

1. By influencing Starling forces in the peripheral capillary bed
2. By affecting the central blood volume and thus the distribution of the ECF
3. By alterations in renal sympathetic nerve activity.

It is now well established that alterations of the input of cardiopulmonary receptors induce changes in renal sympathetic activity.[58] Thus, a decrease in blood volume can alter the afferent input from the cardiopulmonary region with a resultant increase in sympathetic nerve activity.[30] Such an increase in sympathetic tone could contribute to the antinatriuresis of cirrhosis in one of several ways. First, such an alteration could effect a redistribution of blood flow within the kidney resulting in increased net reabsorption of filtrate.[59] Furthermore, studies by DiBona and co-workers provide evidence for a direct tubular effect of the renal sympathetic nerves on renal sodium handling in the absence of alterations in renal hemodynamics.[60] It is thus conceivable that an increase in renal sympathetic activity, possibly due to a diminished "effective" volume, may contribute in several ways to the observed sodium retention of cirrhosis. It should be noted, however, that unanimity is lacking regarding the role of the renal nerves on renal sodium handling. Lifschitz[61] observed that a renal denervation failed to blunt either the antinatriuretic response to hemorrhage or the natriuretic response to volume expansion in awake dogs. Additional studies are needed to better define the precise role of the renal nerves in the physiologic control of urinary sodium excretion.

Despite such theoretical considerations, there are few data bearing on the importance of alterations in the sympathetic nervous system in cirrhotic man. Furthermore, the available information regarding the role of the adrenergic nervous system in cirrhosis is conflicting. Mashford et al.[62] assessed the response to the administration of tyramine and norepinephrine in a group of cirrhotic patients. In contrast to the "normal" blood pressure response to tyramine in a group of normotensive cirrhotic patients, cirrhotic patients with hypotension, oliguria, and progressive azotemia manifested a diminished blood-pressure response to tyramine injection. The authors interpreted their results to suggest impaired tissue norepinephrine stores and altered sympathetic nervous system function in some cirrhotic patients.

In contrast, studies by Epstein et al.[63] have raised the possibility that adrenergic activity may be enhanced in cirrhosis. In the course of assessing intrarenal hemodynamics in cirrhotic patients, these investigators observed that the intrarenal infusion of the alpha adrenergic blocker, phentolamine, failed to increase renal blood flow significantly in cirrhotic patients. While the failure of phentolamine to enhance renal perfusion was interpreted initially as indicating the absence of increased renal adrenergic tone, caution is required in reaching such a conclusion. It should be underscored that a majority of patients who received an intrarenal infusion of phentolamine manifested transient hypotension with significant reductions in arterial blood pressure. It is conceivable that such concomitant alterations in systemic hemodynamics would have masked any phentolamine-induced enhancement of renal perfusion.

In summary, it is apparent that the available data regarding the role of the sympathetic nervous system are fragmentary and inconclusive. Clearly, more direct indices of autonomic activity are required to determine whether a diminished effective volume with concomitant increase in sympathetic activity, both in the kidney and in other areas, contributes to the sodium retention of cirrhosis. Future studies utilizing one of several newly developed sensitive radioenzymatic assays for catecholamines should provide further insight into the alterations in autonomic activity which supervene in cirrhotic man.

The preceding four sections exemplify the mechanisms which may contribute to the deranged sodium homeostasis of cirrhosis. It is apparent that the renal sodium retention of advanced liver disease is a complex pathophysiologic constellation with numerous diverse causes. Figure 4 represents an attempt to integrate these diverse findings and to summarize some of the mechanisms whereby these diverse hormonal mediators may act in concert to induce sodium retention.

IV. WHY ASCITES PREDOMINATES IN LIVER DISEASE

The preceding four sections have reviewed a number of hormonal and neural mechanisms which may mediate the abnormal salt retention in cirrhosis. It is quite apparent that these determinants may be similar to those which are operative in other edematous states such as congestive heart failure and nephrosis. Why, then, do cirrhotic patients manifest their fluid retention so frequently as ascites? There are a number of factors which appear to play a role in determining why the peritoneal sac constitutes the site of accumulation of this retained fluid.

A. Portal Hypertension

Fibrosis and regenerating nodules in the cirrhotic liver create a postsinusoidal block to blood flow through the liver. As a result, the hydrostatic pressure increases behind this functional block in sinusoids, and more distantly, in the portal vein and its tributaries. Fluid enters the peritoneal cavity by transudation from the splanchnic capillary bed. The high pressure in the sinusoids may also force increased amounts of fluid to transude into perisinusoidal spaces, and this fluid is then drained by hepatic lymphatics which converge onto the hilus of the liver. This heightened transport of lymph is reflected in enlarged hilar nodes and distended lymphatic channels, which are often strikingly apparent at surgery. However, even this enhanced transport may be insufficient to drain off the excessive amounts of lymph transuding from sinusoids; the surplus fluid may actually "weep" from the liver surface and enter the peritoneal cavity. This fluid is high in protein content and may be responsible for the exudative ascites found in about 20% of cirrhotic patients.

B. Hypoalbuminemia

A second factor which contributes to the leakage of fluid into the peritoneal cavity is the decline in plasma oncotic pressure caused by a lowered serum albumin concentration. This feature of cirrhosis provided the basis for the former practice of treating

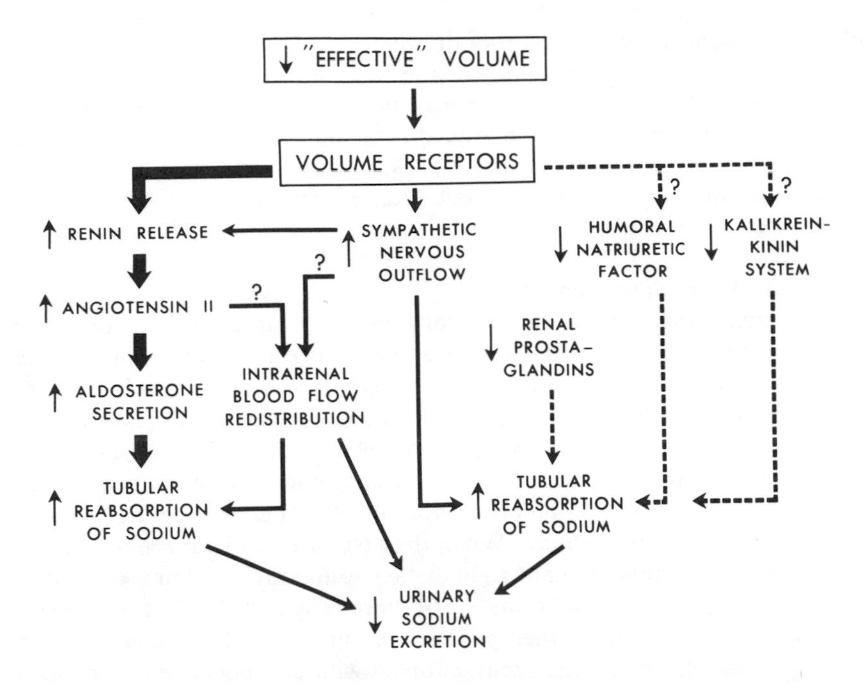

FIGURE 4. Schematic drawing of possible mechanisms whereby a diminished "effective" volume results in sodium retention. The heavy arrows indicate pathways for which evidence is available. The dashed lines represent proposed pathways, the existence of which remains to be established. It is clear that the diverse mediators interact to promote sodium retention. (From Epstein, M., *Gastroenterology*, 76, 622, 1979. With permission.)

patients with albumin infusion, which often proved unsuccessful. In most instances, hypoalbuminemia is the consequence of decreased hepatic synthesis of albumin. It should be noted, however, that this is not the sole cause of the hypoalbuminemia; some cirrhotic patients, in fact, synthesize normal amounts of albumin. It has been suggested that albumin distribution may be impaired, so that rather than remaining in the vascular system, much albumin is consigned to the peritoneal cavity.

C. Other Factors

In addition to portal hypertension and low serum albumin (which clearly play roles in the genesis of ascites), other less understood factors may also be involved. Included among these factors are the peritoneal membrane abnormalities which may retard the normal cycling of fluid between the peritoneal cavity and the vascular compartment.

V. THE MANAGEMENT OF ASCITES AND EDEMA IN LIVER DISEASE

Ascites is associated with many unwanted side effects in patients with liver disease.[64,65] Clearly, the accumulation of marked ascites is associated with significant discomfort in afflicted patients. Some observers have proposed a causal relationship between ascites on the one hand and both high portal pressure and gastroesophageal reflux on the other hand.[64] According to this formulation, ascites would enhance the possibility of variceal bleeding by favoring both rupture of varices and reflux with a resultant erosion of the varices. While such a relationship has not been clearly established, such formulations underscore the general clinical notion that ascites per se is detrimental and requires relief. Furthermore, it has been suggested that ascites is the sine qua non of spontaneous bacterial peritonitis.

While ascites is indeed the " . . . root of much evil",[64] the decision to relieve ascites with diuretic agents is not clear-cut. On the one hand, several studies suggest that diuretic therapy in the cirrhotic patient may be associated with a substantial risk of adverse effects.[66-68] While it may be argued that some of these surveys are outdated and thus may be unrepresentative, recent prospective reports from a number of drug surveillance programs suggest that diuretic-induced complications still constitute formidable problems even today.[68]

A. A Rational Approach to Management

The initial goal of any treatment program should be an attempt to obtain spontaneous diuresis by consistent and scrupulous adherence to a well-balanced diet with rigid dietary sodium restriction (250 mg per day). It should be emphasized that the sodium intake prescribed for cardiac patients (1200 to 1500 mg daily) is not sufficiently restrictive for the cirrhotic patient who continues to gain weight on such a regimen. While the frequency with which such dietary management successfully relieves ascites is unsettled, such a diet should be prescribed for all patients since it is impossible to predict which patients will respond. When the response to dietary management is inadequate, or when the imposition of rigid dietary sodium restriction is not feasible due to cost or unpalatability of the diet, the use of diuretic agents may be considered.

The rational basis of diuretic therapy lies in an understanding of the mechanism(s) and sites of action of the diuretic agent, coupled with an understanding of the pathophysiology of sodium retention in cirrhosis. Since the attributes and efficacy of the varying diuretic agents have been reviewed in detail elsewhere,[69,70] I will focus solely on therapeutic considerations which are unique to the cirrhotic patient. When diuretics are used, the therapeutic aim is a slow and gradual diuresis not exceeding the capacity for mobilization of ascitic fluid. Shear et al.[71] have demonstrated that ascites absorption averages about 300 to 500 mℓ/day during spontaneous diuresis, and has as its upper limits 700 to 900 mℓ/day. Thus, any diuresis which exceeds 900 mℓ/day (in the ascitic patient without edema) must perforce be mobilized at the expense of the plasma compartment with resultant volume contraction.

Finally, the dangers of diuretic-associated hypokalemia should be emphasized. Since total body potassium depletion is frequently associated with cirrhosis, the use of any diuretic that acts proximally to the distal potassium-secretory site may result in profound hypokalemia. Because of the frequently observed temporal relationship between diuretic therapy and the induction of hepatic encephalopathy, and the probability that the enhanced renal ammonia production of hypokalemia may be related to the encephalopathy,[72] great care should be exercised in monitoring potassium derangements in the cirrhotic patient receiving diuretics.

B. Role of LeVeen Shunt

One of the recent advances in the management of the patient with ascites and edema formation is the availability of the LeVeen peritoneovenous (P-V) shunt.[73] Since the underlying abnormality in patients which ascites and sodium retention is not solely an excess of fluid, but rather a maldistribution of ECF, much attention has focused on developing procedures which might redistribute body fluids between compartments, so that the central compartment is replenished at a time when ascites is decreasing. Earlier attempts at autogenous reinfusion of ascites have proven too cumbersome a technique to constitute a useful form of therapy. In 1974, LeVeen and associates resolved many of the technical problems associated with maintaining shunt patency by developing a one-way valve which is activated by a pressure gradient.[73] The experience to date has clearly indicated that the insertion of the P-V shunt can eventuate in significant clinical responses in patients who have been previously crippled by intractable ascites. Immediate and dramatic increases in renal blood flow, GFR, and renal salt and water excretion have been reported following shunt insertion.[74,75] The success of the new valve in facilitating continuous P-V shunting of ascitic fluid and the technical simplicity of its insertion has resulted in its widespread acceptance. However, there is increasing awareness that the widespread utilization of the P-V shunt has been attended by a wide array of complications.[75,76] While we await the results of randomized clinical trials which will establish indications for P-V shunting, we should utilize P-V shunting with restraint.[76] Some observers have reported an operative mortality of 20% with morbid events occurring in over 60% of shunted patients.[75,77] It is my opinion that at present the shunt should be reserved only for those ascitic patients who are truly refractory to a regimen consisting of moderate doses of diuretics with concomitant restriction of sodium intake.

VI. SUMMARY

In summary, the renal sodium retention of advanced liver disease is a complex pathophysiologic constellation with numerous and diverse causes. Although our understanding of the many ways whereby liver dysfunction adversely affects renal sodium handling is incomplete, it is apparent that the past several years have witnessed much progress in the delineation of the abnormalities of deranged sodium homeostasis which characterize advanced liver disease. Advances in the measurement of renal vasoactive hormones that have recently been demonstrated to affect renal hemodynamics and renal sodium handling, including renal prostaglandins and the kallikrein-kinin system, are providing a basis for more complete understanding of the mechanisms that promote sodium retention in cirrhosis. The availability of new radioimmunoassay techniques for the determination of renal prostaglandins and the kallikrein-kinin system should allow further testing of recent provocative suggestions regarding participation of these hormones in mediating sodium retention. Finally, it has been suggested that the peritoneovenous shunt may provide a rational and effective means of managing ascites. Unfortunately, this procedure is accompanied by formidable morbidity and mortality. Clearly, the immediate challenge is the carrying out of carefully designed, randomized clinical trials to identify those patients who may benefit most from such a procedure.

REFERENCES

1. **Papper, S. and Vaamonde, C.,** Renal and electrolyte abnormalities in liver disease: a listing, in *The Kidney in Liver Disease,* Epstein, M., Ed., Elsevier North-Holland, New York, 1978, 33.
2. **Brown, J.,** A remarkable account of a liver, appearing glandulous to the eye, *Philos. Trans. R. Soc. London,* 15, 1266, 1685.
3. **Eisenmenger, W. J., Blondheim, S. H., Bongiovanni, A. M., and Kunkel, H. G.,** Electrolyte studies on patients with cirrhosis of the liver, *J. Clin. Invest.,* 29, 1491, 1950.
4. **Faloon, W. W., Eckhardt, R. D., Cooper, A. M., and Davidson, C. S.,** The effect of human serum albumin, mercurial diuretics and a low sodium diet on sodium excretion in patients with cirrhosis of the liver, *J. Clin. Invest.,* 28, 595, 1949.
5. **Goldman, R.,** Studies in diurnal variation of water and electrolyte excretion: nocturnal diuresis of water and sodium in congestive cardiac failure and cirrhosis of the liver, *J. Clin. Invest.,* 30, 1191, 1951.
6. **Jones, R. A., McDonald, G. O., and Last, J. H.,** Reversal of diurnal variation in renal function in cases of cirrhosis with ascites, *J. Clin. Invest.,* 31, 326, 1952.

7. **Papper, S. and Rosenbaum, J. D.,** Abnormalities in the excretion of water and sodium in "compensated" cirrhosis of the liver, *J. Lab. Clin. Med.,* 40, 523, 1952.
8. **Klingler, E. L., Jr., Vaamonde, C. A., Vaamonde, L. S., Lancestremere, R. G., Morosi, H. J., Frisch, E., and Papper, S.,** Renal function changes in cirrhosis of the liver, *Arch. Intern. Med.,* 125, 1010, 1970.
9. **Epstein, M., Pins, D. S., Schneider, N., and Levinson, R.,** Determinants of deranged sodium and water homeostasis in decompensated cirrhosis, *J. Lab. Clin. Med.,* 87, 822, 1976.
10. **Epstein, M.,** Renal Sodium handling in cirrhosis, in *The Kidney in Liver Disease,* Epstein, M., Ed., Elsevier North-Holland, New York, 1978, 35.
11. **Farnsworth, E. B. and Krakusin, J. S.,** Electrolyte partition in patients with edema of various origins: qualitative and quantitative definition of cations and anions in hepatic cirrhosis, *J. Lab. Clin. Med.,* 33, 1545, 1948.
12. **Gabuzda, G. J.,** Cirrhosis, ascites, and edema: clinical course related to management, *Gastroenterology,* 58, 546, 1970.
13. **Bosch, J., Arroyo, V., Rodes, J., Bruguera, M., and Teres, J.,** Compensacion espontanea de la ascitis en la cirrosis hepatica, *Rev. Clin. Esp.,* 133, 441, 1974.
14. **Papper, S. and Saxon, L.,** The diuretic response to administered water in patients with liver disease. II. Laennec's cirrhosis of the liver, *Arch. Intern. Med.,* 103, 750, 1959.
15. **Vaamonde, C. A.,** Renal water handling in liver disease, in *The Kidney in Liver Disease,* Epstein, M., Ed., Elsevier North-Holland, New York, 1978, 67.
16. **Chaimovitz, C., Rochman, J., Eidelman, S., and Better, O. S.,** Exaggerated natriuretic response to volume expansion in patients with primary biliary cirrhosis, *Am. J. Med. Sci.,* 274, 173, 1977.
17. **Sherlock, S. and Scheuer, P. J.,** The presentation and diagnosis of 100 patients with primary biliary cirrhosis, *N. Engl. J. Med.,* 289, 674, 1973.
18. **Wilkinson, S. P., Arroyo, V., Moodie, H. E., Blendis, L. M., and Williams, R.,** Renal failure and site of abnormal renal retention of sodium in fulminant hepatic failure, *Gut,* 15, 343, 1974.
19. **Conn, H. O.,** The rational management of ascites, in *Progress in Liver Disease,* Vol. 4, Popper, H. and Schaffner, F., Eds., Grune & Stratton, New York, 1972, 269.
20. **Witte, M. H., Witte, C. L., and Dumont, A. E.,** Progress in liver disease: physiological factors involved in the causation of cirrhotic ascites, *Gastroenterology,* 61, 742, 1971.
21. **Gauer, O. H.,** Mechanoreceptors in the intrathoracic circulation and plasma volume control, in *The Kidney in Liver Disease,* Epstein, M., Ed., Elsevier North-Holland, New York, 1978, 3.
22. **Lieberman, F. L., Denison, E. K., and Reynolds, T. B.,** The relationship of plasma volume, portal hypertension, ascites and renal sodium retention in cirrhosis: the overflow theory of ascites formation, *Ann. N.Y. Acad. Sci.,* 170, 202, 1970.
23. **Lieberman, F. L., Ito, S., and Reynolds, T. B.,** Effective plasma volume in cirrhosis with ascites. Evidence that a decreased value does not account for renal sodium retention, a spontaneous reduction in glomerular filtration rate (GFR) and a fall in GFR during drug-induced diuresis, *J. Clin. Invest.,* 48, 975, 1969.
24. **Lieberman, F. L. and Reynolds, T. B.,** Plasma volume in cirrhosis of the liver: its relation to portal hypertension, ascites and renal failure, *J. Clin. Invest.,* 46, 1297, 1967.
25. **Levy, M.,** Observations on renal function and ascites formation in dogs with experimental portal cirrhosis, in *The Kidney in Liver Disease,* Epstein, M., Ed., Elsevier North-Holland, New York, 1978, 131.
26. **Levy, M., Wexler, M. J., and McCaffrey, C.,** Sodium retention in dogs with experimental cirrhosis following removal of ascites by continuous peritoneovenous shunting, *J. Lab. Clin. Med.,* 94, 933, 1979.
27. **Schedl, H. P. and Bartter, F. C.,** An explanation for and experimental correction of the abnormal water diuresis in cirrhosis, *J. Clin. Invest.,* 39, 248, 1960.
28. **Vlahcevic, Z. R., Adam, N. F., Jick, H., Moore, E. W., and Chalmers, T. C.,** Renal effects of acute expansion of plasma volume in cirrhosis, *N. Engl. J. Med.,* 272, 387, 1965.
29. **Tristani, F. E. and Cohn, J. N.,** Systemic and renal hemodynamics in oliguric hepatic failure. Effect of volume expansion, *J. Clin. Invest.,* 46, 1894, 1967.
30. **Epstein, M.,** Renal effects of head-out water immersion in man: implications for an understanding of volume homeostasis, *Physiol. Rev.,* 58, 529, 1978.
31. **Levinson, R., Epstein, M., Sackner, M. A., and Begin, R.,** Comparison of the effects of water immersion and saline infusion on central hemodynamics in man, *Clin. Sci. Mol. Med.,* 52, 343, 1977.
32. **Epstein, M., Pins, D. S., Arrington, R., DeNunzio, A. G., and Engstrom, R.,** Comparison of water immersion and saline infusion as a means of inducing volume expansion in man, *J. Appl. Physiol.,* 39, 66, 1975.
33. **Epstein, M., Levinson, R., Sancho, J., Haber, E., and Re, R.,** Characterization of the renin-aldosterone system in decompensated cirrhosis, *Circ. Res.,* 41, 818, 1977.

34. **Epstein, M., Levinson, R., and Loutzenhiser, R.,** Effects of water immersion on renal hemodynamics in normal man, *J. Appl. Physiol.,* 41, 230, 1976.
35. **Schroeder, E. T., Anderson, G. H., Jr., and Smulyan, H.,** Effects of a portacaval or peritoneovenous shunt on renin in the hepatorenal syndrome, *Kidney Int.,* 15, 54, 1979.
36. **Chaimovitz, C., Szylman, P., Alroy, G., and Better, O. S.,** Mechanism of increased renal tubular sodium reabsorption in cirrhosis, *Am. J. Med.,* 52, 198, 1972.
37. **Saruta, R., Saito, I., Nakamura, R., and Oka, M.,** Regulation of aldosterone in cirrhosis of the liver, in *The Kidney in Liver Disease,* Epstein, M., Ed., Elsevier, New York, 1978, 271.
38. **Coppage, W. S., Jr., Island, D. P., Cooner, A. E., and Liddle, G. W.,** The metabolism of aldosterone in normal subjects and in patients with hepatic cirrhosis, *J. Clin. Invest.,* 41, 1672, 1962.
39. **Vecsei, P., Dusterdieck, G., Jahnecke, J., Lommer, D., and Wolff, H. P. D.,** Secretion and turnover of aldosterone in various pathological states, *Clin. Sci.,* 36, 241, 1969.
40. **Rosoff, L., Jr., Zia, P., Reynolds, T., and Horton, R.,** Studies of renin and aldosterone in cirrhotic patients with ascites, *Gastroenterology,* 69, 698, 1975.
41. **Chonko, A. M., Bay, W. H., Stein, J. H., and Ferris, T. F.,** The role of renin and aldosterone in the salt retention of edema, *Am. J. Med.,* 63, 881, 1977.
42. **Wernze, H., Spech, H. J., and Muller, G.,** Studies on the activity of the renin-angiotensin-aldosterone system (RAAS) in patients with cirrhosis of the liver, *Klin. Wschr.,* 56, 389, 1978.
43. **Zipser, R. D., Hoefs, J. C., Speckart, P. F., Zia, P. K., and Horton, R.,** Prostaglandins: modulators or renal function and pressor resistance in chronic liver disease, *J. Clin. Endocrinol. Metab.,* 48, 895, 1979.
44. **Zusman, R. M., Axelrod, L., and Tolkoff-Rubin, N.,** The treatment of the hepatorenal syndrome with intrarenal administration of prostaglandin E_1, *Prostaglandins,* 13, 819, 1977.
45. **Wernze, H., Muller, G., and Goerig, M.,** Relationship between prostaglandin (PGE_2 and $PGF_{2\alpha}$) and sodium excretion in various stages of chronic liver disease, in *Advances in Prostaglandin and Thromboxane Research,* Samuelsson, B., Ramwell, P. W., and Paoletti, R., Eds., Raven Press, New York, in press.
46. **Epstein, M., Lifschitz, M., and Preston, S.,** Augmentation of renal prostaglandin E (PGE) in decompensated cirrhosis (DC). Implications for renal function, *Kidney Int.,* 16 (Abstr.), 920, 1979.
47. **Robertson, R. P.,** Prostaglandin E (PGE)-induced generation of PGA in man: mechanisms for systemic PGE action, *Clin. Res.,* 24, 157A, 1976.
48. **Arieff, A. I. and Chidsey, C. A.,** Renal function in cirrhosis and the effects of prostaglandin A_1, *Am. J. Med.,* 56, 695, 1974.
49. **Frolich, J. C., Williams, W. M., Sweetman, B. J., Smigel, M., Carr, K., Hollifield, J. W., Fleisher, S., Nies, A. S., Frisk-Holmberg, M., and Oates, J. A.,** Analysis of renal prostaglandin synthesis by competitive protein binding assay and gas chromatography-mass spectrometry, in *Advances in Prostaglandin and Thromboxane Research,* Vol. 1, Samuelsson, B. and Paoletti, R., Eds., Raven Press, New York, 1976, 65.
50. **McGiff, J. C. and Itskovitz, H. D.,** Prostaglandins and the kidney, *Circ. Res.,* 33, 479, 1973.
51. **Boyer, R. D., Zia, P., and Reynolds, T. B.,** Effect of indomethacin and prostaglandin A_1 on renal function and plasma renin activity in alcoholic liver disease, *Gastroenterology,* 77, 215, 1979.
52. **Stein, J. H., Congbalay, R. C., Karsh, D. L., Osgood, R. W., and Ferris, T. F.,** The effect of bradykinin on proximal tubular sodium reabsorption in the dog: evidence for functional nephron heterogeneity, *J. Clin. Invest.,* 51, 1709, 1972.
53. **Nasjletti, A., Colina-Chourio, J., and McGiff, J. C.,** Disappearance of bradykinin in the renal circulation of dogs: effects of kininase inhibition, *Circ. Res.,* 37, 59, 1975.
54. **Margolius, H. S., Horwitz, D., Pisano, J. J., and Keiser, H. R.,** Relationships among urinary kallikrein mineralocorticoids and human hypertensive disease, *Fed. Proc.,* 35, 203, 1976.
55. **Levinsky, N. G.,** The renal kallikrein-kinin system, *Circ. Res.,* 44, 441, 1979.
56. **Bagdasarian, A., Lahiri, B., Talamo, R. C., Wong, P., and Coleman, R. W.,** Immunochemical studies of plasma kallikrein, *J. Clin. Invest.,* 54, 1444, 1974.
57. **Wong, P. Y., Talamo, R. C., and Williams, G. H.,** Kallikrein-kinin and renin-angiotensin systems in functional renal failure of cirrhosis of the liver, *Gastroenterology,* 73, 1114, 1977.
58. **Thames, M. D.,** Neural control of renal function: contribution of cardiopulmonary baroreceptors to the control of the kidney, *Fed. Proc.,* 37, 1209, 1977.
59. **Stein, J. H., Boonjarern, S., Wilson, C. B., and Ferris, T. F.,** Alterations in intrarenal blood flow distribution. Methods of measurement and relationship to sodium balance, *Circ. Res.,* 32 and 33 (Suppl. 1), 61, 1973.
60. **DiBona, G. F.,** Neurogenic regulation of renal tubular sodium reabsorption, *Am. J. Physiol.,* 233, F73, 1977.
61. **Lifschitz, M. D.,** Role of the renal nerves in renal sodium reabsorption in conscious dogs, *Clin. Sci. Mol. Med.,* 54, 567, 1978.

62. **Mashford, M. L., Mahon, W. A., and Chalmers, T. C.,** Studies of the cardiovascular system in the hypotension of liver failure, *N. Engl. J. Med.,* 267, 1071, 1962.
63. **Epstein, M., Berk, D. P., Hollenberg, N. K., Adams, D. F., Chalmers, T. C., Abrams, H. L., and Merrill, J. P.,** Renal failure in the patient with cirrhosis. The role of active vasoconstriction, *Am. J. Med.,* 49, 175, 1970.
64. **Conn, H. O.,** Diuresis of ascites: fraught with or free from hazard, *Gastroenterology,* 73, 619, 1977.
65. **Conn, H. O. and Fessel, J. M.,** Spontaneous bacterial peritonitis in cirrhosis. Variations on a theme, *Medicine,* 50, 161, 1971.
66. **Sherlock, S.,** Ascites formation in cirrhosis and its management, *Scand. J. Gastroenterol.,* 7(Suppl.), 9, 1970.
67. **Greenblatt, D. J. and Koch-Wester, J.,** Adverse reactions to spironolactone. A report from the Boston collaborative drug surveillance program, *JAMA,* 225, 40, 1973.
68. **Spino, M., Sellers, E. M., Kaplan, H. L., Stapleton, C., and MacLeod, S. M.,** Adverse biochemical and clinical consequences of furosemide administration, *Can. Med. Assoc. J.,* 118, 1513, 1978.
69. **Burg, M. B.,** Mechanisms of action of diuretic drugs, in *The Kidney,* Brenner, B. M. and Rector, F. C., Eds., W. B. Saunders, Philadelphia, 1976, 373.
70. **Linas, S. L., Anderson, R. J., Miller, P. D., and Schrier, R. W.,** Rational use of diuretics in cirrhosis, in *The Kidney in Liver Disease,* Epstein, M., Ed., Elsevier North-Holland, New York, 1978, 313.
71. **Shear, L., Ching, S. and Gabuzda, G. J.,** Compartmentalization of ascites and edema in patients with hepatic cirrhosis, *N. Engl. J. Med.,* 282, 1391, 1970.
72. **Gabuzda, G. J. and Hall, P. W., III,** Relation of potassium depletion to renal ammonium metabolism and hepatic coma, *Medicine,* 45, 481, 1966.
73. **LeVeen, H. H., Christoudias, G., Ip, M., Luft, R., Falk, G., and Grosberg, S.,** Peritoneo-venous shunting for ascites, *Ann. Surg.,* 180, 580, 1974.
74. **Kinny, M. J., Wapnick, S., Ahmed, N., Ip, M., Grosberg, S., and LeVeen, H. H.,** Cirrhosis, ascites, and impaired renal function: treatment with the LeVeen-type chronic peritoneal-venous shunt, in *The Kidney in Liver Disease,* Epstein, M., Ed., Elsevier North-Holland, New York, 1978, 349.
75. **Blendis, L. M., Greig, P. D., Langer, B., Baigrie, R. S., Ruse, J., and Taylor, B. R.,** The renal and hemodynamic effects of the peritoneovenous shunt for intractable hepatic ascites, *Gastroenterology,* 77, 250, 1979.
76. **Epstein, M.,** The LeVeen shunt for ascites and hepatorenal syndrome, *N. Engl. J. Med.,* 302, 628, 1980.
77. **Epstein, M.,** Role of the peritoneovenous shunt in the management of ascites and the hepatorenal syndrome, in *The Kidney in Liver Disease,* 2nd ed., Epstein, M., Ed., Elsevier North-Holland, New York, in press.

Chapter 6

RENAL EDEMA

Anthony W. Czerwinski and Francisco Llach

TABLE OF CONTENTS

I. INTRODUCTION

This chapter describes three common renal syndromes which can result in edema. In the first section, the primary emphasis is directed toward those renal mechanisms which develop in the organism's attempt to maintain salt and water homeostasis in chronic renal failure. The second section is concerned with nephrotic edema and reviews both the hypoalbuminemia and the compensatory mechanisms involved in attempts of the system to maintain an effective intravascular volume. The final section reviews edema in the acute nephritic syndrome, e.g., acute glomerulonephritis. In this syndrome, primary glomerular failure results in a failure to excrete ingested salt and water.

II. RENAL FAILURE EDEMA

A. Introduction

Edema is common in patients with severe renal failure. The normal adult consumes approximately 180 mEq of sodium daily (4.14 g sodium, 10.53 g sodium chloride), but can maintain sodium balance with an intake of 10 mEq per day or with an intake as great as 1000 mEq per day. If we assume an unchanging glomerular filtration rate (GFR) and that modulation of sodium balance only involves the kidney, consider a hypothetical example in Table 1. With a sodium intake of 180 mEq/day, a serum sodium concentration of 140 mEq/ℓ and a GFR of 120 mℓ/min, the normal adult would filter 24,192 mEq of sodium daily, reabsorb more than 99% of the filtered load, and excrete 0.78% (fractional excretion of Na, i.e., FE_{Na}) of the filtered sodium. When the sodium intake is reduced to 10 mEq/day, the FE_{Na} decreases to 0.04% and when the sodium intake is increased to 1000 mEq/day, the FE_{Na} increases to 4.13. If a patient with renal disease has a sodium intake of 180 mEq/day, filtration rate of 22 mℓ/min, and a maximum FE_{Na} of 4.0%, sodium balance is barely maintained and any increase in sodium intake or reduction in GFR would result in a state of positive sodium balance. Sodium and concomitant water retention would result in an increase in plasma volume, an increase in capillary hydrostatic pressure, and a decrease in plasma oncotic pressure; this would favor the efflux of fluid from intravascular to interstitial compartments. With the continued accumulation of sodium and water, clinical edema would occur. However, if the FE_{Na} could be increased to 100%, then the sodium balance could be maintained until the filtration rate decreased to less than 1 mℓ/min.

In the following pages, studies related to the regulation of renal sodium excretion and the compartmental distribution of sodium and water in renal failure will be reviewed. Although this review centers on studies in chronic renal failure, much of what is presented applies to acute renal failure.

B. Pathophysiology

1. Compartmental Distribution of Sodium and Water

Early studies[1-3] suggested that patients with chronic renal failure had an increased total body water, extracellular fluid volume (ECFV), and total exchangeable sodium. Blumberg et al.[4] reported their studies on eight well-nourished, physically active uremic patients with C_{cr} 4 to 22 mℓ/min. In this study, ECFV and total exchangeable sodium were normal when related to lean body mass. A later study[5] of a heterogeneous group of 77 patients with $C_{cr} < 10$ mℓ/min also noted that the uremic patient demonstrated an increase in total body water, ECFV, and exchangeable sodium. This investigator noted that the changes in sodium and water distribution were similar to those seen with protein-calorie malnutrition. A recent study[6] compared the extracellular and blood volumes of 31 patients with renal failure, C_{cr} 2 to 14 mℓ/min, with those of 15

Table 1
EXAMPLE OF THE RELATIONSHIP BETWEEN SODIUM INTAKE, FILTERED SODIUM, AND EXCRETED SODIUM

	Sodium[a] intake (mEq/d)	GFR[b] (mℓ/min)	Filtered[c] sodium (mEq/d)	Excreted sodium (mEq/d)	Fractional[d] excretion sodium (%)
Normal patient	10	120	24,192	10	0.04
	180	120	24,192	180	0.74
	1,000	120	24,192	1,000	4.13
Renal failure patient	180	22	4,435	180	4.06
	180	0.89	180	180	100

[a] mEq/d — milliequivalents per day.
[b] GFR — glomerular filtration rate.
[c] Filtered sodium — GFR × serum sodium conc mEq/mℓ × 1,440 min per day. It is assumed that serum sodium concentration is 140 mEq/ℓ.
[d] Fractional excretion sodium (FE_{Na}) — excreted sodium ÷ filtered sodium.

patients with hypertension and 11 healthy controls. This study reported an increase in interstitial volume in uremic patients and an increase in blood volume and interstitial volume in uremic hypertensive patients. These studies suggest that the patient wtih severe renal failure has an increased interstitial fluid volume, but certain questions regarding methodology are unanswered: i.e., do patients with renal failure handle radioisotopes in the same manner as healthy patients; what is the error of measurement; and how should the data be standardized?

2. Renal Mechanisms

Nickel et al.[1] studied eight edema-free patients, with C_{cr} of 2 to 50 mℓ/min, while taking a sodium intake of 70 to 105 mEq/day and again after reducing the sodium intake to 18 to 25 mEq/day. These investigators reported that the renal response to sodium deprivation was slower than in patients with normal renal function and that this was associated with a decrease in filtration rate and an apparent increase in renal tubular reabsorption. Slatopolsky et al.[7] noted that patients with chronic renal failure seldom developed edema or sustained sodium deficits during normal sodium intake. Seventeen edema-free patients (C_{cr} of 1.2 to 63 mℓ/min) ingesting 60 and 120 mEq of sodium daily were studied. The study showed that FE_{Na} increased as the GFR decreased, and that when the sodium intake was increased, there was a further increase in fractional sodium excretion, the increase being greatest in those patients with the most severely depressed glomerular filtration (Figure 1). The increase in fractional sodium excretion, associated with an increased sodium intake, occurred without a significant alteration in the GFR and, thus, involved a decrease in tubular reabsorption. The responsiveness of the control system was tested by extracellular volume expansion and cuff occlusion of the legs; in the former instance, the FE_{Na} increased and in the latter, the FE_{Na} decreased. Since the changes in FE_{Na} occurred while large doses of 9α-fluorohydrocortisone were administered, these investigators concluded that the control system involved a nonmineralocorticoid regulator of tubular reabsorption. Four years later, Wilkinson et al.[8] reported studies in eight edema-free patients with C_{cr} of 3.8 to 24 mℓ/min. These patients, on a diet containing 300 mEq of sodium, had an increased blood pressure, GFR, and fractional excretion of sodium. When the sodium intake was reduced to 10 mEq/day, the blood pressure, GFR, and fractional excretion of sodium decreased. The above studies indicate that both glomerular and tubular mechanisms are involved in renal adaptation and suggest that the usual day to day altera-

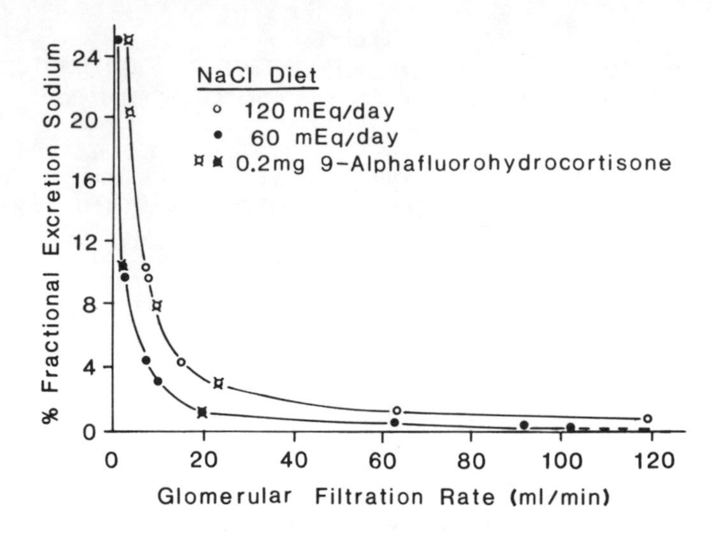

FIGURE 1. Effect of dietary sodium and glomerular filtration rate on fractional excretion of sodium. (From Slatopolsky, E., Elkan, I. O., Weerts, C., and Bricker, N. S., *J. Clin. Invest.*, 47, 521, 1968. With permission.)

tions in sodium excretion primarily occur as in the normal through changes in tubular reabsorption, while with extreme variations in sodium intake, alterations in GFR assume increasing importance.

The question of whether the tendency of the diseased kidney for "salt losing" represented a functional adaptation or the effect of irreversible nephron damage was answered in 1977[9] (Table 2). During this study, five patients with a C_{cr} of 5.2 to 16 mℓ/min had a graded reduction in sodium intake. At the completion of the study, the urinary sodium excretion was reduced to between 4.6 and 9.4 mEq/24 hr and the fractional excretion of sodium was reduced to 0.5 to 1.1%. Similar results were obtained by another investigator[10] in 13 patients with polycystic kidney disease. These studies show that the increased FE_{Na} in chronic renal failure represents a functional adaptation.

The area of nephron involved in this adaptation has been studied using micropuncture techniques. Bank and Aynedjian[11,12] using pyelonephritic rats, concluded that the functional adaptation primarily affected the distal tubule. A later study[13] in rats with an infarcted remnant kidney shows that the functional adaptation involves both an increase in single nephron glomerular filtration and a decrease in fractional proximal reabsorption. Wen et al.[14] then studied the effect of graded volume expansion, 3% and 10% of body weight, in dogs with infarcted remnant kidneys. These investigators concluded that the functional change affected both the proximal and distal tubule. Finally, Webber et al.[15] studied rats with either intact or remnant kidneys and one of three sodium intakes, i.e., 3.0, 1.0, or 0.13 mEq Na/24 hr. Micropuncture studies were done 1 to 2 weeks after institution of the diet. When compared to rats with intact kidneys, the animals with the remnant kidney had an increase in single nephron GFR and a decrease in fractional proximal reabsorption. Furthermore, the remnant kidney animals receiving three different sodium intakes had similar rates of filtration and proximal reabsorption, and each dietary subgroup had approximately the same absolute amount of sodium delivered to the distal tubule. Since the animals were in sodium balance and since the urinary excretion of sodium reflected the intake of sodium, the final adjustments determining the amount of sodium excreted occurred within the distal tubule. These studies indicate that the entire nephron is involved in the functional

Table 2
THE EFFECTS OF GRADUAL SODIUM RESTRICTION IN FIVE CHRONIC RENAL FAILURE PATIENTS[a]

	Initial 3 days of study	Last day of study
Sodium intake (mEq/d)	145 ± 50[b]	5.0 ± 1.3
Weight (kg)	72.9 ± 4.2	72.3 ± 3.6
Plasma volume (mℓ/kg)	46.8 ± 1.3	43.9 ± 1.5
Plasma urea nitrogen (mg/dℓ)	71.4 ± 9.5	79.1 ± 15.2
C inulin (mℓ/min)	10.5 ± 1.9	10.5 ± 1.6
Urine sodium (mEq/d)	138 ± 44	6.6 ± 0.9
FE_{Na} (%)	10.1 ± 2.8	0.5 ± 0.1

[a] Data derived from Reference 9.
[b] Mean ± SEM.

adaptation, that coarse adjustments in sodium excretion occur by increasing the single nephron GFR and decreasing fractional proximal sodium absorption, and that the fine adjustments, which determine the amount of sodium excreted, occur within the distal nephron.

3. Afferent Mechanisms

The afferent pathways modulating sodium excretion in chronic renal failure have not been studied. Studies in animals and man using alterations in sodium intake, saline and albumin infusions, changes in posture, and cuff occlusion of the legs suggest that the same mechanisms operative in normal animals and man, to sense and preserve the "effective" extracellular fluid volume ("effective" ECFV), continue to operate in patients with renal impairment.

4. Effector Mechanisms

a. Oncotic and Hydrostatic Pressure

The role of physical factors as effectors of proximal tubular absorption in severe renal failure is uncertain. One study[16] using micropuncture techniques demonstrated a decrease in proximal fractional reabsorption within 1 hr of nephrectomy, but detectable alterations in peritubular hydrostatic or oncotic pressure were not present. A second study,[17] using clearance techniques and infusion of iso-oncotic saline, dopamine, and vasodilators, suggests that peritubular hydrostatic and oncotic pressure can continue to regulate proximal tubular reabsorption in chronic renal failure.

b. Natriuretic Factor

Schultze et al.[18] studied the mechanisms controlling sodium excretion in dogs on a constant sodium intake after unilateral nephrectomy and 80% infarction of the contralateral kidney. In this model, nephrectomy produced an increase in FE_{Na} from the remnant kidney, which persisted despite mechanical reduction of the glomerular filtration and an excess of exogenous mineralocorticoids. Since the increase in FE_{Na} occurred prior to a significant increase in impermeable solutes and because the rise in FE_{Na} could be prevented by a proportionate reduction in sodium intake, these investigators argued that additional factors must control sodium excretion. A subsequent study,[19] using similar methods and either a constant sodium intake or a proportionate reduction in sodium intake, demonstrated that the serum of dogs on a constant sodium intake contained an extract which was natriuretic when injected into rats. In the interim, Bourgoignie et al.[20] identified a serum extract in patients with severe renal impairment that, when injected into rats, produced a natriuresis and an increase in FE_{Na} without measurable change in inulin or para-aminohippurate clearances. While not isolated, the substance had an estimated molecular weight between 500 and 3000 and was not inactivated by boiling, freezing, or incubation with pronase or chymotrypsin. A later study,[21] using the same methods, demonstrated a similar extract in the urine of 17 uremic patients, C_{cr} <20 mℓ/min and in 1 uremic patient with the nephrotic syndrome. The extract was not present in 7 other nephrotic patients or 14 healthy adult controls. One patient of particular interest was studied twice, once when volume depleted, and a second time after volume repletion. Natriuretic activity was not detected with volume depletion but was present when repleted. The authors concluded that the natriuretic substance was not retained in serum because of a failure in renal excretion, but rather that this factor was physiologically important in the maintenance of sodium homeostasis. In vitro studies[22,23] using isolated proximal tubules and collecting ducts have demonstrated that natriuretic extract inhibits sodium transport when applied to the peritubular surface. Furthermore, urine extracts decrease net sodium transport and are associated with an increase in intracellular sodium concentration in the toad bladder preparation.[24] This implies that the natriuretic extract acts at the peritubular membrane and inhibits sodium transfer from cell to interstitial fluid. Thus, these studies suggest the existence of additional natriuretic factors; however, their identity and site of origin is unknown. Furthermore, the relationship between the natriuretic factor of Bricker and the large molecular weight natriuretic factor of Godon has not been defined.

Using the remnant kidney model, Fine et al.[25] injected standardized amounts of urine natriuretic extract into the renal artery of (1) nonuremic rats with two normal kidneys; (2) nonuremic rats with one normal kidney; and (3) uremic rats after contralateral nephrectomy. The extract induced natriuresis in all groups; however, the natriuretic response was greatest when injected into the uremic rats. This suggests that uremia somehow "magnifies" the renal response to natriuretic extract.

c. Renin-Aldosterone

Cope and Pearson[26] studied aldosterone kinetics in 12 patients with blood urea nitrogen concentrations of 80 to 355 mg/dℓ and demonstrated an increased aldosterone secretion rate in six patients. As part of the study, urinary aldosterone excretion was measured and the absolute levels of excretion were lower than anticipated and did not necessarily coincide with either the plasma aldosterone concentration or aldosterone secretion rate. Two years later, Gold et al.[27] reported that seven of nine patients with C_{cr} <20 mℓ/min had an increased aldosterone secretion rate while consuming a diet containing 100 mEq of sodium daily. Schrier and Regal[28] measured urinary aldosterone in nine patients with C_{cr} of 4.3 to 72 mℓ/min, while on a normal sodium intake and again after sodium deprivation. These investigators showed that urinary aldosterone excretion increased with sodium deprivation and that this was associated with a decrease in urinary sodium. They then gave spironolactone, a competitive aldosterone inhibitor, and demonstrated an increase in urine sodium. A similar study[29] was performed in eight normokalemia patients, C_{cr} 8 to 33 mℓ/min, measuring plasma renin activity and plasma aldosterone concentrations. This study demonstrated an increased plasma renin activity and plasma aldosterone in five patients while ingesting 120 to 145 mEq Na^+ daily and a further increase in renin activity and aldosterone concentration with a sodium intake of 10 to 20 mEq per day. Furthermore, these investigators showed that appropriate changes in renin activity occurred with change in posture and that spironolactone increased urine sodium and decreased urine potassium excretion. In summary, aldosterone concentrations are frequently increased in patients with chronic renal failure and the reviewed studies indicate that aldosterone continues to act as a physiologic regulator of sodium reabsorption.

5. Summary — Pathophysiology

Earlier it was implied that, without adaptation, sodium accumulation was inevitable, given a normal sodium intake, when the GFR decreased below 22 mℓ/min. While this can occur with severe renal impairment, i.e., C_{cr} <10 mℓ/min, the bulk of the evidence indicates the kidney can adapt by increasing the fractional excretion of sodium per residual nephron. The adaptation involves an increase in single nephron GFR and a decrease in tubular sodium reabsorption in both the proximal and distal tubule. At present, defined effectors include an unknown natriuretic factor and aldosterone; however, it is anticipated that additional mechanisms will be identified. In addition, it is possible that the sensitivity of individual nephrons to a variety of effectors could be increased and this magnification of response would contribute to the maintenance of sodium homeostasis. The interplay between effectors has not been studied and whether a subgroup of patients with inadequate adaptive mechanisms exists is unknown.

Therefore, when edema occurs in the course of renal failure, it represents either a sudden excess intake of sodium and water, a failure of adaptive mechanisms, or it reflects a fall in GFR to such low levels that the increase in FE_{Na} is no longer adequate as compensation. Regardless of the mechanism, when sodium excretion is less than sodium intake, sodium retention occurs and this results in an expansion of intravascular volume. The expanded intravascular volume is associated with an increase in hydrostatic pressure and by dilution with a decrease in plasma oncotic pressure. These factors favor the movement of fluid from intravascular to interstitial compartments, and when sufficient sodium retention occurs, edema is apparent.

C. Treatment

The appearance of edema in a patient with a creatinine clearance of greater than 10 mℓ/min should suggest that either the edema is related to another problem, e.g., congestive heart failure, or that the patient has an unusually large sodium intake, either oral or parenteral. If another cause of edema can be identified, this should be treated as indicated; however, if the sodium intake is excessive, this should be managed by a judicious reduction in sodium intake.[30] If clinically significant edema persists, diuretics may be necessary; however, since the patient with renal failure is frequently less responsive to diuretics than patients with normal renal function, a potent loop diuretic, furosemide, is frequently indicated. Furosemide should be started in relatively small doses, 40 mg once daily, and increased as needed until the desired response is obtained. When diuretic therapy is inadequate in controlling severe peripheral edema or pulmonary congestion, then intervention with dialysis, peritoneal or hemodialysis, may be necessary.

III. NEPHROTIC EDEMA

A. Introduction

The nephrotic syndrome is a clinical entity of diverse origin characterized by proteinuria, hypoalbuminemia, and edema. The usual explanation of the edema is that the fundamental abnormality in this syndrome is an increase in glomerular permeability resulting in a marked increase in urine protein excretion.[31,32] Since the proteinuria originates from an increased filtration of serum proteins and because albumin is relatively small when compared to other serum proteins, the predominent protein excreted in nephrotic urine is albumin. Severe albuminura results in hypoalbuminemia and this, in turn, alters the Starling capillary dynamics to favor the egress of fluid from the intravascular to interstitial compartment. The increase in interstitial fluid increases lymphatic flow[33] but when this is inadequate to compensate for the flux of sodium and water into the interstitial space, then clinically evident edema occurs. The egress of fluid from the intravascular compartment causes a decrease in intravascular volume and activates mechanisms causing the kidney to retain sodium, an attendant anion, and water. This further dilutes serum albumin and favors the additional transfer of fluid into the interstitial compartment. The decrease in serum albumin facilitates glomerular filtration and it would be anticipated that the filtration fraction would increase. In addition, the hypoalbuminemia results in decrease in peritubular oncotic pressure and this decreases the proximal tubular reabsorption of sodium and water. Thus, according to the usual explanation, the excess renal retention of sodium and water occurs predominantly through distal tubule mechanisms.[34] In the following pages, studies related to the pathophysiology and treatment of nephrotic edema will be reviewed.

B. Pathophysiology

1. Mechanisms Related to the Hypoalbuminemia

Central to the pathophysiology of nephrotic edema is hypoalbuminemia. The major mechanism causing the hypoalbuminemia is the renal loss of albumin. However, some patients with severe albuminuria have relatively normal serum albumin concentrations while other patients with modest albuminuria have severe hypoalbuminemia.[35] This implies that in addition to urinary excretion, other mechanisms, for example, decreased synthesis, increased catabolism, or even gastrointestinal losses, may be causal in the genesis and maintenance of the decreased serum albumin. These nonrenal mechanisms remain controversial.

The decrease in serum albumin and in intracapillary oncotic pressure causes an alteration in Starling forces favoring the egress of fluid from the intravascular to interstitial compartments. Since the fall in plasma oncotic pressure is general in distribution, the accumulation of interstitial fluid would be expected to occur throughout the body. However, the pattern of edema is influenced by intravascular hydrostatic pressure. Thus, it occurs early in dependent areas.

While nephrotic edema can be the direct result of the hypoalbuminemia, two lines of evidence indicate that additional factors must be involved. The first relates to the rare syndrome of analbuminemia. In this syndrome, albumin synthesis is markedly reduced, serum albumin concentrations are frequently less than 1.0 g/dℓ and edema is absent or minimal.[36-38] A second line of evidence suggesting that factors in addition to hypoalbuminemia are causal in nephrotic edema relates to measurements of blood volume. The loss of fluid from the intravascular space would be expected to produce hypovolemia. While many patients with established nephrotic syndrome have a decreased blood volume, other patients have a normal or increased blood volume[39-43] (Table 3).

While this does not preclude hypoalbuminemia as the primary factor initiating nephrotic edema, it does suggest that other mechanisms are important in pathogenesis.

2. Renal Mechanisms

The GFR, renal plasma flow, and filtration fraction, when measured in nephrotic patients or experimental nephrotic animals, demonstrates heterogeneity of function. Coggins,[34] in explaining the mechanism of edema, selected a relatively homogenous group of nephrotic patients with a normal or increased GFR and an increase in filtration fraction. While this simplifies the description of mechanisms, it does not include a large group of nephrotic patients with abnormalities of renal function,[39,44,45] i.e., a decrease in GFR. Thus, any description of renal mechanisms must consider this diversity of function.

The ability of nephrotics to excrete a sodium load depends on the stage at which disease is studied. Early in the disease, when edema is accumulating, an infusion of

Table 3
BLOOD VOLUME MEASUREMENTS IN PATIENTS WITH THE NEPHROTIC SYNDROME

Method of measurement	Number of patients with normal blood volume[a]			Ref.
	<10%	±10%	>10%	
^{51}Cr—RBC, RISA[b]	40	22	10	57
RISA	9	10	3	58
RISA	2	20	8	38
RISA	7	3	5	59
^{51}Cr—RBC	3	10	2	60
Total (%)	61(40)	65(42)	28(18)	

[a] Estimated normal blood volume when not edematous.
[b] ^{51}Cr—RBC,^{51} Chromium-labeled red blood cells; RISA, radio iodine labeled serum albumin.

hypertonic saline is associated with impaired renal sodium excretion.[46,47] Later, when sodium balance is achieved, sodium excretion is appropriate, providing that overall renal function is not seriously impaired. The ability to excrete a water load was studied by Gur et al.[48] This study demonstrated an impairment in water excretion during the nephrotic phase, with return of normal water excretion when remission was induced.

Discrete nephron function has been studied using micropuncture techniques in rats, 4 days to 15 weeks after the induction of the nephrotic syndrome with puromycin or a variety of immunologic methods.[32,47,49,50,51-53] Generally, the studies have shown a reduced single nephron GFR and a reduction in filtered sodium. All studies, except one, demonstrate a decrease in absolute and fractional proximal tubular sodium and water reabsorption. Since the excreted urine in many of the nephrotic animals had a decreased volume, sodium concentration, and sodium excretion, these studies imply that enhanced tubular sodium and water absorption primarily involves the distal nephron. The results of Kuroda et al.[53] differ in that while absolute proximal sodium and water reabsorption was decreased, the fractional proximal reabsorption of sodium and water was increased. A possible explanation for this difference is that the latter study was performed during a period of avid sodium accumulation, while many of the other studies were performed after sodium balance was achieved.

Evidence supporting the postulate that the time of study and stimulus for sodium accumulation can be important is derived from two studies. The study of von Baeyer[32] demonstrated sodium retention during the 2 to 7 days after induction of disease, but thereafter, sodium excretion returned toward pretreatment levels. Further evidence is derived from the study of Kuschinsky et al.[54] These investigators studied proximal tubular reabsorption during minimal volume expansion, after acute volume expansion and after 9 to 17 days of chronic volume expansion. Acutely volume expanded animals demonstrated a decrease in proximal fractional reabsorption; however, with chronic volume expansion, proximal fractional reabsorption returned toward normal.

Clearance studies in patients are conflicting. Grausz et al.[55] studied five patients with stable nephrotic syndrome during tubular blockade by diuretics and isooncotic volume expansion and concluded that proximal tubular reabsorption was reduced. On the other hand, Gur et al.[48] while investigating renal concentrating and diluting ability in six patients concluded that proximal tubular reabsorption was increased.

3. Effector Mechanisms

a. Oncotic and Hydrostatic Pressure

Effector mechanisms stimulating sodium and water retention in nephrotic edema are incompletely studied. Most investigators imply that the decreased proximal tubular reabsorption is related to the decrease in peritubular oncotic pressure.[49,56,57] However, other studies[32,53] suggest a poor correlation between calculated peritubular oncotic pressure and proximal tubule fractional reabsorption. The study of Kuroda et al.[53] showed an increased proximal tubular fractional reabsorption despite a decrease in calculated peritubular oncotic pressure. After intraluminal nephron albumin microinfusion studies, these investigators concluded that intraluminal albumin induced an increase in fractional reabsorption by slowing tubular transit time and increasing the intraluminal hydrostatic pressure. One study[58] supports the contention that transit time is a regulating variable, but further confirmation is necessary.

b. Renin-Aldosterone

Aldosterone has been implicated in the pathogenesis of nephrotic edema and some nephrotic patients have shown an increased urine aldosterone excretion,[59] plasma aldosterone concentration and aldosterone secretion rate.[60] Seven nephrotic patients and five control subjects on daily diets containing 40 mEq Na^+ and 60 mEq K^+ were studied by Medina et al.[61] When compared to the control subjects, the nephrotic patients demonstrated an increased plasma renin activity, a normal or low concentration or renin substrate, and modestly increased plasma aldosterone. They concluded that the increase in aldosterone was related to the increase in plasma renin, and that in some instances, the decreased concentration of renin substrate could limit the secretion of aldosterone. An earlier study[62] showed that renin was not excreted inappropriately in urine and that the renin activity was not related to the amount of proteinuria or measured creatinine clearance. Chonko et al.[63] measured plasma renin and aldosterone in ten nephrotics 2 days after they started a low sodium diet (10 mEq Na per day) and again after they ingested approximately 300 mEq of sodium daily. With sodium restriction, plasma renin activity was increased and in some patients, this was associated with an increased plasma aldosterone. On the high sodium intake, only two of ten patients had a suppression of plasma renin and aldosterone, despite documented increased in weight, increased edema, and increased urine sodium excretion. This study implies that renin production stimulates aldosterone secretion, but that in addition to volume depletion and urinary sodium excretion, other stimuli must influence renin secretion in the nephrotic syndrome. Meltzer et al.[64] studied 16 nephrotic patients. Patients with high renin, high aldosterone nephrosis tended to be hypovolemic, had minimal histopathologic abnormalities, and were corticosteroid responsive. Patients with low aldosterone nephrosis were hypervolemic, had more severe histopathologic abnormalities, and were steroid resistant. The limitations in this study were that the study groups were small, the groups were dissimilar in age and renal function, and it is uncertain at what stage in the disease patients were studied. Although the data were not presented, these investigators also noted that while high renin levels could not be suppressed with saline, albumin infusions resulted in a decrease in plasma renin activity. Another study[65] suggests that plasma renin may not be useful prognostically and that other mechanisms, in addition to changes in overall renal hemodynamics and aldosterone, are involved in the pathogenesis of nephrotic edema. In that study, ten adults with normal or hypervolemic nephrotic syndrome were studied while nephrotic and during corticosteroid induced remission. Plasma renin was normal or low during the nephrotic phase and did not change substantially with remission.

While these studies indicate that the renin-aldosterone system can be an effector stimulating nephrotic renal sodium retention, they also indicate that nephrotic edema can be present with normal or suppressed plasma aldosterone. Furthermore, it should be remembered that when normal subjects are given aldosterone they do not develop edema and within 1 or 2 weeks escape from the sodium-retaining effects. While hypovolemia can be a major stimulus to renin secretion, other as yet undefined mechanisms may be involved in the increased renin secretion of nephrotic patients.

c. Other Effectors

Other effector systems have been investigated in a less systematic fashion. Godon[66] showed that the urine of salt-loaded normal rats contained a natriuretic substance, 10,000 to 50,000 mol wt, and that this substance was absent from the urine of salt-loaded nephrotic rats. Subsequently, a similar substance was identified in the plasma of salt-loaded normal animals. Two reports[21,67] in humans suggest that the edematous nephrotic patient lacks plasma natriuretic activity. Another potential effector was studied by Oliver et al.[68] These investigators noted a significantly increased urinary norepinephrine excretion and demonstrated an inverse relationship between urine norepinephrine and urine sodium excretion. How and whether this is important in the renal retention of sodium in nephrosis is uncertain. The role of antidiuretic hormone in the nephrotic syndrome has not been defined. As previously noted,[48] nephrotic children have a decreased capacity to form a dilute urine and while this suggests an ADH effect, only one report[69] using a bioassay method suggests that ADH concentrations are increased. Thus, the precise role of ADH as an effector in the pathogenesis of nephrotic edema is uncertain.

4. Affector Mechanisms

The afferent stimuli resulting in activation of effector mechanisms and sodium retention is just beginning to be unraveled. Initially, it was assumed that hypoalbuminemia produced hypovolemia and that this increased renal sodium retention. Support for this view was derived from the fact that albumin infusions frequently resulted in a natriuresis and a decrease in clinically apparent edema.[70] Measurements of blood volume have confused the issue by showing that many nephrotic patients had normal or increased measured blood volume. Present data, using water immersion to the neck, suggests that there is a decrease in central "effective" ECFV and that this initiates mechanisms influencing renal sodium reabsorption.[71,72]

5. Summary — Pathophysiology of Nephrotic Edema

Collating the many clinical and experimental observations suggests the following descriptions of the pathogenesis of nephrotic edema. Glomerular disease of diverse etiology results in significant proteinuria. Proteinuria, particularly albuminuria, when large in amount is the principal factor responsible for hypoalbuminemia. Hypoalbuminemia results in a decreased plasma oncotic pressure and this favors the egress of fluid from the vascular to interstitial compartments. Intravascular hypovolemia initiates mechanisms causing the renal retention of sodium, anions, and water. Early in the disease, sodium reabsorption increases in both the proximal and distal tubules, but in later phases the regulation of sodium reabsorption primarily involves the distal tubule. Known effector mechanisms include proximal tubular oncotic and hydrostatic pressure, the presence or absence of circulating natriuretic factors, aldosterone, and antidiuretic hormone. The dynamics of the process and the interplay between various effector mechanisms have not yet been studied. The renal retention of sodium and water increases the intravascular volume and this contributes to the maintenance of nephrotic edema. In the chronic state, intravascular volume can be decreased, normal, or increased, depending on how the body senses the central effective blood volume.

C. Treatment

Treatment of nephrotic edema could best be accomplished by treatment of the primary glomerular disease and curtailing the proteinuria. Unfortunately, this is frequently not possible and thus, ancillary methods, e.g., assuring an adequate nutritional intake to allow maximum albumin synthesis, limiting sodium and water intake, and the mobilization of edema using diuretics are necessary. While any diuretic may be effective, some of the experimental data would suggest that a diuretic with a distal site of action may be preferred. While an elevated aldosterone secretion is found in many nephrotic patients, it must be remembered that a significant number of patients have a low intravascular volume and that in these patients, a more potent diuretic may be necessary. Edema mobilization should proceed slowly with the intent of reducing the body weight of adults 1 to 2 kg each day. Complications of diuretic therapy can include volume depletion,[73] hyponatremia,[74] and possibly the depletion of other cations, e.g., potassium and magnesium. While theoretically attractive, albumin infusion is not indicated, since the effect is brief, the majority of the infused albumin is excreted in the urine, and the cost is prohibitive.

IV. NEPHRITIC EDEMA

A. Introduction

Acute glomerulonephritis is a clinical syndrome characterized by the abrupt onset of urinary abnormalities, hypertension, and edema. Multiple hypotheses have been put forth to explain the edema. One early hypothesis was a general increase in capillary permeability with the loss of plasma proteins into the interstitial space and that this resulted in a disruption of normal Starling forces and edema. Evidence to the contrary included a decrease rather than an increase in packed cell volume and the presence of low concentrations of protein in nephritic edema fluid.[75,76] Many patients with acute glomerulonephritis have cardiomegaly, pulmonary congestion, and an elevated venous pressure, and on this basis, congestive heart failure was suggested as the cause of nephritic edema. However, most nephrotic patients have a normal or increased cardiac output and this increases appropriately with demand.[77-79] Thus, present evidence indicates that heart failure is not a primary causal mechanism but that cardiomegaly and an increase in intravascular volume results from a primary renal retention of sodium and water. Renal salt and water retention produces an increase in plasma volume,[80] an increase in intravascular hydrostatic pressure, and by dilution, a decrease in plasma oncotic pressure.[81] These changes upset the normal Starling forces and edema occurs.

B. Pathophysiology

1. Renal Mechanisms

Earle et al.[82] using clearance techniques repeatedly over a period of months, studied eight patients with acute glomerulonephritis and four patients with an apparent exacerbation of glomerulonephritis. A decrease in GFR, a normal or minimal decrease in renal plasma flow, and a decrease in filtration fraction was observed. Tubular function, as assessed by clearance techniques, was less impaired than glomerular function, and the fractional excretion of sodium was variable, although fractional excretion and total excretion of sodium increased with the onset of diuresis. It was observed that diuresis and natriuresis could occur without a concomitant change in GFR and it was implied that tubular mechanisms might be involved in salt retention of acute glomerulonephritis. Other studies have demonstrated the same patterns of renal dysfunction;[83,84] however, these studies do not indicate the mechanism of sodium retention or the nephron site that is responsible for sodium retention.

Although studies of experimental glomerulonephritis do not necessarily relate to human nephritis, the study of animals can give insight into the possible mechanisms involved. Blantz and Wilson[85] studied nephrotoxic serum nephritis in rats within 1 hour of injection. In this model, the single nephron GFR and renal plasma flow were decreased, there was an increase in glomerular hydrostatic pressure, a decrease in plasma oncotic pressure, an increase in average effective filtration pressure, and a decrease in urinary sodium excretion. Although the mechanism causing the increase in glomerular capillary pressure was not elucidated, the calculated afferent and efferent glomerular arteriolar resistance was increased, and it was noted that these changes were similar to those seen during an infusion of angiotensin II. Since electron microscopy demonstrated that the glomerular capillary lumens were patent, the investigators concluded the primary cause for the decrease in glomerular filtration was an abrupt decrease in glomerular hydraulic permeability. The relationship between single nephron glomerular filtration, filtration pressure, permeability, and capillary area is

$$\text{Single nephron glomerular filtration} = \text{effective filtration pressure} \times \text{glomerular permeability coefficient}$$

$$\text{Glomerular permeability coefficient} = \text{hydraulic permeability} \times \text{capillary surface area}$$

Another study[86] using neutral dextran indicates that the functional glomerular capillary surface area is reduced within 5 to 15 days of inducing nephrotoxic serum nephritis. Wen et al.[87] studied dogs before and 1 to 2 hr after inducing nephrotoxic serum nephritis. In this model, nephritis was associated with a decrease in total glomerular filtration rate, filtration fraction, urine volume, urine sodium excretion, and fractional excretion of sodium. Micropuncture demonstrated a decrease in single nephron glomerular filtration rate with a proportional reduction in proximal tubular reabsorption, i.e., no change in proximal fractional reabsorption when compared to control. Since the excreted urine had a decreased sodium content, it was concluded that distal sodium absorption was inappropriate, and they suggested that increased distal fractional absorption was related to the decreased delivery of fluid into the distal nephron. Maddox et al.[88,89] studied histologically mild glomerulonephritis in rats. Clearance studies demonstrated an increase in total renal plasma flow, a decrease in filtration fraction, and no significant change in filtration rate. Micropuncture studies 5 to 16 days after disease induction and while volume expanded showed a decrease in proximal tubule reabsorption. This study suggested that tubular mechanisms are operative in the maintenance of sodium homeostasis within one week of onset of disease. The study also demonstrated an increase in effective filtration pressure within the glomerulus, and the investigators concluded that this was why the glomerular filtration rate was normal.

The question of whether the severity of disease and time of study were important variables in the mechanism of renal sodium homeostasis was studied by Wagnild and Gutman.[90] They induced a mild to moderate unilateral glomerulonephritis in one group of dogs and a more severe unilateral nephritis in a second group of dogs. Early in the disease (average time of study 6 days) dogs with mild nephritis had a decreased FE_{Na} from the diseased kidney and increased the FE_{Na} in response to systemic volume expansion. Three of these dogs were studied again 10 to 59 days after injection and at that time, their basal FE_{Na} was increased and they demonstrated an exaggerated increase in FE_{Na} with volume expansion. Dogs with severe disease (average time of study 5 days) had an increased basal FE_{Na} and an exaggerated natriuresis after volume expansion. These studies indicate that both time of study and severity of disease are important variables to be considered when studying sodium homeostasis in glomerulonephritis. In addition, the study indicates that the mechanisms of adaptation can operate locally within one kidney without significantly affecting the function of the opposite kidney.

Whether this indicates a locally released effector mechanism or an altered nephron sensitivity to systemic effectors is uncertain. Further evidence suggesting that the mechanisms of compensation change with time is drawn from the study of Allison et al.[91] These investigators produced an acute proliferative glomerulonephritis with nephrotoxic serum and a chronic membranous glomerulopathy by inducing immune complex formation. Animals were studied using histologic, clearance, and micropuncture techniques at the times shown in Table 4. During micropuncture, all animals were hydropenic and all were given exogenous aldosterone and vasopressin. Nephritic rats had an impressive heterogeneity of nephron structure and function and this variance was greatest in the animals with chronic nephritis. Rats with acute nephritis had a proportionate reduction in total GFR and mean single nephron glomerular filtration, while animals with chronic nephritis and a reduced total GFR had a relatively normal mean single nephron filtration rate. While this could be related to sampling, another possibility is that with chronic disease the reduction in total nephron mass is associated with an increased single nephron GFR in remaining functional nephrons. Absolute proximal tubular reabsorption varied directly with the single nephron GFR and despite he heterogenity of glomerular function, glomerulo-tubular balance was maintained.

The animal studies indicate that changes in nephron function during the evolution of glomerulonephritis represent a dynamic process. During the initial few days, glomerular filtration decreases, tubular function is generally unaltered, and in the presence of a constant sodium intake, sodium retention occurs. Later in the disease, there is an increase in fractional sodium excretion and this is associated with an adaptive decrease in tubular sodium reabsorption. Still later, with the loss of functioning nephrons, the remaining nephrons develop an increase in filtration rate, in addition to the decrease in tubular sodium reabsorption. While this is a plausible description of the renal mechanisms, the process controlling the adaptation has not been studied and whether this applies to patients with glomerulonephritis is unknown.

2. Effector and Affector Mechanisms

Alterations in glomerular physical forces have been reviewed in the section on renal mechanisms and will not be repeated. The role of altered physical factors in glomerulonephritis and how they affect proximal tubular function has been studied. Maddox et al.[89] measured hydrostatic pressure and oncotic pressure, calculated from measurements of protein concentration, in peritubular capillaries and found a significant increase in peritubular hydrostatic and a decrease in peritubular oncotic pressure. On the basis of this study, they suggest that physical forces represented a major determinant of absolute proximal tubular reabsorption. Allison et al.[91] showed a decrease in proximal peritubular oncotic pressure in both acute and chronic nephritis; however, these investigators raised the question of how this could explain the heterogeneity of proximal tubular function considering the number of anastomoses between peritubular capillaries. Thus, while physical forces continue to influence proximal tubular function their exact role is not defined.

Godon[92] using the nephrotoxic serum model, demonstrated that with progressive saline loading, nephritic rats had a less than optimal decrease in proximal sodium reabsorption. Salt-loaded normal rats[66] had a plasma natriuretic factor, between 10,000 and 50,000 mol wt, which was absent from plasma of nephritic rats. If the natriuretic factor was injected in either normal or nephritic animals, a natriuresis ensued. On the basis of these studies, Godon argued that one mechanism explaining the sodium retention of glomerulonephritis is the absence of natriuretic factor. He has subsequently reported that the substance is absent in patients during the edematous phase of acute glomerulonephritis; however, with diuresis, the substance can be identified.[93] The exact nature and significance of this factor is unknown.

Table 4
CLEARANCE DATA IN RATS WITH EXPERIMENTAL GLOMERULONEPHRITIS[a]

	Normal control rats	Acute nephrotoxic nephritis	Chronic immune complex nephritis
Period of study (weeks)		1—5	20—64
GFR (mℓ/min)[b]	1.96 ± 0.5[c]	0.71 ± 0.4	1.04 ± 0.4
SNGFR (nℓ/min)	40.6 ± 6.0	19.4 ± 8.4	43.3 ± 12.6
RPF (mℓ/min)	7.97 ± 2.7	7.42 ± 2.7	9.18 ± 4.1
Filtration fraction (%)	27 ± 5	11 ± 5	14 ± 5
FE_{Na} (%)	0.06 ± 0.07	0.19 ± 0.09	0.14 ± 0.09

[a] Data from Reference 91.
[b] GFR = glomerular filtration rate; SNGFR = single nephron glomerular filtration rate; RPF = renal plasma flow; FE_{Na} = fractional excretion filtered sodium.
[c] Mean ± SD

Studies of plasma renin activity in acute glomerulonephritis have shown that renin activity is low or normal during the edematous phase and increases with diuresis.[94,95] Powell et al.[95] noted a significant inverse relationship between measured plasma renin activity and percent body weight loss with diuresis. While these studies seemingly indicate that renin does not play a significant role in the sodium retention of acute glomerulonephritis, the question has been raised as to whether the renin activity is inappropriately high for the degree of volume expansion.

Afferent mechanisms have not been studied in acute glomerulonephritis, but the available data suggest that the body is reacting as if there was an increase in "effective" ECFV.

3. *Summary of Pathophysiology*

Patients with acute glomerulonephritis on a normal sodium intake develop salt and water retention because of a failure in renal excretion. The primary mechanism causing this failure is an abrupt decrease in GFR related to a decrease in hydraulic permeability and a loss in functional capillary surface area. With salt and water retention, the plasma volume increases, capillary hydrostatic pressure increases, capillary oncotic pressure falls, interstitial fluid accumulates, and if enough salt and water are retained, edema occurs. An early glomerular compensatory mechanism is an increase in afferent and efferent glomerular arteriolar resistance and an increase in effective filtration pressure. The cause of the increased arteriolar resistance has not been defined, but this could involve humoral or neurogenic mechanisms. The increase in effective filtration pressure results in an increase in GFR. In most instances, tubular function is initially normal and because of the decrease in glomerular filtration, there is an increase in distal tubular fractional sodium absorption. Later, the tubules develop an increased capacity to reject a larger portion of the filtered sodium, and the fractional excretion of sodium increases. With recovery, glomerular and tubular mechanisms return to normal, diuresis occurs, plasma volume decreases, and edema is reabsorbed. If recovery does not occur, the mechanisms described in the section on renal failure edema become operative so as to maintain normal sodium homeostasis. The effector and affector mechanisms responsible for those functional changes have been only incompletely studied.

C. Treatment

The initial therapeutic maneuver should be a reduction in sodium and water intake in an attempt to minimize edema accumulation. In most patients, this is sufficient and the edema usually subsides with the onset of diuresis. However, in an occasional patient, pulmonary congestion and volume-dependent hypertension necessitate the use of other measures. In these patients, a loop diuretic can be tried, but if this is ineffective, then peritoneal dialysis to remove the excess salt and water is indicated.

REFERENCES

1. **Nickel, J. J., Lowrance, P. B., Leifer, E., and Bradley, S. E.,** Renal function, electrolyte excretion and body fluids in patients with chronic renal insufficiency before and after sodium deprivation, *J. Clin. Invest.,* 32, 68, 1953.
2. **Moore, F. D., Edelman, I. S., Olney, J. M., James, A. H., Brooks, L., and Wilson, G. M.,** Body sodium and potassium, *Metabolism,* 3, 334, 1954.
3. **Edelman, I. S., Leibman, J., O'Meara, M. P., and Birkenfeld, L. W.,** Interrelations between serum sodium concentration, serum osmolality and total exchangeable sodium, total exchangeable potassium, and total body water, *J. Clin. Invest.,* 37, 1236, 1958.
4. **Blumberg, A., Nelp, W. B., Hegstrom, R. M., and Schribner, B. H.,** Extracellular volume in patients with chronic renal disease treated for hypertension by sodium restriction, *Lancet,* 2, 69, 1967.
5. **Coles, G. A.,** Body composition in chronic renal failure, *Quart. J. Med.,* 41, 25, 1972.
6. **Cangiano, J. L., Ramierz-Muzo, O., Ramirez-Gonzales, R., Trevino, A., and Campos, J. A.,** Normal renin uremic hypertension, *Arch. Intern. Med.,* 136, 17, 1976.
7. **Slatopolsky, E., Elkan, I. O., Weerts, C., and Bricker, N. S.,** Studies on the characteristics of the control system governing sodium excretion in uremic man, *J. Clin. Invest.,* 47, 521, 1968.
8. **Wilkinson, R., Leutscher, J. A., Dowdy, A. J., Gonzales, C., and Nakes, G. W.,** Studies on the mechanism of sodium excretion in uremia, *Clin. Sci.,* 42, 711, 1972.
9. **Danovitch, G., Bourgoignie, J., and Bricker, N. S.,** Reversibility of the "salt losing" tendency of chronic renal failure, *N. Engl. J. Med.,* 296, 14, 1977.
10. **Martinez-Maldonado, M., Yium, J. J., Suki, W. N., and Eknoyan, G.,** Electrolyte excretion in polycystic kidney disease: interrelationship between sodium, calcium, magnesium, and phosphates, *J. Lab. Clin. Med.,* 90, 1066, 1977.
11. **Bank, N. and Aynedjian, H. S.,** Individual nephron function in experimental pyelonephritis. I. Glomerular filtration rate and proximal tubular sodium potassium and water reabsorption, *J. Lab. Clin. Med.,* 68, 728, 1966.
12. **Bank, N. and Aynedjian, H. S.,** Individual nephron function in experimental pyelonephritis. II. Distal tubular sodium and water reabsorption and the concentrating defect, *J. Lab. Clin. Med.,* 68, 728, 1966.
13. **Schultze, R. G., Weisser, F., and Bricker, N. S.,** The influence of uremia on fractional sodium reabsorption by the proximal tubule, *Kidney Int.,* 2, 59, 1972.
14. **Wen, S. F., Wong, N. L. M., Evanson, R. L., Lockhart, E. A., and Dirks, J. H.,** Micropuncture studies of sodium transport in the remnant kidney of the dog, *J. Clin. Invest.,* 52, 386, 1973.
15. **Weber, H., Linn, K. Y., and Bricker, N. S.,** Effect of sodium intake on single nephron glomerular filtration rate and sodium reabsorption in experimental uremia, *Kidney Int.,* 8, 14, 1975.
16. **Allison, M. E. M., Lipham, E. M., Lassiter, W. E., and Gottschalk, C. W.,** The acutely reduced kidney, *Kidney Int.,* 3, 354, 1973.
17. **Wagnild, J.P., Gutman, F. D., and Rieselbach, R. E.,** Influence of hydrostatic and oncotic pressure on sodium reabsorption in the unilateral pyelonephritic dog kidney, *Clin. Sci. Mol. Med.,* 47, 367, 1974.
18. **Schultz, R. G., Shapiro, H. S., and Bricker, N. S.,** Studies on the control of sodium excretion in experimental uremia, *J. Clin. Invest.,* 48, 869, 1969.
19. **Schmidt, R. W., Bourgoignie, J. J., and Bricker, N. S.,** On the adaptation in sodium excretion in chronic uremia, *J. Clin. Invest.,* 53, 1736, 1974.
20. **Bourgoignie, J. J., Hwang, K. H., Espinel, C., Klahr, S., and Bricker, N. S.,** A natriuretic factor in the serum of patients with chronic uremia, *J. Clin. Invest.,* 51, 1514, 1972.

21. **Bourgoignie, J. J., Hwang, K. H., Ipakchi, E., and Bricker, N. S.,** The presence of a natriuretic factor in the urine of patients with chronic uremia, *J. Clin. Invest.*, 53, 1559, 1974.
22. **Grantham, J. J., Irwin, R. L., Qualizza, P. B., Tucker, D. R., and Whittier, F. C.,** Fluid secretion in isolated proximal straight renal tubules, *J. Clin. Invest.*, 52, 2441, 1973.
23. **Fine, L. G., Bourgoignie, J. J., Hwang, K. H., and Bricker, N. S.,** On the influence of the natriuretic factor from the urine of patients with chronic uremia on the bioelectric properties and sodium transport of the isolated mammalian collecting tubule, *J. Clin. Invest.*, 58, 590, 1976.
24. **Kaplan, M. A., Bourgoignie, J. J., Rosecan, J., and Bricker, N. S.,** The effects of the natriuretic factor from uremic urine on sodium transport, water and electrolyte content, and pyruvate oxidation by the isolated toad bladder, *J. Clin. Invest.*, 53, 1568, 1974.
25. **Fine, L. G., Bourgoignie, J. J., Weber, H., and Bricker, N. S.,** Enhanced end-organ responsiveness of the uremic kidney to the natriuretic factor, *Kidney Int.*, 10, 364, 1976.
26. **Cope, C. L. and Pearson, J.,** Aldosterone secretion in severe renal failure, *Clin. Sci.*, 25, 331, 1963.
27. **Gold, E. M., Kleeman, C. R., Ling, S., Yawata, M., and Maxwell, M.,** Sustained aldosterone secretion in chronic renal failure, *Clin. Res.*, 13, 135, 1965.
28. **Schrier, R. W. and Regal, G. M.,** Influence of aldosterone on sodium, water, and potassium metabolism in chronic renal disease, *Kidney Int.*, 1, 156, 1972.
29. **Berl, T., Katz, F. H., Henrich, W. L., Detorrente, A., and Schrier, R. W.,** Role of aldosterone in the control of sodium excretion in patients with advanced chronic renal failure, *Kidney Int.*, 14, 228, 1978.
30. **Talbot, N. B., Crawford, J. D., Kerrigan, G. A., Hillman, D., Bertucio, M., and Terry, M.,** Application of homeostatic principles to the management of nephritic patients, *N. Engl. J. Med.*, 255, 655, 1956.
31. **Knutson, D., Daugharty, T. M., and Brenner, B. M.,** Determinants of glomerular filtration in experimental glomerulonephritis in the rat, *J. Clin. Invest.*, 55, 305, 1975.
32. **Von Baeyer, H., Van Liew, J. B., Klassen, J., and Boylan, J. W.,** Filtration of protein in the antiglomerular basement membrane nephritic rat, *Kidney Int.*, 10, 425, 1976.
33. **Zweifach, V. W.,** Capillary filtration and mechanisms of edema formation, *Pfluegers Archiv. Gesamte Physiol. Menschen Tiere*, 336 (Suppl.), 81, 1972.
34. **Coggins, C. H.,** Nephrotic and nephritic edema, in *Contemporary Issues in Nephrology*, Vol. 1A, Brenner B. M. and Stein, J. R., Eds., Churchill Livingstone, London, 1978, 118.
35. **Peters, J. P., Bruckman, F. S., Eisenman, A. S., Hald, P. N., and Wakeman, A. M.,** The plasma proteins in relation to blood hydration, *J. Clin. Invest.*, 10, 941, 1931.
36. **Waldmann, T. A., Gordon, R. S., Jr., and Rosse, W.,** Studies on the metabolism of the serum proteins and lipids in a patient with analbuminemia, *Am. J. Med.*, 37, 960, 1964.
37. **Freeman, T.,** Analbuminemia — a study of albumin and transferrin metabolism, in *Physiology and Pathophysiology and Plasma Protein Metabolism*, Birke, G., Norberg, R., Plantin, L. O., Eds., Pergamon, London, 1969, 75.
38. **Weinstock, J. V., Kawanishi, H., and Sisson, J.,** Morphologic, biochemical, and physiologic alterations in a case of idiopathic hypoalbuminemia, *Am. J. Med.*, 67, 132, 1979.
39. **Jensen, H., Rossing, N., Andersen, S. B., and Jarnum, S.,** Albumin metabolism in the nephrotic syndrome in adults, *Clin. Sci.*, 33, 445, 1967.
40. **Yamauchi, H., Hooper, J., Jr., and McCormack, K.,** Blood volume and fainting in nephrosis, *Clin. Res.*, 18, 195, 1960.
41. **Oliver, W. J.,** Physiologic responses associated with steroid-induced diuresis in the nephrotic syndrome, *J. Lab. Clin. Med.*, 62, 449, 1963.
42. **Garnett, E. S. and Webber, C. E.,** Changes in blood volume produced by treatment in the nephrotic syndrome, *Lancet*, 2, 798, 1967.
43. **Eisenberg, S.,** Blood volume in persons with the nephrotic syndrome, *Am. J. Med. Sci.*, 255, 320, 1968.
44. **Hopper, J., Jr., Ryan, P., Lee, J. C., and Rosenau, W.,** Lipoid nephrosis in 31 adult patients, *Medicine*, 49, 321, 1970.
45. **Cameron, J. S., Turner, D. R., Ogg, C. S., Sharptone, P., and Brown, C. B.,** The nephrotic syndrome in adults with "minimal change" glomerular lesions, *Quart. J. Med.*, 43, 461, 1974.
46. **Metcoff, J., Nakasone, M., and Rance, C. P.,** On the role of the kidney during nephrotic edema, *J. Clin. Invest.*, 33, 665, 1954.
47. **Lubowitz, H., Mazumdar, D. C., Kawamura, J., Crosson, J. T., Weisser, F., Rolf, D., and Bricker, N. S.,** Experimental glomerulonephritis in the rat: structural and functional observations, *Kidney Int.*, 5, 356, 1974.
48. **Gur, A., Adefuin, P. Y., Siegel, N. J., and Hayslett, J. P.,** A study of the renal handling of water in lipoid nephrosis, *Ped. Res.*, 10, 197, 1976.
49. **Oken, D. E., Cotes, S. C., and Mende, C. W.,** Micropuncture study of tubular transport of albumin in rats with aminonucleoside nephrosis, *Kidney Int.*, 1, 3, 1972.

50. **Landwehr, D. M., Carvalho, J. S., and Okeñ, D. E.,** Micropuncture studies of the filtration and absorption of albumin by nephrotic rats, *Kidney Int.,* 11, 9, 1977.
51. **Rocha, A., Marcondes, M., and Malnic, G.,** Micropuncture study in rats with experimental glomerulonephritis, *Kidney Int.,* 3, 14, 1973.
52. **Bernard, D. B., Alexander, E. A., Couser, W. G., and Levinsky, N. G.,** Renal sodium retention during volume expansion in experimental nephrotic syndrome, *Kidney Int.,* 14, 478, 1978.
53. **Kuroda, S., Aynedjian, H. S., and Bank, N.,** A micropuncture study of renal sodium retention in nephrotic syndrome in rats, *Kidney Int.,* 16, 561, 1979.
54. **Kuschinsky, W., Wahl, M., Wunderlich, P., and Thurau, K.,** Different correlations between plasma protein concentration and proximal fractional reabsorption in the rat during acute and chronic saline infusion, *Pfluegers Arch.,* 321, 102, 1970.
55. **Grausz, H., Lieberman, R., and Earley, L. E.,** Effect of plasma albumin on sodium reabsorption in patients with nephrotic syndrome, *Kidney Int.,* 1, 47, 1972.
56. **Brenner, B. M. and Troy, J. L.,** Postglomerular vascular protein concentration: evidence for a causal role in governing fluid reabsorption and glomerulotubular balance by the renal proximal tubule, *J. Clin. Invest.,* 50, 336, 1971.
57. **Ichikawa, I., Badr, K. F., Troy, J. L., and Brenner, B. M.,** Mechanisms of renal salt retention in experimental nephrotic syndrome, *Kidney Int.,* 16, 783, 1979.
58. **Andreoli, T. E., Schafer, J. A., and Troutman, S. L.,** Perfusion rate-dependence of transepithelial osmosis in isolated proximal convoluted tubules, *Kidney Int.,* 14, 263, 1978.
59. **Luetscher, J. A., Jr. and Johnson, B. B.,** Observations on the sodium retaining corticord (aldosterone) in the urine of children and adults in relation to sodium balance and edema, *J. Clin. Invest.,* 33, 1441, 1954.
60. **Vecsei, P., Dusterdieck, C., Jahnecke, J., Lommer, D., and Wolff, H. P.,** Secretion and turnover of aldosterone in various pathological states, *Clin. Sci.,* 36, 241, 1969.
61. **Medina, A., Davies, D. L., Brown, J. J., Fraser, R., Lever, A. F., Mallick, N. P., Morton, J. J., Robertson, J. J. S., and Tree, M.,** A study of the renin-angiotensin system in the nephrotic syndrome, *Nephron,* 12, 233, 1974.
62. **Pessina, A. C., Hulme, B., Rapelli, A., and Peart, W. S.,** Renin excretion in patients with renal disease, *Circ. Res.,* 27, 891, 1970.
63. **Chonko, A. M., Bay, W. H., Stein, J. H., and Ferris, T. F.,** The role of renin and aldosterone in the salt retention of edema, *Am. J. Med.,* 63, 881, 1977.
64. **Meltzer, J. I., Keim, H. J., Laragh, J. H., Sealey, J. E., Kung-Ming, J., and Chien, S.,** Nephrotic syndrome vasoconstriction and hypervolemic types indicated by renin-sodium profiling, *Ann. Intern. Med.,* 91, 688, 1979.
65. **Mees, E. J. D., Roos, J. C., Boer, P., Yoe, O. H., and Simatupong, T. A.,** Observations on edema formation in the nephrotic syndrome in adults with minimal lesions, *Am. J. Med.,* 67, 378, 1979.
66. **Godon, J. P.,** Sodium and water retention in experimental glomerulonephritis: urinary natriuretic material, *Nephron,* 14, 382, 1975.
67. **Kruck, F. and Kramer, J. H.,** Third factor and edema formation, *Contr. Nephrol.,* 13, 12, 1978.
68. **Oliver, W. J., Kelsch, R. C., and Chandler, J. P.,** Demonstration of increased catecholamine excretion in the nephrotic syndrome, *Proc. Soc. Exp. Biol. Med.,* 125, 1176, 1967.
69. **Wilson, H. E. C. and Muirhead, A.,** Evidence of a pitressin-like substance in the tissue fluids in nephrosis, *Acta Pediat.,* 45, 77, 1956.
70. **Eder, H. A., Lauson, H. D., Chinard, F. P., Grief, R. L., Cotzias, G. C., and Van Slyke, D. D.,** A study of the mechanisms of edema formation in patients with the nephrotic syndrome, *J. Clin. Invest.,* 33, 636, 1954.
71. **Brown, C. D., Hirsch, S. R., Adler, A., Feirroth, M. V., Friedman, E. A., and Berlyne, G. M.,** Water emersion induced diuresis in the nephrotic syndrome, *Kidney Int.,* 16, 927, 1979.
72. **Epstein, M.,** Renal effects of head-out water immersion in man, *Physiol. Rev.,* 58, 529, 1978.
73. **Yamauchi, H. and Hopper, J., Jr.,** Hypovolemic shock and hypotension as a complication in the nephrotic syndrome, *Ann. Intern. Med.,* 60, 242, 1964.
74. **Llach, F. and Czerwinski, A. W.,** Hyponatremia, in *Sodium,* Papper, S., Ed., CRC Press, Boca Raton, Fla., 1982, 237.
75. **Roscoe, M. H.,** Biochemical and haematological changes in type I and type II nephritis, *Quart. J. Med.,* 19, 161, 1950.
76. **Warren, J. V., and Stead, E. A., Jr.,** The protein content of edema fluid in patients with acute glomerulonephritis, *Am. J. Med. Sci.,* 208, 618, 1944.
77. **Morita, Y.,** Hemodynamic and electrolyte studies in acute glomerulonephritis, *Clin. Res. Proc.,* 5, 206, 1957.
78. **Bradley, R. D., Jenkins, B. S., and Branthwaite, M. A.,** Myocardial function in acute glomerulonephritis, *Cardiovas. Res.,* 5, 223, 1971.

79. **Binak, K., Sirmaci, N., Ucak, D., and Harmanci, N.,** Circulatory changes in acute glomerulonephritis at rest and during exercise, *Br. Heart J.,* 37, 833, 1975.
80. **Eisenberg, S.,** Blood volume in patients with acute glomerulonephritis as determined by radioactive chromium tagged red cells, *Am. J. Med.,* 27, 241, 1959.
81. **Earle, D. P.,** Physiologic abnormalities in acute glomerulonephritis, in *Acute Glomerulonephritis,* Metcoff, J., Ed., Little, Brown, Boston, 1967, chap. 20.
82. **Earle, D. P., Farber, S. J., Alexander, J. D., and Pellegrino, E. D.,** Renal function and electrolyte metabolism in acute glomerulonephritis, *J. Clin. Invest.,* 30, 420, 1951.
83. **Bradley, S. E., Bradley, G. P., Tyson, C. J., Curry, J. J., and Blake, W. D.,** Renal function in renal disease, *Am. J. Med.,* 9, 766, 1950.
84. **Brun, C., Hilden, T., and Raaschou, F.,** Physiology of diseased kidney, *Acta Med. Scand.,* Suppl. 234, 71, 1949.
85. **Blantz, R. C. and Wilson, C. B.,** Acute effects of antiglomerular basement membrane antibody on the process of glomerular filtration in the rat, *J. Clin. Invest.,* 58, 899, 1976.
86. **Chang, R. L. S., Deen, W. M., Robertson, C. R., Bennett, C. M., Glassock, R. J., and Brenner, B. M.,** Permselectivity of the glomerular capillary wall, *J. Clin. Invest.,* 57, 1272, 1976.
87. **Wen, S. F. and Wagnild, J. P.,** Acute effect of nephrotoxic serum on renal sodium transport in the dog, *Kidney Int.,* 9, 243, 1976.
88. **Maddox, D. A., Bennett, C. M., Deen, W. M., Glassock, R. J., Knutson, D., Daugharty, T. M., and Brenner, B. M.,** Determinants of glomerular filtration in experimental glomerulonephritis in the rat, *J. Clin. Invest.,* 55, 305, 1975.
89. **Maddox, D. A., Bennett, C. M., Deen, W. M., Glassock, R. J., Knutson, D., and Brenner, B. M.,** Control of proximal tubule fluid reabsorption in experimental glomerulonephritis, *J. Clin. Invest.,* 55, 1315, 1975.
90. **Wagnild, J. P. and Gutmann, F. D.,** Functional adaption of nephrons in dogs with acute progressing to chronic experimental glomerulonephritis, *J. Clin. Invest.,* 57, 1575, 1976.
91. **Allison, M. E. M., Wilson, C. B., and Gottschalk, C. W.,** Pathophysiology of experimental glomerulonephritis in rats, *J. Clin. Invest.,* 53, 1402, 1974.
92. **Godon, J. P.,** Evidence of increased proximal sodium and water in experimental glomerulonephritis, *Nephron,* 21, 146, 1978.
93. **Godon, J. P.,** Origin of edemas during human glomerulonephritis, *Kidney Int.,* 11, 290, 1977.
94. **Birkenhager, W. H., Schalekamp, M. A. D. H., Schalekamp-Kuyken, M. P. A., Kolsters, C., and Krauss, X. H.,** Interrelations between arterial pressure, fluid volumes and plasma renin concentration in the course of acute glomerulonephritis, *Lancet,* 1, 1086, 1970.
95. **Powell, H. R., Rotenberg, E., Williams, A. L., and McCredie, D. A.,** Plasma renin activity in acute post streptococcal glomerulonephritis and the haemolytic-uraemic syndrome, *Arch. Dis. Child.,* 49, 802, 1974.

Chapter 7

IDIOPATHIC EDEMA

David C. Kem

TABLE OF CONTENTS

I. DEFINITION AND DIFFERENTIAL DIAGNOSIS

Idiopathic edema, by its name, indicates that there is no known cause to the excessive fluid retention.[1,2] Such common conditions as subtle congestive heart failure, renal disease, hepatic disease with hypoproteinemia, and other less common etiologies (see Table 1) must be considered and excluded before this diagnosis can be established. Several unique features of idiopathic edema, however, suggest that it is different from edema of a known etiology. Idiopathic edema rarely occurs in males and occurs in females only after puberty. The edema may be aggravated in the premenstrual period, but this is variable and this condition seems to be different from the premenstrual tension and weight gain syndrome found in many women, which resolves following completion of the cycle. A strong postural component is present and is responsible for the exaggerated diurnal weight gain of greater than 2.5 lb (1.1 kg).

II. SYMPTOMS AND SIGNS

The edema is diffusely distributed, but worsens in dependent portions of the body with upright posture. The edema is not always pitting, and the patient's perception of diffuse swelling and discomfort are frequently greater than the physician would suspect on examination alone. Abdominal bloating may be a prominent complaint and non-articular rheumatism may also be present. Mental symptoms include headaches, tension, and "slowness". Anxiety and other psychological complaints may coexist and be amplified by exacerbations of the weight gain as well as by frustration in not finding a diagnosis and effective treatment. Blood pressure is usually normal but orthostatic hypotension may be prominent if diuretic therapy is aggressively pursued. Many patients note a marked increase in thirst in the afternoon concurrent with a significant decrease in urine output. Sleep and recumbency are thus interrupted by nocturia and the diuresis may return the patient's weight to or near normal. This pattern of weight gain may occur daily or irregularly over a 3- to 7-day period during which a gradual build up of weight will be followed by a massive diuresis.

III. PATHOPHYSIOLOGY

Edema formation results from the derangement of:

1. The capillary hydrostatic pressure
2. Decreased plasma oncotic pressure
3. Increased oncotic pressure of the extracellular fluid, or
4. By changes in capillary wall permeability

Streeten has reviewed the evidence for and against involvement of each in producing idiopathic edema.[2]

There is little evidence to support an increase in capillary hydrostatic pressure in idiopathic edema. Gill[3] reported an unusual patient who had subtle congestive heart failure which was not evident until catheterization revealed an elevated end-diastolic pressure and responded to digitalization. Hemodynamic studies in small groups of patients have failed to find similar evidence in idiopathic edema.

Gill et al.[4] have also implicated mild hypoproteinemia of unknown etiology which might lead to edema through the decreased plasma oncotic pressure. Their measurements have not been confirmed by others, suggesting that there might be significant differences in patient selection from a heterogeneous group. Thibonnier et al.[5] have reported slightly lower plasma proteins in these patients than in their normal subjects, but these changes are not sufficient to be causative of this condition.

Table 1
DIFFERENTIAL DIAGNOSIS OF EDEMA

Common

Congestive heart failure
Nephrotic syndrome
Renal insufficiency
Hepatic cirrhosis
Venous insufficiency
Lymphatic obstruction
Hypoproteinuria
Premenstrual edema syndrome

Less Common

Allergic and inflammatory states
Hypothyroidism
Hyperthyroidism

The oncotic pressure of interstitial fluid may be increased by selective transudation of protein, but this apparently is an uncommon event.[2,6] Abnormal changes in capillary permeability to both water and solute appear to be the most likely single factor contributing to idiopathic edema. Streeten now believes that rapid transudation of fluid into the legs causes a series of compensatory events which may lead to exacerbation of the symptoms.

The possibility that hormonal changes are important in genesis of the condition has been much discussed. Early observations of a consistent increase in aldosterone secretion have not been confirmed. Abnormalities in ADH metabolism, renal filtration, and sodium clearance have also been proposed and some evidence supporting each possibility has been presented.

Streeten et al.[2] have been impressed with the effect of posture and have divided their patients into groups dependent on their renal response to upright posture and to oral salt and water loading. They reported that 26 of 31 patients had an orthostatic increase in weight that was greater than normal during upright posture. Sixteen of these 26 patients with an orthostatic component retained significantly more sodium during orthostasis than did the control subjects, while the others tended to retain water but not salt. Plasma aldosterone but not renin activity was elevated in 7 of the 16 (44%) sodium retainers, but only during the upright posture. Since urinary K^+ was increased in these seven patients during upright posture, Streeten et al. have suggested that the elevated aldosterone levels could be of some significance in the sodium retention. More importantly, measurements of inulin and PAH clearance were strikingly abnormal in the patients with orthostatic sodium retention but were normal in the others. It would appear likely therefore that the abnormal increase in sodium retention is due to a posture-mediated drop in GFR and renal plasma flow, and the change in aldosterone, when it occurs, is secondary to volume-mediated changes in production and metabolic clearance of the hormone.

The patients with increased water retention but normal sodium excretion following upright posture were suspected of having an intermittent and abnormal increase in ADH. This hypothesis was strengthened by demonstrating that the majority of these patients markedly increased water clearance when given alcohol concurrent with their water loading. No data were provided, however, for the effects of ethanol on water clearance in normal control subjects, so there is incomplete evidence for "normalization" of the H_2O clearance. Thibonnier et al.[5] have reported an abnormal increase in urinary ADH during upright posture, but not during recumbency, in 10 subjects with idiopathic edema. Water loading decreased the urinary ADH only in the supine position, but not in the upright posture.

The evidence for a decrease in plasma volume during upright posture in both orthostatic water as well as salt retainers is strengthened by the observations that leg compression by wrapping with bandages, use of a G suit, or immersion in water will prevent the posture-mediated changes and provoke a diuresis.

Deficient catecholamine levels, especially of dopamine, have been reported[7,8] but no confirmatory data are available. Recent reports that dopaminergic stimulation inhibits aldosterone production in man are of interest, but use of agents with dopaminergic activity have not been universally successful to date.

In summary, abnormalities in several systems affecting edema have been observed, but none have proven to be conclusive for causing this syndrome. The most plausible sequence, however, is a defect in capillary permeability which is aggravated by upright posture and leads to a transiently decreased plasma volume. Reactive and secondary increases in plasma aldosterone and/or ADH are concurrent with a decrease in renal filtration and plasma clearance. These changes are more exaggerated than those occurring in normal subjects, and the reason for this difference in patients with idiopathic edema is unknown.

IV. DIAGNOSIS

A careful history and physical examination are of great help in ruling out edematous states with a potentially definable cause. Attention to the timing, location, and onset of the swelling is important. The presence of intense thirst and oliguria, especially during upright posture in the afternoon, may help separate idiopathic edema from the other conditions.

Documentation of edema is important and may be difficult unless the patient is examined in the late afternoon. In many cases, asking the patient to weigh in the morning and in the evening just prior to sleep or recumbency will permit one to assess the magnitude of water retention and diuresis. A chart of two to four weeks duration is generally adequate for such purpose.

The physical examination should include measurement of recumbent and upright blood pressure and pulse in order to look for volume depletion or autonomic abnormalities. The screening laboratory studies generally include measurement of serum protein concentration, electrolytes, and tests of hepatic and renal function. Further testing is generally not necessary unless indicated by specific findings in the history or physical examination. A chest X-ray is of potential value for cardiac decompensation, but ultrasound measurement of left ventricular thickness appears to be the best noninvasive test short of radionuclide studies or catheterization.

V. TREATMENT

The management of idiopathic edema should always be accomplished within the context that the symptons, although unpleasant and unwanted, will not produce an effect on life expectancy and only mildly incapacitate the patient.

Once the diagnosis appears established, reassurance is critically important for the patient and will need to be reinforced over a period of time. Frequently only mild salt restriction to 50 to 80 mEq/day, reduction of standing during the day, and avoiding exacerbating factors may be all that is necessary for those with mild or modest symptoms. When symptoms are the worst, a mild diuretic may be used intermittently to minimize the fluctuations of weight. The availability of relief by using intermittent diuretics also provides the patient with some reassurance that she has control over the symptoms. It must be pointed out, however, that too vigorous diuretic therapy can enhance secondary aldosteronism and lead to rebound swelling, or produce volume depletion, postural hypotension, and hypokalemia.

The best time to use a diuretic is in the evening, and especially so if the patient can remain recumbent during this time. Hydrochlorothiazide is the best choice since it is of intermediate duration and usually acts rapidly to avoid nocturia. Furosemide may be a better choice because of its more rapid onset and shorter duration of effect, but the possibility of fluid and electrolyte abnormalities is significantly greater. If hypokalemia occurs, combining 50 to 75 mg per day of aldactone with a diuretic may decrease the effects of the secondary aldosteronism. There is less experience with triamterene and amiloride, but these should be effective. Oral potassium supplements might be used, but never in combination with the just-mentioned potassium sparing agents.

Leg compression with tight-fitting stockings has clearly been helpful in minimizing the symptoms, but there is still some accumulation of fluid during the day. This leads to unacceptable discomfort, and although this is potentially the best way to manage the condition, it also seems to be the least tolerated by the patient.

Sympathomimetic agents have been tried with variable success. Dextroamphetamine in doses of 10 to 25 mg/day has been advocated by Streeten[2] for resistant cases and some success has been noted. The mechanism is not clear, and although a decrease in aldosterone secretion has been noted, this is unlikely to be its primary mode of action.

Bromoergocryptine, a dopaminergic agonist, has been tried on several occasions and initial success was encouraging. More recently, this agent has been reported to provide minimal effect on the weight changes and it should continue to be considered an experimental drug for this condition.

Management of idiopathic edema is complicated by the inability of the patient to obtain complete relief and by the inability of the physician to provide a "real diagnosis". In the face of demanding patients and breach of trust from both the patient and physician, it is not uncommon to have a patient "shop" for a diagnosis or cure. This must be recognized as part of the syndrome and attempts should be made to reassure the patient of the difficulties and appropriate expectations that she might have. Finally, although the diagnosis has been firmly established in the past, it is important to remember that intercurrent disease might be present and periodic reevaluation is in order.

REFERENCES

1. **Thorn, G. W.,** Cyclical edema, *Am. J. Med.,* 23, 507, 1957.
2. **Streeten, D. H. P.,** Idiopathic edema: pathogenesis, clinical features, and treatment, *Metabolism,* 27, 353, 1978.
3. **Gill, J. R., Jr., Mason, D. T., and Bartter, F. C.,** "Idiopathic" edema resulting from occult cardiomyopathy, *Am. J. Med.,* 38, 475, 1965.
4. **Gill, J. R., Jr., Waldmann, T. A., and Bartter, F. C.,** Idiopathic edema. I. The occurrence of hypoalbuminemia and abnormal albumin metabolism in women with unexplained edema, *Am. J. Med.,* 52, 444, 1972.
5. **Thibonnier, M., Marchetti, J., Corvol, P., Menard, J., and Milliez, P.,** Abnormal regulation of antidiuretic hormone in idiopathic edema, *Am. J. Med.,* 67, 67, 1979.
6. **Coleman, M., Horwith, M., and Brown, J. L.,** Idiopathic edema. Studies demonstrating protein-leaking angiopathy, *Am. J. Med.,* 49, 106, 1970.
7. **Kuchel, O., Cuche, J. L., Buu, N. T., Guthrie, G. P., Jr., Unger, T., Nowaczynski, W., Boucher, R., and Genest, J.,** Catecholamine excretion in "idiopathic" edema: decreased dopamine excretion, a pathogenic factor?, *J. Clin. Endocrinol. Metab.,* 44, 639, 1977.
8. **Gill, J. R., Jr., Cox, J., Delea, C. S., and Bartter, F. C.,** Idiopathic edema. II. Pathogenesis of edema in patients with hypoalbuminemia, *Am. J. Med.,* 52, 452, 1972.

Hypertension

Chapter 8

SODIUM AND HYPERTENSION

Edward D. Frohlich and Franz H. Messerli

TABLE OF CONTENTS

I. INTRODUCTION

In probably no area of health care delivery has there been greater controversy than in the relationship between dietary factors and cardiovascular illness. However, in no area of cardiovascular diseases has a relationship been as liberally documented as it has been with dietary sodium intake and elevated arterial pressure.[1,2] Nevertheless, controversy rages and arguments continue as to whether the sodium ion participates actively in the pathogenesis of systemic arterial hypertension.[3]

Sodium and hypertension have been related on a number of levels. Data have been offered both clinically and experimentally for the association; and this chapter is concerned with documentation that supports the fundamental assertion that the sodium ion is inextricably interwoven into the mosaic of hypertension clinically, pathophysiologically, and therapeutically. It remains to be shown if this relationship is etiological. But, regardless of such a demonstration, it seems that the strong recommendations from many quarters of an enlightened preventive therapeutic approach have justification and merit.[4]

This discussion provides supporting physiological and other data for the above thesis, and the reader is left with the task of arriving at (or rejecting) the broad conclusion thereby implied. Thus, if there is an intrinsic defect in sodium metabolism in patients with hypertension and if this defect is made manifest by excessive sodium intake, it should follow that a program of sodium restriction may provide the most rational means for preventing or modifying the pandemic expression of this disease.

II. EPIDEMIOLOGY

Ever since the first scientific observation relating sodium excess to elevated arterial pressure was recorded by Ambard and Beaujard in 1904[5] and the later emphasis by Allen,[6] investigators have explored this relationship through many varied disciplines. Perhaps one of the strongest of the bases supporting this relationship has been provided by the time-consuming and tedious population studies by epidemiologists. These studies initially provided a direct relationship between the quantity of sodium intake in the diet of large regionalized populations and the prevalence of hypertension (Figure 1). Later, these studies demonstrated more sophisticated associations based upon dietary patterns, cultural differences, and racial or ethnic factors. It would seem that one common denominator for each of these factors is the sodium intake of these varied populations.

A. Populations with Low Sodium Intake

Arterial hypertension is virtually absent in primitive cultures, and arterial pressure does not rise with age as in so-called westernized or industrialized populations.[7] These observations suggest that some environmental factor may be influencing the course of arterial pressure throughout life and therefore the development of hypertension. Several possibilities come to mind that could account for the relative absence of high blood pressure in these primitive people: (1) overall health status of primitive man may be so affected with more frequent maladies, e.g., malnutrition, parasitic infections, anemia, and other chronic debilitating diseases; (2) primitive people generally lack obesity and are physically more active; (3) primitive man is less exposed to stress, noise, and pollution, and lives at a slower pace; (4) a variety of other so-called ''benefits of civilization are absent (such as television, soft drinks, ''gas-guzzlers'', refrigerators, chewing gum, etc.); (5) dietary habits, especially sodium intake, are different in primitive populations than in westernized man. However, food shortage, poor sanitation, and reduced access to modern medical care are all more common in nonwesternized societies.

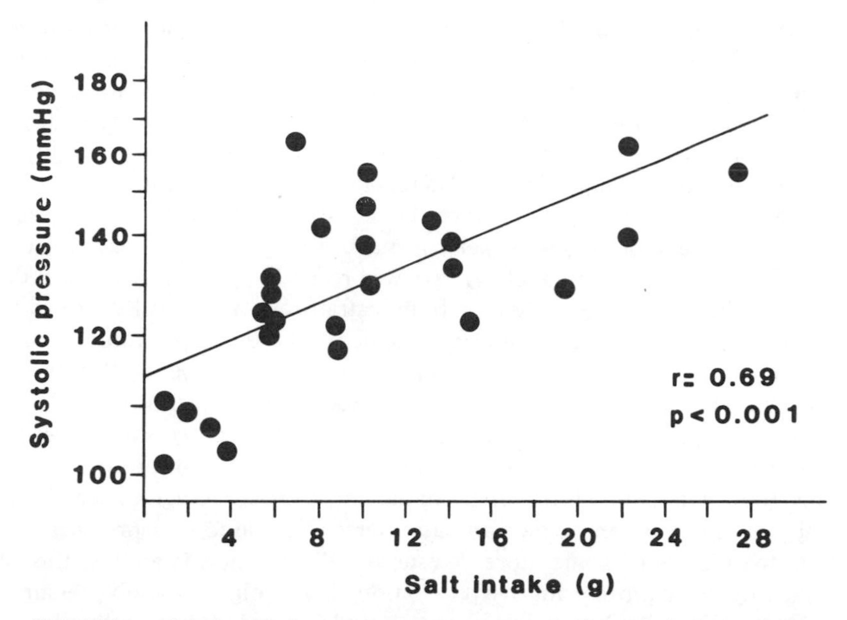

FIGURE 1. Relationship between systolic pressure and salt intake in 27 populations drawn from the literature. (From Glieberman, L., *Ecol. Food Nutr.*, 2, 143, 1973. With permission.)

Therefore, malnutrition, chronic illnesses, and debility are certainly more prevalent in primitive populations. Indeed, arterial pressure has been found to be consistently lower in patients with splenomegaly compared to subjects with a normal spleen within such tribes.[8,9] Nonetheless, athletic performance as measured by oxygen consumption during treadmill testing has been shown to be excellent in Masai men.[10] Also, no physical signs of malnutrition were found by Page in the study of a Solomon Island population.[11] This report clearly showed that tribal people may be in excellent physical condition and yet have no evidence of hypertension or blood pressure increase with age.

It becomes a somewhat more difficult task to examine the second possibility, the relationship between obesity and hypertension, which might explain the normal blood pressure in lean, physically active primitive man. Hypertension is more prevalent in obese people; and normotensive obese subjects are at a greater risk of developing hypertension that those who are nonobese.[12-15] Furthermore, weight reduction has been associated with a fall in arterial pressure regardless of sodium intake.[16] Primitive man is usually lean and his variation in body weight minimal, and blood pressure and weight have not been found to be correlated in most of these populations studied. For example, Page and his associates found no correlation between body weight and systolic pressure in the Solomon Island Study.[11] Similarly, Prior and his colleagues compared two Polynesian populations with different degrees of "westernization". They found no hypertension in the more primitive group but the one that was partially westernized had a substantially higher prevalence of hypertension; and the difference could not be explained on the basis of body weight.[17] Thus, in some primitive societies, body weight remains constant or even falls with age and is not related to blood pressure.[5,17,18] Even when so-called primitive populations undergo westernization, weight changes are usually small when compared to the weight differences observed within our civilization.[19-21] Hence, the lack of a correlation between weight and arterial pressure is not surprising when one considers the rather limited degree of "obesity" observed.[18] Moreover, there is evidence to indicate that the relative absence of hypertension in primitive populations may also be related to their leaner body habitus and, perhaps, more active lifestyle.

The third possibility, that the absence of hypertension may be caused by lack of some "negative" aspects of western civilization (e.g., stress, noise, tension, pollution, and urban living), has been a most difficult consideration to evaluate scientifically. The fourth possibility, that certain more "positive" (or attractive) aspects of our civilization (e.g., affluence, conveniences, exposure to news media, etc.) influence our blood pressure pattern throughout life, also remains open to question.

One strong point made by several investigators is that the lack of hypertension in primitive societies may be due to a lack (or marked reduction) of dietary sodium intake. This hypothesis has been carefully reviewed recently by Freis who concluded that of the various changes brought about with industrialization (or, as he termed it, "acculturization"), by far the most important seemed to be the remarkable increased intake of dietary sodium.[4] His conclusions were substantiated further by the observations of Lowenstein who studied two neighboring tribes in the Amazon basin.[22] The tribe that had been converted to Catholicism and used salt and coffee showed an increase in arterial pressure with age, whereas the other tribe, remaining primitive and using plant ashes (i.e., potassium chloride) as condiment, showed no age-dependent pattern of arterial pressure. Similarly, Page and associates also found that pressure increased with age in the women of some more "westernized" Solomon Islanders; this correlation was lacking in the more primitive populations.[11] The highest blood pressure values and greatest prevalence of hypertension were found in a group that traditionally boiled their vegetables in sea water and consequently had the highest sodium intake.[11] Furthermore, hypertension was observed and arterial pressures significantly increased with age in nomads of the Qash'qai pastoral tribe of southern Iran whose daily urinary sodium excretion averaged 186 mEq.[23] However, although this nomadic tribe lives at a relatively low level of acculturation, their body weight showed no tendency to increase with age.[23] In contrast, no age-dependent trend of arterial pressure was observed by Oliver in the Yanomamo Indians, an unacculturated tribe inhabiting the tropical rain forests of northern Brazil and southern Venezuela. In fact, the blood pressure levels averaged 100/64 mmHg in the over-50-year-old subjects who excreted as little as 1 mEq of urinary sodium per day. This remarkably low sodium diet was associated with an aldosterone excretion rate that exceeded values commonly found in patients with primary aldosteronism.[24]

B. Secondary Hypertension

Secondary hypertension can be found in up to 5% of all hypertensive patients in the U.S. Since not even a single patient with hypertension had been found, in very careful and extensive studies of some low dietary sodium ingesting cultures, it follows that secondary hypertension does not seem to exist in these populations either. This puzzling extrapolation may be explained by the general experience that, similar to essential hypertension, a minimal amount of sodium may be necessary for the pathogenesis of secondary hypertension. In other words, although renal artery stenosis, renal parenchymal disease, etc. may exist in primitive man, the sequence of events leading to hypertension does not occur, perhaps because of sodium depletion.

C. Blood Pressure Changes with Westernization

Page has recently reviewed the changes occurring when more primitive people migrate into a westernized or industrialized environment.[25] Invariably sodium intake rises and so does blood pressure. Probably the most clear-cut and instructive study in this regard was by Shaper and collaborators and was concerned with Somburn nomads from Kenya, who belong to a population that normally shows no rise in blood pressure throughout life and no hypertension.[26] In these natives daily salt intake has been estimated to average 2 to 3 g. When the young men enlisted in the army, their food habits were changed drastically from milk and meat diet to one of maize meal having a high (approximately 16 g) sodium content with a high carbohydrate-to-protein content ratio. Shaper observed an increased body weight and skin-fold thickness during their first year of military service without any change in arterial pressure.[26] However, during the second and third years, blood pressure climbed progressively at a time when body weight and skin-fold thickness actually decreased. Although in this example the association between dietary sodium and incidence of hypertension was striking various other changes (e.g., exposure to stress and yelling) associated with military life might have influenced the pathogenesis of hypertension.

D. Sodium Intake in Industrialized Populations

Numerous partially contradictory reports have concerned sodium intake and prevalence of hypertension in societies with a so-called "normal" sodium intake.[27-28] However, lack of correlation between arterial pressure and sodium excretion within a specific population with relatively little variation of dietary sodium should not be too surprising. First, Dahl has suggested that salt may exert its main hypertensionogenic effect during early childhood, as it does in salt-susceptible rats.[2] Hence, subsequent high or low sodium intake may not influence any further chance for development of hypertension. Secondly, too narrow a range of dietary sodium, even with a broad range of arterial pressures, will yield a statistically insignificant relationship. Moreover, daily sodium intake is subject to great fluctuation, and excessive sodium intake does not usually have an immediate effect on arterial pressure, at least not in patients who are on an otherwise "normal" diet. Neverthless, the early work of Sasaki shows good correlation of sodium intake with mortality from strokes in northern Japan where a high sodium intake is common.[39] We have recalculated the data of Sasaki, based on the age-adjusted death rate per 100,000 men from strokes in 46 prefectures, and found the prevalence of hypertension ranged from 30 to 40% in the inhabitants of the prefectures with the highest sodium intake (Figure 2). Not surprisingly, arterial pressure correlated best with sodium intake among the different populations with greatly varying dietary habits (Figure 1).

How is it possible that some individuals may heavily use the salt shaker throughout life without developing hypertension? The relationship demonstrated for bronchogenic cancer and cigarette smoking may throw some light on this question. Approximately 20% of men will develop bronchogenic carcinoma after 40 years of smoking more than two packages per day, but 80% will not. It is obvious that the absence of bronchogenic carcinoma in the 80% group argues as little against a causal relationship between smoking and lung cancer as the fact that many people who excessively salt their food remain normotensive throughout life. Thus, a genetic susceptibility to hypertension with sodium excess may exist in human beings similar to the maintenance of normal pressure in Dahl's salt-resistant rats despite a high salt intake and pressure increases in his salt-sensitive rats even after a short exposure to sodium excess.[2] Kawasaki et al. increased the daily sodium intake of 19 patients with essential hypertension from 9 to 240 mEq.[40] Nine of these subjects were sensitive and showed an 18 mmHg (average) increase in pressure; the others showed only a 4 mmHg increase in arterial pressure. Hence, it seems justifiable to conclude that human beings, similar to some rats, vary in their sodium sensitivity and that perhaps this characteristic may be inherited.

III. PHYSIOLOGICAL CONSIDERATIONS

A. Sodium Requirements

Salt appetite in animals has been reviewed recently in a careful study by Dethier.[41] It seems that carnivore animals usually get enough sodium by eating the flesh of other

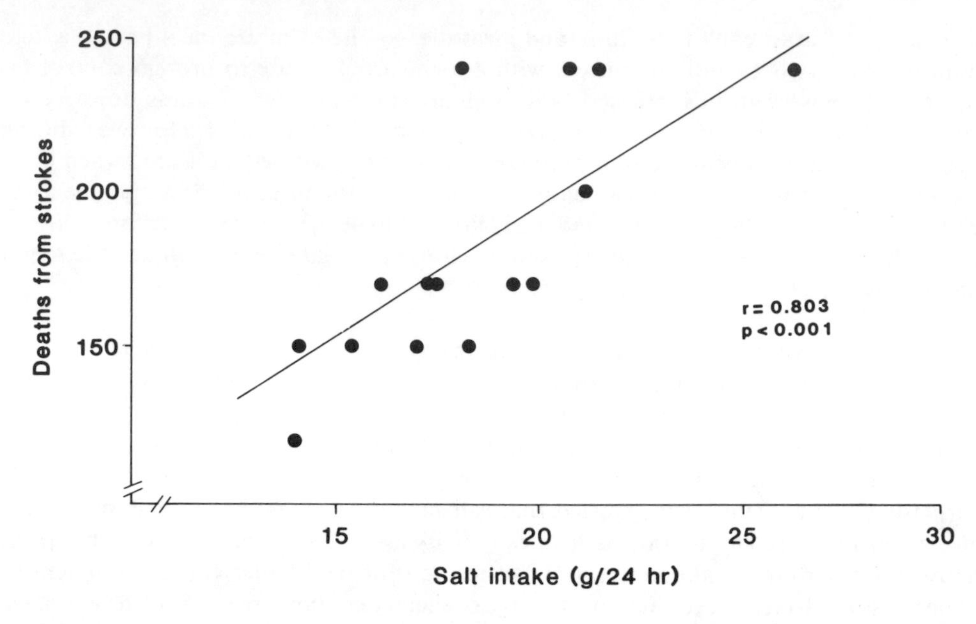

FIGURE 2. Relationship between salt intake and death from strokes (age-adjusted per 100,000 men) in 46 prefectures of northern Japan.

animals; herbivores, however, may need additional sodium. Blair-West has correlated avid sodium appetite of animals with high aldosterone levels and low urinary sodium excretion.[42]

Sodium chloride is a condiment that was cherished and used in various cultures whenever available; and literature is replete with references to salt craving in man. However, there is good evidence that this taste and craving is acquired in early childhood and related to the properties of salt as a condiment rather than to its physiologic effects.[1] Sodium and chloride are the principal ions in the extracellular space, thereby regulating volume (and pressure) of the extracellular and intracellular fluid. In this context, 1 g of salt commonly accompanies 125 $m\ell$ of isotonic water. Thus, salt intake of 12 g/day could implicate a 1.5 ℓ fluctuation of the body fluid spaces. Although fluid and ionic shifts are buffered, at least to some extent, by the intracellular compartment, fluctuations in the extracellular fluid space caused by a "normal" salt intake are still considerable. It is very difficult to estimate minimal daily sodium requirements in man since sodium is lost in sweat, stool, and urine, in amounts that vary with climate, body temperature, and coexisting diseases, medications, and diets.

B. Sodium Losses

With excessive sweating, sodium loss through skin may be impressive, especially in subjects receiving a "normal" sodium intake. However, as sodium is restricted, a marked reduction in the sodium content of sweat occurs. Patients can be maintained on a diet of 100 to 200 mg of sodium per day without adding extra salt when sweating during hot spells in summer.[43] Similarly, healthy persons can maintain sodium balance even when sweating fluid volumes of 5 to 9 ℓ/day while receiving a salt intake varying between 1.9 and 20 g.[44] Moreover, most populations ingesting low sodium diets live in tropical or subtropical climates exposed to heat; and although these tribal people are physically very active, they maintain sodium balance even when urinary sodium excretion averages only 1 mEq/24 hr.[24]

Fecal sodium losses are very small and do not vary much with sodium intake. Dahl has shown that, regardless of sodium intake, only about 10 to 125 mg is excreted daily in the stool.[45] In contrast to this route of excretion, urinary sodium grossly reflects sodium intake in the absence of heavy perspiration. As already mentioned, in those cultures following very low sodium diets, daily urinary sodium losses may be as low as 1 mEq.[24] In "westernized" patients with good renal function daily urinary sodium excretion may range from 6 to 30 mg with a sodium intake of 50 mg/day.[45]

From the above discussion, it can be calculated that the obligatory daily sodium losses (and hence the sodium requirements) most likely should not exceed 200 mg. Nonobligatory requirements are created by: increased sweating, periods of rapid growth, pregnancy, lactation, and in critical situations such as hemorrhage, tissue injury, or gastrointestinal electrolyte losses. Under these unusual circumstances, effective circulatory volume can be maintained by adrenal steroid induced mobilization of additional sodium from the skeleton.[42]

C. Salt Appetite

When considering the minimal daily sodium requirements to maintain sodium balance, it becomes obvious that the salt appetite of most human beings is inappropriately high, suggesting that the craving for salt may be an acquired habit. Children show a dislike for salty food when first presented with it; however, baby food manufacturers, as well as caring mothers, have been adding variable amounts of sodium to baby food to enhance its taste. Indeed, even in recent years when food companies have withheld sodium from baby foods, mothers have added salt to the specially prepared diets. Consequently, a baby may get 10 to 20 times more sodium than he actually needs.[2,46] Since essential hypertension probably originates in early childhood, the sodium intake during this period of life may be crucial in determining whether a sodium-sensitive subject will subsequently develop hypertension regardless of dietary sodium intake later in life. After the childhood period this sodium "intoxication" continues in the form of the obvious foods and condiments (e.g., peanuts, potato chips, popcorn, pickles) as well as the surreptitious additives and preservatives, (e.g., glutamate, ascorbate, citrate, nitrite). Conceivably, our bodies become adapted to chronic high-sodium intake by various mechanisms and subsequent reduction of this may even produce emotional symptoms.[47] These, however, are generally mild and transient and subside after a few days with a low-sodium intake. Dahl has shown that when a patient remained on a low-sodium diet (250 mg/day), addition of only 1 g of salt to this daily intake was noticeable; and the patient became adapted very quickly to the higher level.[45] In contrast, a patient ingesting a so-called "normal" amount of salt (10 to 20 g/day) did not realize when the daily intake was increased by 5 to 10 g.[45] Thus, a patient's discriminatory ability with regard to salt depends on the total amount in the diet. If he or she ingests a "normal" amount of sodium, the addition of 5 or even 10 g of salt may go entirely unnoticed.

D. Clinical and Experimental

Studies concerned with the threshold for the perception of saltiness in patients or experimental animals with hypertension have been conflicting.[48-53] The genetically (spontaneously) hypertensive rat demonstrates a preference for salt; and will consume more salt and water than its normotensive control.[51,52] Schechter demonstrated no difference between hypertensive and normotensive subjects.[53] However, when hypertensive patients and normotensive subjects could freely choose between distilled water or water with sodium chloride, the patients consumed about four times as much salt and twice as much fluid as did the normotensive subjects. This difference may be explained to some extent by the observations of the so-called exaggerated natriuresis in hypertensive patients (*vide infra*). An increased arterial pressure is associated with an increased renal sodium excretion;[54] hence, an increased sodium preference in patients with hypertension could conceivably reflect a greater urinary sodium loss.

How is it, then, that wild animals, which, unlike man, are not manipulated by advertising of the food industry and show instinctive intelligence with regard to diet, often show a profound salt preference and make long treks to salt licks? This holds true only for herbivores which may experience overt salt deficiency.[41] With few exceptions, all plants have a very low sodium content, especially away from the sea. The farther away an area is from natural salt sources, the lower the sodium content in the naturally growing food of that area.[41] Omnivores and carnivores may also be observed in the environment of salt licks, not because of their craving for salt, but because of an enhanced chance to find some herbivores. Milk cows, on the other hand, have a fairly high sodium loss in milk and, as every farmer knows, a lack of salt interferes with their milk output and, when extreme, may even lead to growth deficiencies and other illnesses.

E. Sodium vs. Chloride and Other Ions

In 1928 Addison found that, "Potassium salt regularly produced a decline in blood pressure, while sodium salt just as regularly produced a rise."[55] Subsequently, it was suggested by several investigators that the sodium ion may not be the only culprit in the pathogenesis of hypertension, but that lack of potassium or permissible or protective effects of calcium, magnesium, cadmium, etc. should be taken into account.[1] Langford et al. have shown that hypertensive subjects from lower socioeconomic rural areas seem to have a significantly higher urinary sodium-to-calcium and sodium-to-potassium excretion ratio than comparable normotensive subjects.[56] These findings suggest that a high potassium or high calcium intake may have some protective effect and may prevent, or at least slow down, the development of hypertension in the presence of a "normal" sodium intake. Similarly, high potassium intake in the diet has been shown partially to protect salt-sensitive rats from the deleterious effects of sodium.[57] Moreover, Bartorelli has shown that acute infusion of potassium is natriuretic and may even lower arterial pressure in hypertensive patients.[58] Furthermore, Sasaki and co-workers found that the intake of large amounts of apples by farmers of the Akita prefecture had a distinct blood pressure lowering effect.[59] A high calcium intake promotes sodium excretion and an intravenous calcium load increases arterial pressure, primarily through an increased vascular resistance.[60-62] Moreover, chronic hypercalcemia disorders such as hyperparathyroidism, sarcoidosis, and vitamin D intoxication are associated with hypertension, and arterial pressure may return to normal values after correction of the underlying abnormality.[60,61,63] These data are supported by a large body of information demonstrating a direct vasoconstrictor effect of the calcium ion, a vasodilating effect of magnesium, and both effects with potassium (dilator in lower doses and constrictor in high doses).[64-66]

Experimental data have shown that adrenalectomized rats have a greater preference for sodium salts of chloride, nitrate, sulfate, acetate, and iodide than nonadrenalectomized control rats.[67] Obviously, the taste preference focuses upon the sodium ion of these sodium salts with the preference for sodium appearing to be an innate response rather than a learned one. It therefore appears that it may be extremely difficult to appease this dietary sodium preference with substitutes containing a salt other than sodium, especially in patients exposed to life-long high dietary sodium intake.

F. Sodium and Chemically Induced Hypertension

Drugs that contain large amounts of sodium or produce sodium retention are particularly apt to expand extracellular fluid volume and perhaps provoke edema (unless an escape phenomenon takes place) or elevate arterial pressure. Moreover, drugs that contain high amounts of sodium or bicarbonate include the sodium salts of the semisynthetic penicillins (carbenicillin, tycrocillin) and various antacids. Other sodium-retaining medications are licorice, carbenoxolone, anabolic steroids, oral contraceptives, and nonsteroidal antiinflammatory drugs such as phenylbutazone or indomethacin. The pathogenesis and mechanism of these sodium and chemically induced forms of hypertension have been reviewed recently.[68]

IV. PHYSIOLOGICAL INTERRELATIONSHIPS

Knowledge progresses in stages: initially, there is a status of being uninformed and misinformed; but as inklings and insights appear from well-controlled studies, the data derived permit formulation of an hypothesis and considerable clearing of early vagueness. However, as further information appears, the simplistic early formulations and concepts give way to more complex theories; and, as further elaboration of fundamental information increases, it usually follows that more information is necessary. Indeed, this is the present state of our knowledge concerning the relationship between the sodium ion and the various systems involved in the pathogenesis and pathophysiology of hypertension. It is apparent that sodium is an important factor that participates in each of the pressor mechanisms; and it therefore follows that sodium plays an important, if not central, role in the pathophysiology of the hypertensive diseases.[69]

A. Renopressor System

By way of brief synopsis, the cells of the juxtaglomerular apparatus of the kidney synthesize renin in response to a variety of stimuli: afferent adrenergic input, volume contraction or expansion, circulating hormones (e.g., angiotensin II and catecholamines), electrolytes (e.g., potassium, calcium) and, of course, sodium concentration.[70,71] Some renin-release mechanisms involve beta-adrenergic receptors within the juxtaglomerular apparatus. Others may respond through intrarenal baroreceptors that are sensitive to the intrarenal arterial (distending) or perfusion pressure.[71-74] The sodium or chemically sensitive receptors are located within the macula densa; when sodium concentration is reduced, renin is released from the kidney.[75-78] The released proteolytic enzyme acts upon a protein substrate synthesized in the liver that is complexed (as angiotensinogen) to angiotensin I. Therefore, the dissociated product, formed from the action of renin, is the relatively inactive and short-lived decapeptide, angiotensin I, that is immediately transformed into an eight-chained polypeptide, angiotensin II. This is accomplished by cleaving of the terminal two peptides of angiotensin II by the converting enzyme (which is most concentrated in the lung). The released octapeptide acts directly upon certain target organs (i.e., the adrenal glomerulosa cells that synthesize the potent retaining steroid, aldosterone, the adrenal medullary cells that store the synthesized catecholamines, the vascular smooth muscle cells, and, also, specific cardiovascular and thirst-regulating centers in the brainstem and forebrain. Alternatively, a septapeptide may be formed that seems to have a more specific action on the adrenal cortical cells. Thus, sodium deprivation initiates activation of a host of physiological mechanisms that serve to restore normal sodium balance or return reduced arterial pressure to previous levels that were associated with a normal sodium intake. This so-called closed-loop servomechanism provides a basis for a logical understanding of the major mechanisms that subserve normal sodium-water-pressure balance. It therefore is clear that sodium restriction normally will stimulate an appropriate release of renin, generation of angiotensin II, secretion of increased aldosterone, and restoration of sodium balance and arterial pressure. However, unexplained at this point is why certain patients with elevated arterial pressure, receiving a diet containing normal amounts of sodium, may have normal, reduced, or elevated levels of plasma renin activity associated with normal aldosterone secretion.

(This subject of the release of renin, generation of angiotensin II, stimulation of aldosterone, and the maintenance of sodium homeostasis would be an appropriate area for a special review in itself. Their role in the normal state has already been discussed in this book and for this reason it was summarized only generally in the foregoing paragraph. Nevertheless, the reader is referred to several excellent texts that summarize this subject in great detail.)[69,79-82]

B. Water Balance

It follows from the foregoing discussion that one major and potent stimulus for maintenance of normal water balance is exerted through the sodium-renin-angiotensin-aldosterone system. Sodium, however, is also implicated in other mechanisms that maintain normal water homeostasis. One is related to the renopressor system since the angiotensin II, generated through sodium deprivation, may stimulate dipsogenic centers in the forebrain that provoke thirst and increased water intake, and thereby restoration of normal water balance.[83-91] In addition, since sodium is the major intravascular cation, reduced circulating sodium concentration (and, hence, plasma osmolality) will also stimulate osmoreceptors that, in turn, will promote release of the antidiuretic hormone (ADH) and thereby also acts to maintain normal sodium and water homeostasis.[92,93] Just precisely what role this mechanism plays in the pathogenesis or in the maintenance of arterial hypertension is unknown at this time.

Nevertheless, although abnormal ADH control has not been demonstrated in patients with primary (essential) or secondary forms of clinical hypertension, there has been some recent evidence of alterations of vasopressin control in several forms of experimental hypertension. Thus, studies in rats with DOCA-salt hypertension have revealed elevated plasma arginine-vasopressin (AVP) levels and urinary excretion rates;[94,95] and elevated AVP levels have also been demonstrated in rats with two-kidney Goldblatt hypertension[96-97] and with genetic spontaneous hypertension.[98,99] More recent studies by Mohring and his colleagues have concluded that AVP (i.e., ADH) does play an important role in the maintenance of the latter form of experimental hypertension.[100] This model is similar to essential hypertension in man[101] in that it bears similar characteristics including genetic predisposition and development in the presence of normal renal and other endocrine functions, and follows a similar hemodynamic pattern in progression of vascular and cardiac disease.[102]

Another as yet poorly defined mechanism regulating sodium, and hence water, homeostasis is that of the "third factor" (or the "natriuretic factor") which may be responsible for additional natriuresis. The normal role of this factor is discussed in detail elsewhere in this volume; and although at this point, its presence and role in patients (or animal models) with hypertension is unknown, it is possible that with its reduced generation (or its absence) sodium and water retention will result, thereby promoting further elevation of arterial pressure.

C. Hormonal Systems

Total body homeostasis, however, is not so simplistic that arterial pressure and sodium balance can be conceived as being solely dependent upon renal or renopressor systems involving only the closed-loop servomechanisms that normalize or maintain an abnormally elevated arterial pressure. A variety of other steroidal hormones (than aldosterone) also may serve to retain sodium by the kidney. To be sure, the most potent sodium-retaining steroid is aldosterone; but others include cortisol, hydroxycortisone, the estrogens and androgens, and weak steroids that may only be present in the fetus or may be extracted from large quantities of urine.[103,104] Surely, other adrenal cortical steroidal hormones will be identified. Evidence for their potential role in sodium homeostasis, and also in elevating arterial pressure, are the clinical observations that hypertension does not occur in the pathological consideration of Addison's disease or following administration of certain pharmacological compounds (e.g., aminoglutethimide) that arrest adrenal steroidal biosynthesis in patients with hypertension having low plasma renin activity and normal aldosterone excretion.[105,106]

At this point, then, we must conclude that the adrenal hormonal systems that affect sodium balance and arterial pressure are incompletely understood. However, there are two newly explored hormonal systems that also function to subserve sodium and pressure homeostasis. These two systems, most likely functioning intrarenally (but possibly also in the vessel wall and brain), are the kallikrein-kinin and prostaglandin-prostaglandin systems.[107] Discussion of the normally functioning interrelationships involving sodium homeostasis and the three intrarenal hormonal systems (i.e., renin-angiotensin and the foregoing two) is found elsewhere in this volume. Data are lacking on the possible alterations that might exist with respect to the role of the prostaglandins in the hypertensive diseases; and, only recently, decreased kallikrein excretion has been demonstrated in patients with essential hypertension[108,109] and in spontaneously and renal hypertensive rats.[110-112] What is known is that: there is a strong intrarenal relationship between the renin-angiotensin, kallikrein-kinin, and prostacyclin-prostaglandin systems;[113] and this relationship most likely involves sodium — at least in part.[114-116] Further information regarding their participation in maintaining normal sodium homeostasis and in mechanisms underlying hypertension should shed further light on the pathophysiology of the primary (essential) and secondary forms of hypertension.

D. Intravascular-Renal Volume Control (Renal Autoregulation)

There is still another physiological mechanism that serves to maintain homeostasis of arterial pressure and sodium and water balance. This is the intrinsic ability of the kidney to preserve fluid balance and maintain control of arterial pressure.[54,117] As postulated, there is an intrinsic intrarenal set-point that "senses" renal perfusion pressure so that when pressure becomes elevated diuresis and natriuresis will result, and when it is reduced, urine volume and sodium excretion will diminish. Just precisely what is involved and how this occurs in providing normal volume and pressure homeostasis by the kidney is unknown; however, it seems to involve an intrinsic renal mechanism.

Guyton has suggested that there is a basic defect in this intrarenal setpoint in hypertension, and, while he has provided much evidence for it in certain experimental models of hypertension, its existence in genetic laboratory models of hypertension and in man with essential hypertension remains an issue of great controversy. Thus, according to Guyton, there is a specific intrarenal defect that involves a higher pressure set-point of the kidney that regulates excretion of intravascular and extracellular fluid volumes. He suggests that, as a result, as volume expands, arterial pressure rises through an increased cardiac output. As part of a total body vascular autoregulatory response, the kidney, now perceiving the higher perfusion pressure and flow, also responds by actively increasing the state of tone of its vascular smooth muscle to the increased flow, thereby increasing renal as well as total peripheral resistance. This increased vascular resistance permits a pressure-diuresis (and natriuresis), restoring the expanded plasma and extracellular fluid volumes to normal, but at the expense of an altered hemodynamic state. This state of established systemic arterial hypertension associated with an increased vascular resistance and normal cardiac output (and organ blood flows) was brought about by total body autoregulation from the foregoing state of a higher cardiac output and normal total peripheral resistance.[82,118,119] This hypothesis has become one of the very attractive and popular concepts for our understanding of the pathogenesis and pathophysiology of essential hypertension — a disease presently of unknown etiology. Further discussion of these hemodynamic considerations will follow, but at this point it should be stated that expanded intravascular and extracellular fluid volumes have not been demonstrated in any form of genetic, experimental, or clinical essential hypertension. Neither has there been shown the necessity for a state of elevated cardiac output and increased total peripheral resistance.[120] However, the published experimental observations are indeed real in certain experimental forms of hypertension;[82] and we believe that these experimental findings together with the other observations in naturally occurring forms of experimental and clinical hypertension are compatible (*vide infra*).

E. Sodium Measurements in Hypertension

The argument that sodium excess produces increased tone of vascular smooth muscle and increased arterial pressure would be much stronger if it were possible to demonstrate that experimental animals and patients with hypertension have increased total body sodium content or increased sodium concentration in vascular smooth muscle. The available evidence is confused and complicated by experimental studies that have involved specific interventions that actively increase (or promote retention of) total body sodium (e.g., administration of agents such as adrenal steroids or surgical extirpation of normally functioning renal parenchymal tissue). Nevertheless, in several appropriately controlled studies, Tobian and his co-workers have demonstrated that associated with the increased arterial pressure in experimental forms of hypertension (e.g., renovascular, but with nephrectomy), the sodium content of the larger arteries was increased.[121-124] Tobian therefore suggested that the increased sodium concentration led to a "waterlogging" phenomenon of the vessel wall that serves to increase resistance to the forward flow of blood.[125]

The major problems with these (and many other) studies is that the experimental animals required artificial means for induction of hypertension involving factors that actively retain sodium. Most importantly, when the vessels were found to have increased amounts of sodium, they were not those that actively participate in increasing the total peripheral resistance[126] which, after all, is the hemodynamic hallmark of hypertension. These studies, performed in the late 1950s and early 1960s, were supplemented by concurrent epidemiological studies (cited above) that demonstrated a high degree of correlation between dietary sodium intake of large populations and the prevalence of hypertension in these populations (*vide supra*). Some of these workers also conducted clinical studies that determined whether total body sodium was, in fact, increased in the patients with hypertension. Thus Dahl and his associates showed that patients with essential hypertension seemed to have an increased total body sodium, as determined by a prolonged biological half-life of radioactivity tagged sodium ions and increased tissue sodium levels.[127] These reports supported earlier investigations that found increased serum sodium concentration,[128,129] expanded total exchangeable body sodium,[130] and increased arterial wall sodium.[121-124] Other workers, however, found conflicting results indicating that total exchangeable body sodium was not increased[130-135] and that there was no significant relationship between height of arterial pressure and serum sodium concentrations.[136-139] One major explanation for this disparity concerns the methods used for determination of total exchangeable sodium: earlier workers used a short-lived isotope, Na^{24} (half-life, 15 hr), whereas others used the more stable ion Na^{22} (half-life, 2.6 years).[131-135] Another explanation for these varying reports may reside in several other factors including: differences in patient diagnostic classifications; administration of drugs during the study; methodological differences for measuring body mass; and differences in the sensitivity of the analytical methods used.

The problem has remained at this level of confusion until rather recently when a variety of studies have demonstrated that administration of excess dietary sodium elevated arterial pressure in: (1) previously normotensive experimental animals that were genetically predisposed to develop hypertension by sodium excess;[140-143] and (2) patients with borderline hypertension having a family history of essential hypertension.[144] Thus, Dahl and his associates had developed a genetically susceptible strain of rats (sodium-sensitive) that, when given a sodium-excess diet, would develop increased arterial pressure. However, rats of the same strain were also selectively bred to be resistant (sodium-resistant) to the hypertensionogenic effect of sodium.[140,141] The mechanism(s) responsible for this increased sensitivity is still incompletely understood although initially it was held that it involved the excess secretion of the adrenal steroid, 18-hydroxydesoxycorticosterone.[145] More recently, Garay and Meyer showed that the increased arterial pressure of genetically hypertensive rats was associated with an altered Na-K-ATPase red blood cell flux system that operated in a similar fashion as in patients with essential hypertension but not as in Goldblatt hypertensive rats or in patients with renovascular hypertension.[146]

F. Nervous System

Just as the relationship between the sodium ion and the renopressor system is neither a simplistic nor an independent system, serving only to maintain arterial pressure, so is the relationship between sodium and the nervous system. Sodium has been implicated in a variety of altered neural mechanisms including: norepinephrine turnover and metabolism,[147-151] storage[148,152-156] and circulating levels,[157,159] and excretion of catecholamines.[158-163] These studies have demonstrated interactions between adrenal steroids and catecholamines in various experimental forms of hypertension and relationship to the renin-angiotensin system including that system within the brain[163] or blood vessel wall.[165] No studies of this nature have yet been reported in man with borderline or established hypertension. However, it is anticipated that with development of more sensitive techniques for measuring the angiotensins, catecholamines, and the characteristics of their respective metabolisms and their receptors, our knowledge of their roles in controlling arterial pressure will also increase. These mechanisms as they relate to underlying normal homeostasic functions are discussed elsewhere in this volume. Nevertheless, from this summary we have some reason to believe that excess sodium will increase norepinephrine turnover after catecholamine metabolism, and perhaps enhance vascular responsiveness (independent of a volume effect), to exogenously administered norepinephrine; sodium excess or deprivation is associated with reciprocal levels of plasma renin activity and circulating catecholamine loads, but this is still not certain in tissues including the renin-angiotensin system of the brain; and that sodium alterations may affect both catecholamine and angiotensin II receptor density and affinity levels to circulating catecholamines.

V. HEMODYNAMIC CONSIDERATIONS

A. Vascular Smooth Muscle

Although the above findings implicate a variety of mechanisms that interact to elevate arterial pressure, other data have suggested direct effects of the sodium ion that serve to increase the tone of the vascular smooth muscle cell itself and the vascular resistance to the forward flow of blood. These effects have been explained through a variety of physiological mechanisms that operate within the vessel wall to increase vascular resistance: through expansion of the extracellular fluid volume and secondarily by inducing a "water-logging" of the vessel's wall; by genetic alterations in the sodium/potassium-ATPase net flux system that controls intracellular Na^+/K^+ gradients; by other factors that alter the sodium ion partition in the smooth muscle cell and hence the transmembrane ionic gradient of sodium; through induction of certain structural changes that alter the response of the vessels to vasoconstrictor or vasodilating stimuli; or by active vasoconstrictor alterations induced locally within the vessel wall by the sodium ion on some of the mechanisms already described above (e.g., autoregulatory, humoral, local "renopressor," or neural) (Table 1). Much of the evi-

Table 1
POSSIBLE MECHANISMS INCREASING VASCULAR SMOOTH MUSCLE TONE SUBSERVED BY AN INCREASED AMOUNT OF SODIUM

1. Structural changes (changing the wall-to-lumen ratio) and hence resistance to blood flow and responsiveness of the vasculature to vasoactive stimuli
   a. Osmotic effect — "waterlogging"
   b. Alteration of the intracellular protein matrix
2. Myogenic response (physical responses of the wall to increased blood flow)
   a. Autoregulation
   b. "Bayliss" response to vascular distension
3. Response of the vascular smooth muscle cell to increased sodium (this mechanism assumes a defect in control of intracellular sodium ion concentration)
   a. Altered protein-synthesizing activity
   b. Defect in the regulation of catecholamine release and uptake
   c. Increased sodium-calcium exchange mechanism
4. Induction of other local mechanisms of vessel tone
   a. Altered responsiveness of receptors to naturally occurring circulating vasoconstrictor and/or vasodilator substances (with vessel wall, heart, brain, kidney)
   b. Altered generation of locally produced vasoactive agents within the blood vessel wall, in the central nervous system, or elsewhere
5. Altered neural function
   a. Increased turnover of norepinephrine at postganglionic adrenergic nerve endings
   b. Altered levels of circulating catecholamines
   c. Stimulation of enhanced adrenergic outflow from the brain

dence has involved pharmacological or surgical induction of a variety of forms of experimental hypertension although, in recent years, greater emphasis has been placed on the role of increased sodium loads in sodium-sensitive genetically predisposed hypertensive animals or man.

1. Structural Changes

With respect to the structural changes, these may increase vascular resistance through an osmotic or "waterlogging" effect whereby sodium gains entrance into vascular smooth muscle or other cellular (or intracellular) material, bringing along with it water;[121,125,166-171] the net effect is a thickening of the vessel wall and an increased wall-to-lumen ratio.[166-173] As an alternative possibility, the increased sodium concentration could also alter the protein matrix of the vascular smooth muscle cell that would also thicken the vessel wall.[167,173-176] The net result of the increased vessel wall-to-lumen ratio is an augmented response of the vessel to constrictor stimuli and an attenuation of the vascular response to vasodilating stimuli.[166,172,177,178] Indeed, this is the characteristic of vessels of most experimental and clinical hypertensive subjects.

2. Myogenic Factors

Another mechanism whereby hypertensive vascular disease may develop may be initiated through intrinsic myogenic factors within the vessel wall that permit its increased resistance to an increased blood flow or distending pressure.[179] Under these circumstances the increased blood flow, vascular distention, or increased perfusion pressure may be associated with an expanded circulatory volume; but, as indicated, these circumstances do not necessarily have to occur in the hemodynamic events associated with all sodium-induced hypertensive states. It is true that an increased cardiac output and expanded intravascular (and extracellular) fluid volumes have been observed in certain forms of experimental hypertension[182] including DOCA hypertension.[180,181] However, in other forms of hypertension associated with only salt excess (or salt excess superimposed upon a genetic mechanism), the expanded volume associated with increased blood flow is not necessarily a prerequisite for later development of an elevated arterial pressure associated with increased total peripheral resistance.[182-186]

3. Altered Membrane Sodium Transport

Other mechanisms involving sodium occur within the vessel wall. In addition to increased intramurally bound sodium that might act through enhanced hydration to stiffen and thicken the vessel wall, sodium ions may alter the transmembrane cellular ionic gradients.[173,187] Sodium is mainly an extracellular ion and its increase within the cell implies an altered sodium permeability of the cell membrane. Transmembrane ionic gradient of sodium usually remains relatively constant over a wide range of physiological conditions; however, under certain pathological conditions (e.g., hypertension), it is conceivable that more sodium ions gain entrance to the intracellular milieu. According to Friedman, this slight increase in intracellular sodium ion concentration could significantly change the protein-synthesizing activity of the vascular smooth muscle cell and, hence, total peripheral resistance directly[167,174,175,189-191] or indirectly through altered catecholamine release and uptake mechanisms.[147-153] Alternatively, Blaustein has suggested that the altered sodium ion electrochemical gradient across the vascular smooth muscle membrane may alter calcium regulation so that the amount of intracellular ionized calcium is increased (by displaced calcium ion in an altered exchange mechanism) thereby increasing vessel wall tension.[192] Since smooth muscle contraction is related directly to the free intracellular calcium ions that are available, the increase in free intracellular calcium produced by the altered sodium-for-calcium ion exchange mechanisms could facilitate the smooth muscle contractile process.

Unfortunately, the smooth muscle cell is relatively inaccessible to experimental studies. However, several recent reports have demonstrated a defect in the movement of ions across the membranes of red blood[146,193-197] and white blood[198] cells of patients with essential hypertension as well as rats with genetic and other experimental hypertension.[146] Not all of the reported abnormalities are reconcilable with Blaustein's hypothesis that suggests an increased intracellular sodium content.[192] Nevertheless, a laboratory distinction between essential and secondary hypertension and normotensive subjects with and without a family history of hypertension has been demonstrated in one study by measuring erythrocyte cation flux. It seems unlikely that these changes resulted from the hypertension or antihypertensive therapy. Whether or not an effective red cell sodium pump is paralleled by similar changes in the arteriolar smooth muscle remains unknown, but this abnormality may provide a useful model of a possible enzymatic defect invovling sodium metabolism in essential hypertension.

4. Humoral Factors

Humoral mechanisms operating via an active sodium-retaining effect of released steroids may participate. Alteration in sodium concentration may initiate the generation of a postulated local renin-angiotensin system within the vessel wall or remotely within the central nervous system.[199] We already have some idea that a sodium-excess diet in rats with spontaneous hypertension fails to provide the normal decrease with aging in angiotensin II receptors within the hindbrain.[164] Moreover, sodium excess may also be associated with increased myocardial beta-adrenergic receptor sites.[200]

Thus evidence has been rapidly accumulating to suggest that dietary sodium excess may be associated with: (1) increased responsiveness of the vessel wall to constrictor stimuli and decreased responsiveness to dilating stimuli; (2) an increased active vessel wall tension that may be mediated through a variety of intracellular vascular smooth muscle events (e.g., altered catecholamine release and/or uptake, altered transmembrane ionic gradients, or an abnormal sodium-calcium exchange mechanism); (3) initiation of enhanced vascular smooth muscle receptor dynamics to other vasoactive agents (e.g., angiotensin II, catecholamines); and (4) alteration of local generation of vasoconstrictor and/or vasodilator compounds (e.g., angiotensin catecholamines, prostaglandins). It seems reasonable to assume that a combination of each of the proposed mechanisms (as well as others) may be initiated in a predisposed animal.

B. Systemic and Regional Hemodynamic Studies

Surprisingly, with the large volume of literature concerned with the role of sodium (or salt) excess in the pathogenesis of hypertension (experimentally or clinically), there has been very little reported concerning the effects of sodium excess, per se, on the hemodynamic hallmark of hypertension, increased vascular resistance. For one reason, if excess sodium is permitted there is a rapid excretion of the acute sodium load in the urine in both normotensive and hypertensive subjects which is exaggerated in untreated patients with hypertension (*vide infra*). Nevertheless, several studies have shown that extremely high sodium loads (1500 mEq/24 hr intravenously) of sodium will markedly increase cardiac output in normotensive subjects and this response may differ in black and white subjects.[201,202] These hemodynamic studies have not been pursued with patient having essential hypertension.

Several studies have been reported concerning the hemodynamic effects of sodium excess in certain secondary forms of experimental hypertension. Thus Guyton and his associates have shown that elevated arterial pressure will result from chronic salt-loading in dogs with removal of as much as 70% of normally functioning renal parenchyma.[203] This hypertension was related to an initial increase in cardiac output (associated with a normal total peripheral resistance) and a later development of more sustained and higher arterial pressure that is associated with a normal cardiac output and increased total peripheral resistance.[204] During the initial stages the elevated pressure and increased cardiac output were associated with an expanded sodium space and intravascular volume and an increased mean circulatory filling pressure.[205] This latter hemodynamic index reflects the pressure and volume within the total circulation when the circulation is transiently arrested;[206, 207] and the postcapillary segment is the major area for vascular capacitance.[208] Similar hemodynamic findings have been reported in the mini-pig with DOCA-induced hypertension.[180] Both of these types of hypertension are secondary either to a surgical or pharmacological means for actively retaining sodium.

More recently, hemodynamic studies have been reported in rats and in man with genetically predisposed hypertension. In these studies there was no evidence of impaired renal excretory function or of an excess of a sodium-retaining steroid hormone. Thus, we have shown that the spontaneously hypertensive rat (SHR) demonstrated a further rise in arterial pressure that was increased still more with the addition of DOCA.[142,143] Arterial pressure was not elevated by salt excess in the normotensive control WKY rat strain. The further increase of arterial pressure in the SHR was associated with a marked increase in total peripheral resistance but a normal cardiac output; in contrast, cardiac output increased as total peripheral resistance fell in the WKY. A similar hemodynamic pattern was observed by Ganguli and associates in strains of rats (Dahl) that were sensitive and resistant to salt hypertension.[209] These studies therefore provided direct evidence that both forms of genetic hypertension were sensitive to sodium excess: in the former (SHR) by further increasing vascular resistance from an already elevated pressure and resistance state, and in the latter by increasing vascular resistance to an abnormal level from a normal pressure and resistance. Most importantly, the control rats for both of these genetically predisposed hypertensive models failed to increase pressure or vascular resistance; in fact, the chronic sodium excess was associated with a fall in total peripheral resistance since arterial pressure remained changed. Other studies (in the Dahl strains) have shown that during excessive salt ingestion, the increased total peripheral resistance was specifically demonstrated in the renal circulation; this renal vasoconstriction was not the consequence of increased neurogenic activity nor due to enhanced vascular responsiveness to angiotensin II or norepinephrine.[210] The kidney of the salt-sensitive hypertensive rat demonstrates reduced sodium excretory capacity[211] associated with low renal papillary plasma flow.[212] These renal circulatory findings are in contrast to earlier findings demonstrating enhanced central nervous system sensitivity to salt[2,3] and an increased neurogenic contribution to hindquarter vasoconstriction[214] in this genetically susceptible strain to salt-induced hypertension.

It is also clear that sodium excess is not essential for the development of all experimental forms of hypertension. In two recent studies, renovascular hypertension was produced in sodium-depleted dogs and rats in whom the height of arterial pressure was similar to control animals receiving a normal sodium intake. Thus, Davis and his associates were able to demonstrate that sodium-depleted dogs increased arterial pressure to the same levels as other dogs made hypertensive either by a chronic perinephritic hull[215] or by renal arterial clipping[216] but receiving normal sodium intake. These findings were corroborated by Trippodo and his associates in sodium-depleted Goldblatt hypertensive rats.[217] Thus, whereas sodium excess may aggravate development of hypertension associated with increased total peripheral resistance in certain genetically predisposed or other experimental hypertensive animals, excessive sodium does not seem to be necessary in other types. Nevertheless, in neither genetic nor two forms of renal hypertension has expanded fluid volume or increased cardiac output been shown to be necessary.

Therefore, these studies have an important bearing on the understanding of mechanisms that have been advocated for the development of sodium-excess hypertension; increased extracellular fluid volume → increased cardiac output → vascular or renal adaptation of the total body and renal circulations in order to reset arterial pressure and sodium and water excretion levels by the kidney, respectively, through increasing vascular resistance (total body or intrarenal autoregulation). These experimental studies have been supported by findings in borderline hypertensive man demonstrating an increased arterial pressure and forearm vascular resistance without demonstrating an increased forearm blood flow.[144]

C. Exaggerated Natriuresis

As already indicated, untreated patients and most experimental animal models of hypertension have demonstrated the phenomenon of exaggerated natriuresis which has been described and confirmed repeatedly in patients with essential hypertension.[218-233] Although a number of explanations have been offered, a precise explanation awaits further elucidation. The abnormality was first demonstrated by Farnsworth as an exaggerated excretion of chloride following a saline infusion only because at that time sodium was not measurable.[218] Later, exaggerated sodium excretion was demonstrated[219] following rapid intravenous infusion of hypertonic saline,[223-229] isotonic sodium chloride-lactate,[231] mannitol,[220] glucose,[232] and water-loading.[225,233] This sodium hyperexcretion persists despite sodium deprivation prior to study[231] and does not seem to be dependent upon sodium loading alone.[229] The natriuresis apparently is unrelated to the glomerular filtration rate[228,230] or to tubular loss of chloride.[227] Earlier studies incriminated a defective concentrating ability of the hypertensive kidney,[234,235] but 56 years later Brodsky and Graubarth postulated a proximal and distal renal tubular defect.[221] Another explanation has implicated augmented tubular rejection of sodium and chloride that is induced by elevated renal vascular or peritubular capillary hydrostatic pressure.[236-239] These observations suggested a possible intrarenal hemodynamic basis underlying this phenomenon; and support for this hypothesis has been demonstrated by elevated wedged renal venous pressure in experimental forms of hypertension and in man with essential hypertension.[239,240] However, several other lines of investigation in normotensive experimental animals have found a depression in fractional sodium reabsorption by the proximal and distal tubules during extracellular fluid volume expansion independent of a renal hemodynamic defect.[227,229,238,241-245] However, most of these studies were in normotensive animals, and still other reports concerned with normotensive man showed a nonspecific excretion of other substances in addition to sodium.[247] Thus, Cannon and his colleagues showed an enhanced excretion of potassium, calcium, chloride, and urate in addition to sodium during and following intravenous infusion of hypertonic saline.[247] It may also be possible that extrarenal factors may be responsible for the enhanced excretion of sodium by the kidneys;[248,249] and this is also supported by the observation that the phenomenon disappears following deconditioning[250] or sympathetic blockade.[251] This suggests that neural mechanisms might be responsible for the abnormality; but alternatively the disappearance could be associated with normalization of arterial pressure. Also supporting the contention of neural factors[250,251] was the work of Ulrych, who demonstrated that the magnitude of the natriuresis was temporally related to the height of cardiac output achieved during the saline infusion.[248] Thus, the patients with hypertension achieved a higher cardiac output that was temporally associated with the exaggerated natriuresis. This could imply that peripheral venoconstriction in the patients with hypertension provoked a shifting intravascular volume to the central circulation thereby augmenting cardiac output. This observation was supported by subsequent studies demonstrating that an elevated cardiac output in patients with hypertension was directly related to the volume of blood redistributed into the cardiopulmonary area and not to an expansion of circulating intravascular volume.[252] These observations suggest that low-pressure volume receptors in the cardiopulmonary area may initiate a reflex-mediated exaggerated urinary excretion of sodium and water. Support for this observation is offered by the brisk diuresis and natriuresis that follows the transient episodes of paroxysmal atrial tachycardia.[253]

D. A Unifying Hypothesis

At this point in a hemodynamic discussion concerning the interrelationship among a variety of factors observed in hypertension, it might be appropriate to offer an integrated hypothesis for their coexistence. This is not intended to provide the final answer for these apparently disassociated phenomena but rather a stimulus for further inquiry. We have already indicated that Guyton[82] and others[254-257] have suggested that early in the development of hypertension intravascular and extracellular fluid volumes may be expanded as a result of an intrarenal defect. As a result, mean circulatory filling pressure and cardiac output become elevated, thereby initiating a second phenomenon of "total body autoregulation" that transforms a high cardiac output — normal total peripheral resistance state to one of normal cardiac output and high total peripheral resistance. We have also indicated that sodium retention and expanded intravascular volume are not mandatory for the development of high vascular resistance hypertension.[215-217] In addition, we have shown that hypertension develops through an unaltered course in SHR rats even when cardiac output is reduced from either conception[184,185] or from weaning.[183] Moreover, at this age (and even until 12 to 14 weeks) during development of sustained hypertension in the SHR, intravascular or extracellular volumes are not expanded[186] and the elevated mean circulatory filling pressure therefore indicates peripheral venoconstriction in this hypertension.[258] Even man with borderline or mild essential hypertension with elevated cardiac output and normal total peripheral resistance demonstrates arteriolar constriction,[259] since when cardiac output and heart rate are normalized acutely by pharmacological means the elevated arterial pressure persists.[260,261] Moreover, the augmented cardiac output and cardiopulmonary volume early in this hypertension provide evidence of venoconstriction in addition to arteriolar constriction.[252] These data, both in experimental and clinical hypertensions, indicate that a state of sodium retention leading to an expanded intravascular (and extracellular) fluid volume and increased cardiac output is not necessary for subsequent development of hypertension provided that a state of peripheral venoconstriction is present. The venoconstriction will therefore increase the mean circulatory filling pressure and venous return to the heart in order to maintain cardiac output associated with arteriolar constriction.

A question then may be asked: just what defect may precipitate the simultaneous hemodynamic events of arteriolar and venular constriction with or without increased cardiac output? Increased adrenergic input to the cardiovascular system is one possible mechanism and, indeed, is also compatible with the altered intrarenal "set-point" postulated since altered neural input to the kidney can affect the renal handling of sodium and water.[262-268] How can these phenomena occur in the setting of altered sodium metabolism? As indicated above, a defect in intracellular sodium regulation (possibly induced by an abnormally low sodium-potassium net flux enzyme system,[193-197] by an abnormality of the Na-Ca exchange system[192] or other currently explored mechanisms) may indirectly serve to alter adrenergic function through altered catecholamine release and uptake as well as receptor action.

Alternatively, de Wardener and MacGregor have proposed that there may be a circulating substance present in patients with essential hypertension that inhibits sodium transport.[268a] This substance, therefore, may favor sodium retention but might also be "saluretic". The nature of this saluretic factor that also inhibits sodium transport remains to be elucidated, and its role in the majority of patients with essential hypertension who have volume contraction must be evaluated. Nevertheless, much remains to be explored, but at this time the multitude of disparate findings seem to be coalescing into a more meaningful understanding. And from this review one thing seems likely: sodium will play a critical role in elucidating our further understanding of the pathogenesis and pathophysiology of the problem.

VI. TREATMENT OF HYPERTENSION

This discussion, dealing with the treatment of hypertension, only concerns the use of low-sodium dietotherapy. Diuretic therapy, the mechanism of action of diuretics, and the proposed antihypertensive mechanisms of these agents are discussed elsewhere in this volume. Discussion of all other antihypertensive agents is not in the scope of this chapter or text.

A. Salt Restriction

Kempner has found that of 500 patients, 62% were improved by a rice-fruit diet, whereas 33% were unchanged and the 5% with the most severe hypertensive cardiovascular disease died.[269] Similar responses to severe sodium restrictions have been observed by others.[270-275] Each of these studies showed that severe sodium restriction significantly reduced arterial blood pressure. However, there was general consensus by most of these confirmatory reports that extreme sodium restriction is not a realistic objective. Indeed, even modest sodium restriction (i.e., by about 50%) has been shown to significantly reduce arterial pressure.[275,276]

B. Prehypertension or Offspring of Hypertensive Parents

In a very careful review, Freis concluded that a reduction of salt intake below 2 g/day would result in the prevention of essential hypertension.[4] He suggested that "it would seem wise for individuals with a family history of essential hypertension to accustom themselves to a truly salt-free diet (less than 1 g of salt or 50 mEq of sodium per day) and to prevent their children from acquiring the habit of eating salty food."[4] Some children of parents (one or both) with hypertension may already manifest abnormalities of the renin-angiotensin-aldosterone system or in the transmembrane sodium-for-potassium flux transport, and a simple clinical test has been proposed to identify potential or hypertension-prone youngsters.[146,196,197] Until a reliable and reproducible procedure is available (or proves to be useful clinically), our judgment as to whether or not salt restriction should be exerted can be based on past clinical observations. Experience has shown that most risk factors for subsequent development of hypertension are blood pressure levels at the upper limit of normal (90 percentile), excessive weight gain, strongly positive family history, inappropriately fast heart rate, and black race.[277-283] Therefore, it seems reasonable that children who have two or more of these risk factors are subjected to a high risk of developing hypertension and should significantly reduce their sodium intake. This should be particularly feasible in the family setting since this is one rational means for treating the entire family predisposed to developing hypertension. Following this practice, it is possible that sodium restriction may be advocated for a person who would not benefit from this approach (i.e., whose genetic background with regard to hypertension may correspond to the salt-resistant rat). Nonetheless, there has been no report of a healthy child who has sustained impaired health because of the sodium restriction proposed by Freis.

C. Borderline Hypertension

The term borderline hypertension has generally been applied to relatively young patients whose ambulatory diastolic pressure fluctuates around 90 mmHg.[284] Their risk of developing established hypertension is two- to threefold greater than a comparable group of age- and sex-matched normotensive subjects.[284] Together with other nonspecific measures such as weight reduction, regular exercise, etc. a sodium-restricted diet may gratify the patient and the physician by postponing need for antihypertensive medication indefinitely or for several years. The younger the patient the more drastically sodium restriction will lower arterial pressure. This measure is inexpensive and harmless, but its success depends, of course, on the patient's insight and compliance. Recent studies in these individuals as well as patients with mild hypertension have demonstrated the value of sodium restriction (associated with weight reduction) not only in reducing arterial pressure but also in slowing heart rate and reducing serum cholesterol levels.[285]

D. Essential Hypertension

Over the past two decades the availability of potent diuretics has made sodium restriction less popular for the treatment of established hypertension. Most physicians and patients prefer the convenience of a once- or twice-a-day diuretic taken with or without dietary sodium restriction. Clearly, diuretic therapy is effective and has become the mainstay for most advocated pharmacologically oriented treatment programs.[286] Nonetheless, it has been well demonstrated that the amount of sodium ingested by some patients (15 to 30 g/day) may completely override the antihypertensive effects of a diuretic.[287] Furthermore, unrestricted sodium intake may predispose the patient to one of the major side effects of the diuretics — hypokalemia.[288] Thus, the more sodium that is filtered across the glomerular membrane, the more will be available at the distal tubular level for exchange with potassium. Further, the contracted plasma volume, the reduced pressure, and the reflex adrenergic stimulation stimulates the juxtaglomerular cells to release renin. This, in turn, generates angiotensin II and promotes aldosterone secretion from the adrenal cortex. An increased aldosterone level enhances sodium reabsorption from the distal tubules and further facilitates the excretion of potassium and hydrogen ions thereby aggravating an already mild state of hypokalemic alkalosis. Modest sodium restriction (e.g., 3 to 5 g salt), on the other hand, will reduce the availability of sodium ions at the level of the distal tubule and will thereby minimize the potassium loss and alkalosis.

Furthermore, modest dietary sodium restriction reduces plasma volume, and so lowers arterial pressure in itself,[274] facilitates better blood pressure control by the thiazide diuretics,[275,276] and enhances the effectiveness of other antihypertensive agents. The third reason to advise patients with established essential hypertension to restrict their sodium intake is an educational one. As already indicated, by this measure the salt shaker will be used less by the entire family, thereby reducing the incidence of hypertension, or at least serving to postpone its development in offspring of hypertensive parents. Although this may be wishful thinking (at this date), the fact that one family member must restrict salt intake will at least suggest to the others that salt is not just a harmless condiment for everybody.

E. Low Sodium Diet

Most sodium-restricted diets are very complex, are prescribed haphazardly, and are unenthusiastically followed.[41] For a palate that has been accustomed to salty foods since infancy, the absence of this condiment is at first distressing; and it may take several weeks to realize that the salty kitchen often overrides the delicious natural flavor of foods. Dahl has suggested the following simple rules for attaining a low sodium diet:[43] (1) never add salt to food during preparation or at the table; (2) avoid milk and milk products; (3) avoid all processed foods except fruits; and (4) examine all labels. As Kaplan has pointed out, the major problem in this regard is the convenience foods such as T.V. dinners, prepared and frozen dishes that continue to replace natural unsalted foods.[289] Moreover, patients who must eat out in restaurants often will find it difficult to follow a low salt regime. Several dietary guides on a low salt diet have been published that we would like to recommend.[290-292]

VII. CONCLUSION

We have presented data that indicate the enhanced prevalence of hypertensive disease in populations that have greater access to increased dietary sodium intake. We have indicated that the sodium ion participates importantly, and perhaps centrally, in each of the pressor mechanisms, and that with its restriction or absence, hypertension is much less likely to become manifest. Moreover, we have shown that sodium is a key factor (one way or the other) in each of the postulated mechanisms underlying essential hypertension. And whether by a low sodium diet or through the major effect of diuretic therapy, sodium restriction plays a key role in antihypertensive therapy. And we believe that by enlightened sodium restriction, particularly in genetically predisposed individuals, hypertensive diseases may be prevented (at best) or at least retarded in development.

REFERENCES

1. **Meneely, G. R. and Battarbee, H. D.**, High sodium-low potassium environment and hypertension, *Am. J. Cardiol.*, 38, 768, 1976.
2. **Dahl, L. K.**, Salt and hypertension, *Am. J. Clin. Nutr.*, 25, 231, 1972.
3. **Pickering, G. W.**, *High Blood Pressure*, Grune & Stratton, New York, 1955.
4. **Freis, E. D.**, Salt and the prevention of hypertension, *Circulation*, 53, 589, 1976.
5. **Ambard, L. and Beaujard, E.**, Causes de l'hypertension arterielle, *Arch. Gen. Med.*, 1, 520, 1904.
6. **Allen, F. M.**, *Treatment of Kidney Diseases and High Blood Pressure*, The Psychiatric Institute, Morristown, 1925.

7. **Glieberman, L.**, Blood pressure and dietary salt in human populations, *Ecol. Food Nutr.*, 2, 143, 1973.
8. **Maddocks, I. and Vines, A. P.**, The influence of chronic infection of blood pressure in New Guinea males, *Lancet*, 2, 262, 1966.
9. **Burns-Cox, C. J.**, Splenomegaly and blood pressure in an Orang-Asli community in West Malaysia, *Am. Heart J.*, 80, 718, 1970.
10. **Mann, G. V., Shaffer, R. D., and Rich, A.**, Physical fitness and immunity to heart disease in Masai, *Lancet*, 2, 1308, 1965.
11. **Page, L. B., Damon, A., and Moellering, R. C., Jr.**, Antecedents of cardiovascular disease in six Solomon Islands societies, *Circulation*, 49, 1132, 1974.
12. **Chiang, B. N., Perlman, L. V., and Epstein, F. H.**, Overweight and hypertension — a review, *Circulation*, 39, 403, 1969.
13. **Tobian, L.**, Hypertension and obesity, *N. Engl. J. Med.*, 298, 46, 1978.
14. **Kannel, W. B., Brand, N., Skinner, J. J., Jr., Dawber, T. R., and McNamara, P. M.**, The relation of adiposity to blood pressure and development of hypertension. The Framingham Study, *Ann. Intern. Med.*, 67, 48, 1967.
15. **Tyroler, H. A., Heyden, S., and Hames, C. G.**, Weight and hypertension: Evans County studies of blacks and whites, in *Epidemiology and Control of Hypertension*, Paul, O., Ed., Stratton Intercontinental, New York, 1975, 177.
16. **Reisin, E., Abel, R., Modan, M., Silverberg, D. S., Eliahou, H. E., and Modan, B.**, Effect of weight loss without salt restriction on the reduction of blood pressure in overweight hypertensive patients, *N. Engl. J. Med.*, 298, 1, 1978.
17. **Prior, A. M., Evans, J. G., Harvey, H. B. P., Davidson, E., and Lindsey, M.**, Sodium intake and blood pressure in two Polynesian populations, *N. Engl. J. Med.*, 279, 515, 1968.
18. **Whyte, H. M.**, Body build and blood pressure of men in Australia and New Guinea, *Aust. J. Exp. Biol.*, 41, 395, 1963.
19. **Hoobler, S. W., Tejada, C., Gusman, M., and Pardo, A.**, Influence of nutrition and "acculturation" on the blood pressure levels and changes with age in the highland Guatemala Indian, *Circulation*, 31 (Suppl. 2), 116, 1963.
20. **Maddocks, I.**, Blood pressure in Melanesians, *Med. J. Aust.*, 1, 1123, 1969.
21. **Prior, I. A. M., Grimely-Evans, J., Harvey, H. P. B., Davidson, F., and Lindsey, M.**, Sodium intake and blood pressure in two Polynesian populations, *N. Engl. J. Med.*, 279, 515, 1968.
22. **Lowenstein, J. W.**, Blood pressure in relation to age and sex in the tropics and subtropics. A review of the literature and an investigation in the tribes of Brazil Indians, *Lancet*, 1, 389, 1961.
23. **Page, L. B., Vandervert, D., and Nader, K., Lubin, N., and Page, J. R.**, Blood pressure, diet and body form in traditional nomads of the Qash Qui tribe, Southern Iran, *Acta Cardiol.*, 33, 102, 1978.
24. **Oliver, W. J., Cohn, E. L., and Neel, J. V.**, Blood pressure, sodium intake, and sodium-related hormones in the Yanomamo Indians, a "no-salt" culture, *Circulation*, 52, 146, 1975.
25. **Page, L.**, Epidemiologic evidence on the etiology of human hypertension and its possible prevention, *Am. Heart J.*, 91, 527, 1976.
26. **Shaper, A. G., Geonard, P. A., and Jones, K. W., and Jones, M.**, Environmental effects on the body build, blood pressure, and blood chemistry of nomadic warriors serving in the army in Kenya, *E. Afr. Med. J.*, 46, 282, 1969.
27. **Ashe, B. I. and Rosenthal, H. O.**, Protein, salt and fluid consumption of 1000 residents of New York, *JAMA*, 108, 1160, 1937.
28. **Dahl, L. K.**, Sodium intake of the American male: implications on the etiology of essential hypertension, *Am. J. Clin. Nutr.*, 6, 1, 1958.
29. **Phear, D. N.**, Salt intake and hypertension, *Br. Med. J.*, 2, 1453, 1958.
30. **Miall, W. E.**, Follow-up study of arterial pressure in the population of a Welsh mining valley, *Br. Med. J.*, 2, 1204, 1959.
31. **Schneckloth, R. E., Corcoran, A. C., Stuart, K. L., and Moore, F. E.**, Arterial pressure and hypertensive disease in a West Indian Negro population. Report of a survey in St. Kitts, West Indies, *Am. Heart J.*, 53, 607, 1962.
32. **Dawber, T. R., Kannel, W. B., Kagan, A., Donabedian, R. K., McNamara, P. M., and Pearson, G.**, Environmental factors in hypertension, in *The Epidemiology of Hypertension*, Stamler J., Stamler, R., and Pullman, R. B., Eds., Grune & Stratton, New York, 1967, 255.
33. **Langford, H. G., Watson, R. L., and Douglas, B. H.**, Factors affecting blood pressure in population groups, *Trans. Assoc. Am. Physicians*, 71, 135, 1968.
34. **Morgan, M., Carney, S., and Wilson, M.**, Interrelationship in humans between sodium intake and hypertension, *Clin. Exp. Pharmacol. Physiol.*, Suppl. 12, 127, 1975.
35. **Berglund, G., Wikstrand, J., Wallenti, I., and Wuehlmsen, L.**, Sodium excretion and sympathetic activity in relation to severity of hypertensive disease, *Lancet*, 1, 324, 1976.

36. Morgan, T., Adam, W., Gillies, A., Wilson, M., Morgan, G., and Carney, S., Hypertension treated by salt restriction, *Lancet*, 1, 227, 1978.
37. Simpson, F. O., Nye, E. R., Bolli, P., Waal-Manning, H. J., Goulding, A. W., Phelan, E. L., de Hamel, F. A., Sterward, R. D. H., Spears, G. F. S., Leek, G. M., and Steward, A. C., The Milton survey. I. General methods, height, weight and 24-hour excretion of sodium, potassium, calcium, magnesium and creatinine, *N. Z. Med. J.*, 87, 379, 1978.
38. Simpson, F. O., Waal-Manning, H. J., Bolli, P., and Spears, G. F. S., Relationship of blood pressure to sodium excretion in a population survey, *Clin. Sci. Mol. Med.*, 55 (Suppl. 4), 373s, 1978.
39. Sasaki, N., The relationship of salt intake to hypertension in the Japanese, *Geriatrics*, 19, 735, 1964.
40. Kawasaki, T., Delea, C. S., Bartter, F. C., and Smith, H., The effect of high-sodium and low-sodium intakes on blood pressure and other related variables in human subjects with idiopathic hypertension, *Am. J. Med.*, 64, 193, 1978.
41. Dethier, V. G., The taste of salt, *Am. Sci.*, 665, 744, 1977.
42. Blair-West, J. R., Coghlan, J. P., Denton, D. A., Nelson, J. F., Orchard, E., Scoggins, B. A., and Wright, R. D., Physiological, morphological and behavioural adaptation to a sodium deficient environment by wild native Australian and introduced species of animals, *Nature*, 217, 922, 1968.
43. Dahl, L. K., Salt intake and hypertension, in *Hypertension: Physiopathology and Treatment*, Genest, J., Koiw, E., and Kuchel, O., Eds., McGraw-Hill, New York, 1977, 755.
44. Conn, J. W., Mechanisms of adaptation to heat, *Adv. Intern. Med.*, 3, 373, 1949.
45. Dahl, L. K., Salt intake and salt need, *N. Engl. J. Med.*, 258, 1152, 1958.
46. Dahl, L. K., Salt in processed baby foods, *Am. J. Clin. Nutr.*, 21, 787, 1968.
47. Kaunitz, H., Causes and consequences of salt consumption, *Nature*, 178, 1141, 1956.
48. Fallis, N., LaSagna, L., and Tetreault, I., Gustatory thresholds in patients with hypertension, *Nature*, 196, 74, 1962.
49. Wotman, S., Mandel, I. D., Thompson, R. H., and Laragh, J. H., Salivary electrolytes and salt taste thresholds in hypertension, *J. Chron. Dis.*, 20, 833, 1967.
50. Henkin, R. I., Taste thresholds in patients with essential hypertension and with hypertension due to primary hyperaldosteronism, *J. Chron. Dis.*, 27, 235, 1974.
51. Catalanotto, F., Schechter, P. J., and Henkin, R. I., Preference for NaCl in the spontaneously hypertensive rat, *Life Sci.*, 11, 557, 1972.
52. McConnell, S. D. and Henkin, R. I., Increased preference for Na^+ and K^+ salts in spontaneously hypertensive (SH) rats, *Proc. Soc. Exp. Biol. Med.*, 143, 1973.
53. Schechter, P. J., Horwitz, D., and Henkin, R. I., Salt preference in patients with untreated and treated essential hypertension, *Am. J. Med. Sci.*, 267, 320, 1974.
54. Selkurt, E. E., Effect of pulse pressure and mean arterial pressure modification on renal hemodynamics and electrolyte and water excretion, *Circulation*, 4, 541, 1951.
55. Addison, W., The uses of sodium chloride, potassium chloride, sodium bromide and potassium bromide in cases of arterial hypertension which are amenable to potassium chloride, *Can. Med. Assoc. J.*, 18, 281, 1928.
56. Langford, H. and Watson, R. L., Electrolytes and hypertension, in *The Epidemiology and Control of Hypertension*, Stamler, J., Stamler, R., and Pullman, T., Eds., Grune & Stratton, Chicago, 1975, 119.
57. Dahl, L. K., Leitl, G., and Heine, M., Influence of dietary potassium and sodium/potassium molar ratios on the development of salt hypertension, *J. Exp. Med.*, 136, 3318, 1972.
58. Bartorelli, C., Gargano, N., and Leonetti, G., Potassium replacement during long-term diuretic treatment in hypertension, in *Antihypertensive Therapy, An International Symposium*, Gross, F., Ed., Springer-Verlag, Berlin, 1966, 422.
59. Mitsushashi, T., Effects of the ingestion of large amount of apples on blood pressure in farmers in Akita prefecture, Igaku to Seibutsugaku, *Hirosaki Med. J.*, 12, 57, 1960.
60. Earll, J. M., Kurtzman, N. A., and Moser, R. H., Hypercalcemia and hypertension, *Ann. Intern. Med.*, 64, 378, 1966.
61. Weidmann, P., Massry, S. G., Coburn, J. W., et al., Blood pressure effects of acute hypercalcemia, *Ann. Intern. Med.*, 76, 741, 1972.
62. Marone, C., Beretta-Piccoli, C., and Weidmann, P., Role of hemodynamics, catecholamines and renin, in acute hypercalcaemic hypertension in man, *Clin. Sci.*, 59 (Suppl. 6), 369 s, 1980.
63. Hellstrom, J., Birke, G., and Edvall, C. A., Hypertension in hyperparathyroidism, *Br. J. Urol.*, 30, 13, 1958.
64. Scott, J. B., Frohlich, E. D., Hardon, R. A., and Haddy, F. J., Na^+, K^+, Ca^{++} and Mg^{++} action on coronary vascular resistance in the dog heart, *Am. J. Physiol.*, 201, 1095, 1961.
65. Frohlich, E. D., Scott, J. B., and Haddy, F. J., Effects of cations on resistance and responsiveness of the renal and forelimb vascular beds, *Am. J. Physiol.*, 203, 583, 1962.
66. Texter, E. C., Lauretta, H. C., Frohlich, E. D., and Chou, C. C. , Effects of major cations on gastric and mesenteric vascular resistances, *Am. J. Physiol.*, 212, 569, 1967.

67. **Fregly, M. J.**, Specificity of the sodium chloride appetite of adrenalectomized rats; substitution of lithium chloride for sodium chloride, *Am. J. Physiol.*, 195, 645, 1958.
68. **Messerli, F. H. and Frohlich, E. D.**, High blood pressure. A side effect of drugs, poisons, and foods, *Arch. Intern. Med.*, 139, 682, 1979.
69. Hypertension Task Force: Current Research and Recommendations from the Task Force Subgroups, Salt and Water, Vol. VIII, Part B, NIH Publ. No. 791630, National Heart, Lung, and Blood Institute, U.S. Department of Health, Education, and Welfare, Bethesda, 1979.
70. **Dustan, H. P., Tarazi, R. C., and Hinshaw, L. B.**, Mechanisms controlling arterial pressure, in *Pathophysiology: Altered Regulatory Mechanisms in Disease,* 2nd ed., Frohlich, E. D., Ed., Lippincott, Philadelphia, 1976, 49.
71. **Freeman, R. H. and Davis, J. O.**, The control of renin secretion and metabolism, in *Hypertension: Physiopathology and Treatment,* Genest, J., Koiw, E., and Kuchel, O. J., Eds., McGraw-Hill, New York, 1977, 210.
72. **Assaykeen, T. A., Clayton, P. L., Goldfien, A., and Ganong, W. F.**, Effect of alpha- and beta-adrenergic agents on the renin response to hypoglycemia and epinephrine in dogs, *Endocrinology,* 87, 1318, 1970.
73. **Winder, N., Chockshi, D. S., Yoon, M. S., and Freedman, A. D.**, Adrenergic receptor mediation of renin secretion, *J. Clin. Endocrinol. Metab.*, 19, 1168, 1969.
74. **Winer, N., Chockshi, D. S., and Walkenhorst, W. G.**, Effects of cyclic AMP, sympathomimetic amines, and adrenergic receptor antagonists on renin secretion, *Circ. Res.*, 29, 239, 1971.
75. **Vander, A. J. and Miller, R.**, Control of renin secretion in the dog, *Am. J. Physiol.*, 207, 537, 1964.
76. **Nash, F. D., Rostorfer, H. H., Bailee, M. D., Wathen, R. L., and Schneider, E. G.**, Relation to renal sodium load and dissociation from hemodynamic changes, *Circ. Res.*, 22, 473, 1968.
77. **Thuran, K., Schnermann, J., Nagel, W., Horster, M., and Wohl, M.**, Composition of tubular fluid in the macula densa segment as a factor regulating the function of the juxtaglomerular apparatus, *Circ. Res.*, 20(2), 79, 1967.
78. **Barajas, L.**, Renin secretion: an anatomical basis for tubular control, *Science,* 172, 485, 1971.
79. **Page, I. H. and Bumpus, F. M., Eds.**, Angiotensin, in *Handbook of Experimental Pharmacology,* Vol. 37, Springer-Verlag, New York, 1974.
80. **Laragh, J. H., Ed.**, Hypertension Manual: Mechanisms, Methods, Management, *York Medical*
81. **Genest, J., Koiw, E., and Kuchel, O., Eds.**, *Hypertension: Physiopathology and Treatment,* McGraw-Hill, New York, 1977.
82. **Guyton, A. C.**, Arterial pressure and hypertension, in *Circulatory Physiology,* Vol. 3, W. B. Saunders, Philadelphia, 1980.
83. **Epstein, A. N.**, The neuroendocrinology of thirst and salt appetite, in *Frontiers in Neuroendocrinology,* Ganong, W. F. and Martini, L., Eds., Raven Press, New York, 1978.
84. **Bonjour, J. P. and Malvin, R. L.**, Stimulation of ADH release by the renin-angiotensin system, *Am. J. Physiol.*, 218, 1555, 1970.
85. **Malvin, R. L.**, Possible role of the renin-angiotensin system in the regulation of diuretic hormone secretion, *Fed. Proc.*, 30, 1383, 1971.
86. **Severs, W. B., Summy-Long, J., and Daniels-Severs, A.**, Angiotensin interaction with thirst mechanisms, *Am. J. Physiol.*, 226, 340, 1974.
87. **Block, S. L., Kucharczyk, J., and Mogenson, G. J.**, Disruption of drinking to intracranial angiotensin by a lateral hypothalamic lesion, *Pharm. Biochem. Behav.*, 2, 515, 1974.
88. **Brody, M. J., Fink, G. D., Buggy, J., Haywood, J. R., Gordon, F. J., and Johnson, A. K.**, The role of the anteroventral third ventricle (AV3V) region in experimental hypertension, *Circ. Res.*, 43(1), 2, 1978.
89. **Ramsay, D. J.**, Beta-adrenergic thirst and its relation to the renin-angiotensin system, *Fed. Proc.*, 37, 2689, 1978.
90. **Hall, J. E., Guyton, A. C., Smith, M. J., Jr., and Coleman, T. G.**, Chronic blockadge of angiotensin II formation during sodium deprivation, *Am. J. Physiol.*, 237, F424, 1979.
91. **Misantone, L. J., Ellis, S., and Epstein, A. N.**, Development of angiotensin-induced drinking in the rat, *Brain Res.*, 186, 195, 1980.
92. **Mouw, D., Bonjour, J. P., Malvin, R. L., and Vander, A.**, Central action of angiotensin in stimulating ADH release, *Am. J. Physiol.*, 220, 239, 1971.
93. **Simonnet, G., Rodriguez, F., Fumoux, F., Czernichow, P., and Vincent, J. D.**, Vasopressin in release and drinking induced by intracranial injection of angiotensin II in monkey, *Am. J. Physiol.*, 237, R20, 1979.
94. **Möhring, J., Möhring, B., Petri, M., and Haack, D.**, Vasopressor role of ADH in the pathogenesis of malignant DOC hypertension, *Am. J. Physiol.*, 232, F260, 1977.
95. **Haack, D., Möhring, J., Möhring, B., Petri, M., and Hackenthal, E.**, Comparative study on development of corticosterone and DOCA hypertension in rats, *Am. J. Physiol.*, 232, F260, 1977.

96. **Möhring, J., Möhring, B., Pctri, M., and Haack, D.**, Plasma vasopressin concentrations and effects of vasopressin antiserum on blood pressure in rats with malignant two-kidney Goldblatt hypertension, *Circ. Res.*, 42, 17, 1978.
97. **Johnston, C. I., Pullau, P. T., and Walter, N. M. A.**, Vasoactive peptides in experimental renal hypertension, *Klin. Wochenschr.*, 56(1), 81, 1978.
98. **Crofton, J. T., Share, L., Shade, R. E., Allen, C., and Tarnowski, D.**, Vasopressin in the rat with spontaneous hypertension, *Am. J. Physiol.*, 235, H361, 1978.
99. **Möhring, J., Kintz, J., and Schoun, J.**, Role of vasopressin blood pressure control of spontaneously hypertensive rats, *Clin. Sci. Mol. Med.*, 55, 247s, 1978.
100. **Möhring, J., Kintz, J., and Schoun, J.**, Studies on the role of vasopressin in blood pressure control of spontaneously hypertensive rats with established hypertension (SHR, stroke-prone strain), *J. Cardiovasc. Pharmacol.*, 1, 593, 1979.
101. **Udenfriend, S. and Spector, S.**, Spontaneously hypertensive rat, *Science*, 176, 1155, 1972.
102. **Frohlich, E. D. and Trippodo, N. C.**, Similarities of genetic (spontaneous) hypertension: man and rat, *Circ. Res.*, 48, 309, 1981.
103. **Nowaczynski, W., Genest, J., and Kuchel, O.**, Aldosterone, deoxycorticosterone, and corticosterone metabolism in essential hypertension, in *Hypertension: Physiopathology and Treatment*, Genest, J., Koiw, E., and Kuchel, O., Eds., McGraw-Hill, New York, 1977, chap. 7.4.
104. **Melby, J. C.**, 18-OH-DOC in clinical and experimental hypertension, in *Hypertension: Physiopathology and Treatment*, Genest, J., Koiw, E., and Kuchel, O., Eds., McGraw-Hill, New York, 1977, chap. 7.6.
105. **Woods, J. W., Liddle, G. W., Stant, E. G., Jr., Michelakis, A. M., and Brill, A. B.**, Effect of an adrenal inhibitor in hypertensive patients with suppressed renin, *Arch. Intern. Med.*, 123, 336, 1969.
106. **Mancheno-Rico, E., Kuchel, O., Nowaczynski, W., Seth, K. K., Sasaki, C., Dawson, K., and Genest, J.**, A dissociated effect of amino-glutethimide on the mineralocorticoid secretion in man, *Metabolism*, 22, 123, 1973.
107. Report of the Hypertension Task Force. Current Research and Recommendations from the Task Force Subgroups, Prostaglandins and Kallikrein-Kinin Vol. 7 (Parts A and B), DHEW Publication No. (NIH) 79-1629, U.S. Department of Health, Education, and Welfare, Bethesda, 1979.
108. **Margolius, H. S., Geller, R. G., de Jong, W., Pisano, J. J., and Sjoerdsma, A.**, Urinary kallikrein excretion in hypertension, *Circ. Res.*, 30(2), 125, 1972.
109. **Lawton, W. and Fitz, A.**, Urinary kallikrein excretion in normal renin essential hypertension, *Circulation*, 56, 856, 1977.
110. **Geller, R. G., Margolius, H. S., Pisano, J. J., and Keiser, H. R.**, Urinary kallikrein in spontaneously hypertensive rats, *Circ. Res.*, 36(1), 103, 1975.
111. **Porcelli, G., Bianchi, G., and Croxatto, H.**, Urinary kallikrein in a spontaneously hypertensive strain of rats, *Proc. Soc. Exp. Biol. Med.*, 149, 983, 1975.
112. **Gulati, O. P., Carretero, O. A., Morino, T., and Oza, N. B.**, Urinary kallikrein and plasma renin during the reversal of renovascular hypertension in rats, *Clin. Sci. Mol. Med.*, 51 (Suppl. 3), 263s, 1976.
113. **McGiff, J. C.**, Interactions of prostaglandins with the kallikrein-kinin and renin-angiotensin systems, *Clin. Sci.*, 59 (Suppl. 6), 105 s, 1980.
114. **Carretero, O. A. and Oza, N. B.**, Urinary kallikrein, sodium metabolism and hypertension, in *Proc. Int. Workshop Conf. Mechanisms Hypertension*, Sambhi, M., Ed., Excerpta Medica, Amsterdam, 1973.
115. **Croxatto, H. R., Huidobro, F., Rojas, M., Roblero, J., and Albertini, R.**, The effect of water, sodium overloading and diuretics upon urinary kallikrein, *Advances in Experimental Medicine and Biology*, 70, 361, 1976.
116. **Gill, J. R., Jr., Frolich, J. C., Bowden, R. E., Raylor, A. A., Keiser, H. R., Seyberth, H. W., Oates, J. A., and Bartter, F. C.**, Bartter's syndrome: a disorder characterized by high urinary prostaglandins and a dependence of hyperreninemia on prostaglandin synthesis, *Am. J. Med.*, 161, 43, 1976.
117. **Shipley, R. E. and Study, R. S.**, Changes in renal blood flow, extraction of insulin, glomerular filtration rate, tissue pressure, and urine flow with acute alterations of renal artery blood pressure, *Am. J. Physiol.*, 167, 676, 1951.
118. **Guyton, A. C., Granger, H. J., and Coleman, T. G.**, Autoregulation of the total systemic circulation and its relation to control of cardiac output and arterial pressure, *Circ. Res.*, 28(1), 93, 1971.
119. **Coleman, T. G., Granger, H. J., and Guyton, A. C.**, Whole-body circulatory autoregulation and hypertension, *Circ. Res.*, 28(1), 76, 1971.
120. **Frohlich, E. D.**, Hemodynamics of hypertension, in *Hypertension: Physiopathology and Treatment*, Genest, J., Koiw, E., and Kuchel, O., Eds., McGraw-Hill, New York, 1977, 15.
121. **Tobian, L. and Binion, J. T.**, Tissue cations and water in arterial hypertension, *Circulation*, 5, 754, 1952.

122. **Tobian, L.**, The electrolytes of arterial wall in experimental renal hypertension, *Circ. Res.*, 4, 671, 1956.
123. **Tobian, L. and Fox, A.**, The effect of norepinephrine on the electrolyte composition of arterial smooth muscle, *J. Clin. Invest.*, 35, 297, 1956.
124. **Redleaf, P. D. and Tobian, L.**, The question of vascular hyperresponsiveness in hypertension, *Circ. Res.*, 6, 185, 1958.
125. **Tobian, L.**, Interrelationship of electrolytes, juxtaglomerular cells in hypertension, *Physiol. Rev.*, 40, 280, 1960.
126. **Constantopoulos, G.**, Cations and norepinephrine content of arteries in hypertension, in *Hypertension: Physiopathology and Treatment*, Genest, J., Koiw, E., and Kuchel, O., Eds., McGraw-Hill, New York, 1977, chap. 8.9.
127. **Dahl, L. K., Smiley, M. G., Silver, L., and Spraragan, S.**, Evidence for a prolonged biological half-life of Na^{22} in patients with hypertension, *Circ. Res.*, 10, 313, 1962.
128. **Holley, H. L., Elliott, H. C., Jr., and Holland, C. M., Jr.**, Serum sodium values in essential hypertension, *Proc. Soc. Exp. Biol. Med.*, 77, 561, 1951.
129. **Albert, D. G., Morita, F., and Iseri, L. T.**, Serum magnesium and plasma sodium levels in essential vascular hypertension, *Circulation*, 17, 761, 1958.
130. **Ross, E. J.**, Total exchangeable sodium in hypertensive patients, *Clin. Sci.*, 15, 81, 1956.
131. **Dole, V. P., Dahl, L. K., Cotzias, G. C., Dziewiathowski, D. D., and Harris, C.**, Dietary treatment of hypertension. II. Sodium depletion as related to the therapeutic effect, *J. Clin. Invest.* , 30, 584, 1951.
132. **Dahl, L. K., Stall, B. G., and Cotzias, G. C.**, Metabolic effects of marked sodium restriction in hypertensive patients: changes in total exchangeable sodium and potassium, *J. Clin. Invest.*, 33, 1397, 1954.
133. **Moore, F. D., Edelman, I. S., Olney, J. M., James, A. H., Brooks, L., and Wilson, G. M.**, Body sodium and potassium. III. Inter-related trends in alimentary, renal and cardiovascular disease: lack of correlation between body stores and plasma concentration, *Metabolism*, 3, 334, 1954.
134. **deGraff, J.**, Inulin space and total exchangeable sodium in patients with essential hypertension, *Acta Med. Scand.*, 156, 337, 1957.
135. **Winer, B. M.**, Studies on the content and distribution of sodium, potassium and water in human hypertension, in *Hypertension, The First Hahnemann Symposium on Hypertensive Disease*, Moyer, J. H., Ed., W. B. Saunders, Philadelphia, 1959, 268.
136. **Mathur, K. S. and Wadhawan, D. N.**, Serum electrolytes estimation at different blood pressure levels, *J. Assoc. Physicians India*, 6, 578, 1958.
137. **Weller, J. M.**, The acid-base balance and sodium distribution of the blood in essential hypertension, *J. Lab. Clin. Med.*, 53, 553, 1959.
138. **Kirkendall, W. M., Peterson, R. D., and Armstrong, M. L.**, Salt loading and exchangeable electrolytes in patients without and with mild hypertension, *J. Lab. Clin. Med.*, 54, 915, 1959.
139. **Levine, B. E., Weller, J. M., and Remington, R. D.**, Serum sodium and potassium in essential hypertension, *Circulation*, 24, 29, 1961.
140. **Dahl, L. K., Heine, M., and Tassinari, L.**, Role of genetic factors in susceptibility to experimental hypertension due to chronic excess salt ingestion, *Nature*, 194, 480, 1962.
141. **Dahl, L. K., Heine, M., and Tassinari, L.**, Effects of chronic salt ingestion. Evidence that genetic factors play an important role in susceptibility to experimental hypertension, *J. Exp. Med.*, 115, 1173, 1962.
142. **Chrysant S. G., Walsh, G. M., and Frohlich, E. D.**, Hemodynamic changes induced by prolonged NaCl and DOCA administration in spontaneously hypertensive rats (SHR), *Angiology*, 29, 303, 1978.
143. **Chrysant, S. G., Walsh, G. M., Kem, D. C., and Frohlich, E. D.**, Hemodynamic and metabolic evidence of salt sensitivity in spontaneously hypertensive rats, *Kidney Int.*, 15, 33, 1979.
144. **Mark, A., Lawton, W., Abbaud, F., Fitz, A., Connor, W., and Heistad, D.**, Effects of high and low sodium intake on arterial pressure and forearm vascular resistance in borderline hypertension. A preliminary report, *Circ. Res.*, 36(1), 194, 1975.
145. **Rapp, J. P. and Dahl, L. K.**, 18-Hydroxycorticosterone secretion in experimental hypertension in rats, *Circ. Res.*, 28(2), 153, 1971.
146. **Garay, R. P. and Meyer, P.**, A new test showing abnormal net Na^+ and K^+ fluxes in erythrocytes of essential hypertensive patients, *Lancet*, 1, 349, 1979.
147. **De Champlain, J., Krakoff, L. R., and Axelrod, J.**, The metabolism of norepinephrine in experimental hypertension in rats, *Circ. Res.*, 20, 136, 1967.
148. **De Champlain, J., Krakoff, L. R., and Axelrod, J.**, Relationship between sodium intake and norepinephrine storage during the development of experimental hypertension, *Circ. Res.*, 23, 479, 1968.
149. **De Champlain, J., Mueller, R. A., and Axelrod, J.**, Turnover and synthesis of norepinephrine in experimental hypertension, *Circ. Res.*, 25, 285, 1969.

150. **Louis, W. J., Krauss, K. R., Kopin, I. J., and Sjoerdsma, A.,** Catecholamine metabolism in hypertensive rats, *Circ. Res.*, 27, 589, 1970.
151. **Nakamura, K., Gerald, M., and Thoenen, H.,** Experimental hypertension of the rat: reciprocal changes of norepinephrine turnover in heart and brain stem, *Naunyn Schmiedebergs Arch. Pharmacol.*, 268, 125, 1971.
152. **Krakoff, L. R., de Champlain, J., and Axelrod, J.,** Abnormal storage of norepinephrine in experimental hypertension in the rat, *Circ. Res.*, 21, 583, 1967.
153. **De Champlain, J., Krakoff, L. R. and Axelrod, J.,** Interrelationship of sodium intake, hypertension, and norepinephrine storage in the rat, *Circ. Res.*, 24(1), 75, 1969.
154. **Doyle, A. E.,** Endogenous catecholamine content of cardiac muscle in sodium loaded and sodium depleted rats, *Lancet*, 1, 1399, 1968.
155. **Kazda, S., Pohlova, I., Bibr, B., and Kockova, J.,** Norepinephrine content of tissues in DOCA hypertensive rats, *Am. J. Physiol.*, 216, 1472, 1969.
156. **Mueller, R. A., Willard, P., and Axelrod, J.,** Alterations in norepinephrine storage in inbred rats made hypertensive by triiodothyronine and sodium chloride, *Pharmacology*, 5, 153, 1971.
157. **Romoff, M. S., Keusch, G., Campese, V. M., Wang, M.-S., Friedler, R. M., Wiedmann, P., and Massry, S. G.,** Effect of sodium intake on plasma catecholamines in normal subjects, *J. Clin. Endocrinol. Metab.*, 48, 26, 1978.
158. **Luft, F. C., Rankin, L. I., Henry, D. P., Bloch, R., Grim, C. E., Weyman, A. E., Murray, R. H., and Weinberger, M. H.,** Plasma and urinary norepinephrine values at extremes of sodium intake in normal man, *Hypertension*, 1, 261, 1979.
159. **Nicholls, M. G., Kiowski, W., Zweifler, A. J., Julius, S., Schork, M. A., and Greenhouse, J.,** Plasma norepinephrine variations with dietary sodium intake, *Hypertension*, 1, 29, 1980.
160. **Gordon, R. D., Kuchel, O., Liddle, G. W., and Island, D. P.,** Role of the sympathetic nervous system in regulating renin and aldosterone production in man, *J. Clin. Invest.*, 46, 599, 1967.
161. **Cuche, J. L., Kuchel, O., Barbeau, A., Boucher, R., and Genest, J.,** Relationship between the adrenergic nervous system and renin during adaptation to upright posture: a possible role for 3,4-dihydroxyphenethylamine (dopamine), *Clin. Sci.*, 43, 481, 1972.
162. **Alexander, R. W., Gill, J. R., Jr., Yamahi, H., Lovenberg, W., and Keiser, H. R.,** Effects of dietary sodium and of acute infusion on the interrelationship between dopamine excretion and adrenergic activity in man, *J. Clin. Invest.*, 54, 154, 1974.
163. **Faucheaux, B., Buu, N. T., and Kuchel, O.,** Effects of saline and albumin on plasma and urinary catecholamines in dogs, *Am. J. Physiol.*, 232, F123, 1977.
164. **Cole, F. E., Blakesley, H. L., Graci, K. A., Frohlich, E. D., and MacPhee, A. A.,** Brain angiotensin II receptor affinity and capacity in spontaneously hypertensive and normotensive rats: effects of dietary NaCl, *Brain Res.*, 190 (1), 272, 1980.
165. **Swales, J. D.,** Local vascular renin activity as a factor in circulatory control, *Cardiovasc. Rev. Rep.*, 1, 309, 1980.
166. **Folkow, B., Grimby, G., and Thulesius, O.,** Adaptive structural changes of the vascular walls in hypertension and their relation to the control of the peripheral resistance, *Acta Physiol. Scand.*, 44, 255, 1958.
167. **Hollander, W., Kramsch, D. M., Farmelant, M., and Madoff, I. M.,** Arterial wall metabolism, in experimental hypertension of coarctation of the aorta of short duration, *J. Clin. Invest.*, 47, 1221, 1968.
168. **Folkow, B.,** The haemodynamic consequences of adapative structural changes of the resistance vessels in hypertension, *Clin. Sci.*, 41, 1, 1971.
169. **Folkow, B., Hallback, M., Lundgren, Y., Sivertsson, R., and Weiss, L.,** Importance of adaptive changes in vascular design for establishment of primary hypertension, studied in man and in spontaneously hypertensive rats, *Circ. Res.*, 32(1), 2, 1973.
170. **Pamnani, M. B. and Overbeck, H. W.,** Abnormal ion and water composition of veins and normotensive arteries in coarctation hypertension in rats, *Circ. Res.*, 38, 375, 1976.
171. **Overbeck, H. W.,** Cardiovascular hypertrophy and ''waterlogging'' in coarctation hypertension. Role of sympathoadrenergic influences and pressure, *Hypertension*, 1, 486, 1979.
172. **Harris, G. S. and Palmer, W. A.,** Effect of increased sodium ion on arterial sodium and reactivity, *Clin. Sci.*, 42, 301, 1972.
173. **Friedman, S. M.,** Sodium in blood vessels, *Blood Vessels*, 16, 2, 1979.
174. **Headings, V. E., Rondell, P. A., and Bohr, D. F.,** Bound sodium in arterial wall, *Am. J. Physiol.*, 199, 783, 1960.
175. **Palaty, V., Gustafson, B., and Friedman, S. M.,** Sodium binding in the arterial wall, *Can. J. Physiol. Pharmacol.*, 47, 763, 1969.
176. **Jones, A. W. and Swain, M. L.,** Chemical and kinetic analysis of sodium distribution in canine lingual artery, *Am. J. Physiol.*, 223, 1110, 1972.

177. **Folkow, B. and Aberg, B.**, The effect of functionally induced changes of wall lumen ratio on the vasoconstrictor response to standard amounts of vasoactive agents, *Acta Physiol. Scand.*, 47, 131, 1959.
178. **Folkow, B. and Heil, E.**, *Circulation,* Oxford University Press, New York, 1971.
179. **Bayliss, W. M.**, On the local reactions of the arterial wall to changes in internal pressure, *J. Physiol. London,* 28, 220, 1902.
180. **Berecek, K. H. and Bohr, D. F.**, Whole body vascular reactivity during the development of desoxycorticosterone acetate hypertension in the pig, *Circ. Res.*, 42, 764, 1978.
181. **Grekin, R. J., Terris, J. M., and Bohr, D. F.**, Electrolyte and hormonal effects of acetate in young pigs, *Hypertension,* 2, 326, 1980.
182. **Pfeffer, M. A. and Frohlich, E. D.**, Hemodynamic and myocardial function in young and old normotensive and spontaneously hypertensive rats, *Circ. Res.*, 32(1), 28, 1973.
183. **Pfeffer, M. A., Frohlich, E. D., Pfeffer, J. M., and Weiss, A. K.**, Pathophysiological implication of the increased cardiac output of young spontaneously hypertensive rats, *Circ. Res.*, 34(1), 235, 1974.
184. **Frohlich, E. D. and Pfeffer, M. A.**, Adrenergic mechanisms in human and SHR hypertension, *Clin. Sci. Mol. Med.*, 48, 225s, 1975.
185. **Pfeffer, M. A., Pfeffer, J. M., Weiss, A. K., and Frohlich, E. D.**, Development of SHR hypertension and cardiac hypertrophy during prolonged beta-blocking therapy, *Am. J. Physiol.*, 232, H639, 1977.
186. **Trippodo, N. C., Walsh, G. M., and Frohlich, E. D.**, Fluid volumes during onset of spontaneous hypertension in rats, *Am. J. Physiol*, 235, H52, 1978.
187. **Freidman, S. M. and Freidman, C. K.**, The energizing role of sodium and potassium in the regulation of vascular smooth muscle tension, in *Electrolytes in Cardiovascular Diseases,* Bajusz, E., Ed., S. Karger, New York, 1965.
188. **Keatinge, W. R.**, Ionic requirements for arterial action potential, *J. Physiol., London,* 194, 169, 1968.
189. **Friedman, S. M.**, A comparison of net fluxes of Li and Na in vascular smooth muscle, *Blood Vessels,* 12, 219, 1975.
190. **Crane, R. K.**, The gradient hypothesis and other models of carrier-mediated active transport, Revue Physiol., *Biochem. Pharmacol.*, 78, 99, 1977.
191. **Friedman, S. M.**, Arterial contractility and reactivity, in *Hypertension: Physiopathology and Treatment,* Genest, J., Koiw, E., and Kuchel, O., Eds., McGraw-Hill, New York, 1977, chap. 8.
192. **Blaustein M. P.**, Sodium ions, calcium ions, blood pressure regulation, and hypertension: a reassessment and a hypothesis, *Am. J. Physiol.*, 232, c165, 1977.
193. **Losse, H., Wehmeyer, N., and Wessels, F.**, Der Wasser — und Elektrolyt — gehalt von Erythorcyten bei arterieller Hypertonie, *Klin. Wochenschr.*, 38, 393, 1960.
194. **Postnov, Y., Orlov, S., Shevchenko, A., and Adler, A.**, Altered sodium permeability, calcium binding, and Na, K-ATPase activity in the red blood cell membrane in essential hypertension, *Pfleugers Arch.*, 371, 263, 1971.
195. **Postnov, Y., Orlov, S., and Pokvdin, N.**, Decrease of calcium binding by red blood cell membrane in spontaneously hypertensive rats and in essential hypertension, *Pfluegers Arch.*, 379, 191, 1979.
196. **Garay, R. P., Elghozi H.-L., Dagher, G., and Meyer, P.**, Laboratory distinction between essential and secondary hypertension by measurement of erythrocyte cation fluxes, *N. Engl. J. Med.*, 302, 769, 1980.
197. **Canessa, M., Adragna, N., Solomon, H. S., Connolly, T. M., and Tosteson, D. C.**, Increased sodium-lithium countertransport in red cells of patients with essential hypertension, *N. Engl. J. Med.*, 302, 772, 1980.
198. **Edmondson, R. P. S., Thomas, R. D., Hilton, P. J., Patrick, J., and Nones, N. E.** Abnormal leucocyte composition and sodium transport in essential hypertension, *Lancet,* 1, 1003, 1975.
199. **Ganten, D., Schelling, P., and Ganten, U.**, Tissue isorenins, in *Hypertension: Physiopathology and Treatment,* Genest, J., Koiw, E., and Kuchel, O., Eds., McGraw-Hill, New York, 1977, chap 6.10.
200. **MacPhee, A. A., Blakesley, H. L., Graci, K. A., Frohlich, E. D., and Cole, F. E.**, Altered cardiac β-adrenergic receptors in SHR rats receiving salt excess, *Clin. Sci.*, 59 (Suppl. 6), 169 s, 1980.
201. **Luft, F. C., Grim, C. E., Fineberg, N., and Weinberger, M. C.**, Effects of volume expansion and contraction in normotensive whites, blacks, and subjects of different ages, *Circulation,* 59, 643, 1979.
202. **Luft, F. C., Rankin, L. I., Bloch, R., Weyman, A. E., Willis, L. R., Murray, R. H., Grim, C. E., and Weinberger, M. H.**, Cardiovascular and humoral responses to extremes of sodium intake in normal black and white men, *Circulation,* 60, 697, 1979.
203. **Douglas, B. H., Guyton, A. C., Langston, J. B., and Bishop, V. S.** Hypertension caused by salt loading. II. Fluid volume and tissue pressure changes, *Am. J. Physiol.*, 207, 669, 1974.
204. **Coleman, T. G. and Guyton, A. C.** Hypertension caused by salt loading in the dog. III. Onset transients of cardiac output and other circulatory variables, *Circ. Res.*, 25, 153, 1969.

205. **Manning, R. D., Jr., Coleman, T. G., Guyton, A. C., Norman, R. A., Jr., and McCaa, R. E.,** Essential role of man circulatory filling pressure in salt-induced hypertension, *Am. J. Physiol.,* 236, R40, 1979.
206. **Guyton, A. C., Polizo D., and Armstrong, G. G.,** Mean circulatory filling pressure measured immediately after cessation of heart pumping, *Am. J. Physiol.,* 179, 261, 1954.
207. **Guyton, A. C., Jones, C. E., and Coleman, T. G.,** Cardiac output and its regulation, in *Circulatory Physiology,* W. B. Saunders, Philadelphia, 1973.
208. **Yamamoto, Y., Trippodo, N. C., Ishise, S., and Frohlich, E. D.,** Total vascular pressure-volume relationship in the conscious rat, *Am. J. Physiol.,* 238, H823, 1980.
209. **Ganguli, M., Tobian, L., and Iwai, J.,** Cardiac output and peripheral resistance in strains of rats sensitive and resistant to NaCl hypertension, *Hypertension,* 1, 3, 1979.
210. **Fink, G. D., Takeshita, A., Mark, A. C., and Brody, M. J.,** Determinants of renal vascular resistance in the Dahl strain of genetically hypertensive rat, *Hypertension,* 2, 274, 1980.
211. **Tobian, L., Lange, J., Azar, S., Iwai, J., Koop, D., Coffee, K., and Johnson, M. A.,** Reduction of natriuretic capacity and renin release in isolated, blood-perfused kidneys of Dahl hypertension-prone rats, *Circ. Res.,* 43 (1), 92, 1978.
212. **Ganguli, M., Tobian, L., and Dahl, L. K.** Low renal papillary plasma flow in both Dahl and Kyoto rats with spontaneous hypertension, *Circ. Res.,* 39, 337, 1976.
213. **Ikeda, T., Tobian, L., Iwai, J., and Goosens, P.,** Central nervous system pressor responses in rats susceptible and resistant to NaCl hypertension, *Clin. Sci. Mol. Med.,* 55 (4), 1, 1978.
214. **Takeshita, A. and Mark, A. L.,** Neurogenic contribution to hindquarter vasoconstriction during high sodium intake in Dahl strain of genetically hypertensive rat, *Circ. Res.,* 43 (1), 87, 1978.
215. **Freeman, R. H., Davis, J. O., and Watkins, B. E.,** Development of chronic perinephritic hypertension in dogs without volume expansion, *Am. J. Physiol.,* 233, F278, 1977.
216. **Davis, J. O., Stephens, G. A., Freeman, R. H., and Deforrest, J. M.,** Changes in cardiac output during the development of renal hypertension in sodium-depleted dogs, *Clin. Sci. Mol. Med.,* 55, 221s, 1978.
217. **Trippodo, N. C., Walsh, G. M., Ferrone, R. A., and Sugan, R. C.,** Fluid partition and cardiac output in volume-depleted Goldblatt hypertensive rats, *Am. J. Physiol.,* 237, H18, 1979.
218. **Farnsworth, E. B. and Barker, M. H.,** Tubular resorption of chloride in essential arterial hypertension; intensive study of one case, *Proc. Soc. Exp. Biol. Med.,* 53, 160, 1943.
219. **Farnsworth, E. B.,** Renal reabsorption of chloride and phosphate in normal subjects and in patients with essential arterial hypertension, *J. Clin. Invest.,* 25, 897, 1946.
220. **Green, D. M., Wedell, H. G., Wald, M. H., and Learned, B.,** The relation of water and sodium excretion to blood pressure in human subjects, *Circulation,* 6, 919, 1952.
221. **Brodsky, W. A. and Graubarth, H. N.,** Excretion of water and electrolytes in patients with essential hypertension, *J. Lab. Clin. Med.,* 41, 43, 1953.
222. **Birchall, R., Tuthill, S. W., Jacobs, W. S., Trautman, W. J., Jr., and Findley, T.,** Renal excretion of water, sodium and chloride, *Circulation,* 7, 258, 1953.
223. **Green, D. N., Johnson, A. D., Bridges, W. C., and Lehmann, J. H.,** Stages of salt exchange in essential hypertension, *Circulation,* 9, 416, 1954.
224. **Thompson, J. E., Silva, T. F., Kinsey, D., and Smithwick, R. H.,** The effect of acute salt loads on the urinary sodium output of normotensive and hypertensive patients before and after surgery, *Circulation,* 19, 912, 1954.
225. **Taquini, A. C., Plesch, S. A., Capris, T. A., and Badano, B. N.,** Some observations on water and electrolyte metabolism in essential hypertension, *Acta Cardiol.,* 11, 109, 1956.
226. **Hollander, W. and Judson, W. E.,** Electrolyte and water excretion in arterial hypertension. I. Studies in non-medically treated subjects with essential hypertension, *J. Clin. Invest.,* 36, 1460, 1957.
227. **Cottier, P. T., Weller, J. M., and Hoobler, S. W.,** Effect of an intravenous sodium chloride load on renal hemodynamics and electrolyte excretion in essential hypertension, *Circulation,* 17, 750, 1958.
228. **Cottier, P. T., Weller, J. M., and Hoobler, S. W.,** Sodium chloride excretion following salt loading in hypertensive subjects, *Circulation,* 18, 196, 1958.
229. **Baldwin, D. S., Biggs, A. W., Golding, W., Hulet, W. H., and Chasis, H.,** Exaggerated natriuresis in essential hypertension, *Am. J. Med.,* 24, 893, 1958.
230. **Hanenson, I. B., Traussky, H. H., Polasky, N., Ransohoff, W., and Miller, B. F.,** Renal excretion of sodium in arterial hypertension, *Circulation,* 21, 498, 1959.
231. **Papper, S., Belsky, J. L., and Bleifer, K. N.,** The response to the administration of an isotonic sodium chloride-lactate solution in patients with essential hypertension, *J. Clin. Invest.,* 39, 876, 1960.
232. **Ek, J.,** The influence of heavy hydration of the renal function in normal and hypertensive man, *Scand. J. Clin. Lab. Invest.,* 7 (Suppl. 19), 1, 1955.
233. **Metzger, R. A., Vaamonde, L. S., Vaamonde, C. A., and Papper, S.,** Renal excretion of sodium durin oral water administration in patients with systemic hypertension, *Circulation,* 38, 955, 1968.

234. **Koranyi, A. V.**, Physiologische und klinische Untersuchungen über den osmotischen Druck thierischer Flussigkeiten, *Z. Klin. Med.*, 33, 1, 1897.
235. **Koranyi, A. V.**, Physiologische und klinische Utersuchungen über den osmotischen Druck thierischer Flussigkeiten, *Z. Klin. Med.*, 34, 1, 1898.
236. **Lewy, J. E. and Windhager, E. E.** Peritubular control of proximal tubular fluid reabsorption in the rat kidney, *Am. J. Physiol.*, 214, 943, 1968.
237. **Koch, K. M., Aynedijan, H. S., and Bank, N.**, Effect of acute hypertension on sodium reabsorption by the proximal tubule, *J. Clin. Invest.*, 47, 1696, 1968.
238. **Martino, J. A. and Early, L. E.**, Relationship between intrarenal hydrostatic pressure and hemodynamically induced changes in sodium excretion, *Circ. Res.*, 23, 371, 1968.
239. **Lowenstein, J., Beranbaum, E. R., Chassis, H., and Baldwin, D. S.**, Intrarenal pressure and exaggerated natriuresis in essential hypertension, *Clin. Sci.*, 38, 359, 1970.
240. **Schalekamp, M. A. D. H., Krauss, X. H., Schalekamp-Kuyken, M. P. A., Kolsteis, G., and Birkenhager, W. H.**, Studies on the mechanism of hypernatriuresis in essential hypertension in relation to measurements of plasma renin concentration, body fluid compartments and renal function, *Clin. Sci.*, 41, 219, 1971.
241. **Livinsky, N. G. and Lalone, R. C.**, The mechanism of sodium diuresis after saline infusion in the dog, *J. Clin. Invest.*, 42, 1261, 1963.
242. **Rector, F. C., Jr., Van Giesen, G., Kil, F., and Seldin, D. W.**, Influence of expansion of extracellular volume on tubular reabsorption of sodium independent of changes in glomerular filtration rate and aldosterone activity, *J. Clin. Invest.*, 43, 321, 1964.
243. **Dirks, J. H., Cirksena, W., and Berliner, R. W.**, The effect of saline infusion on sodium reabsorption by the proximal tubule of the dog, *J. Clin. Invest.*, 44, 1160, 1965.
244. **Courtney, M. A., Hylle, M., Lassiter, W. E., and Gottschald, C. W.**, Renal tubular transport of water, solute and PAH in rats loaded with isotonic saline, *Am. J. Physiol.*, 209, 1199, 1965.
245. **Howard, S. S., Davis, B. B., Knox, F. G., Wright, F. S., and Berlinger, R. W.**, Depression of fractional sodium reabsorption by the proximal tubule of the dog without sodium diuresis, *J. Clin. Invest.*, 47, 1561, 1968.
246. **Buckalew, V. M., Jr., Puschett, J. B., Kintzel, J. E., and Goldberg, M.**, Mechanism of exaggerated natriuresis in hypertensive man: impaired sodium transport in the loop of Henle, *J. Clin. Invest.*, 48, 1007, 1969.
247. **Cannon, P. J., Svahn, D. S., and Demartini, F. E.**, The influence of hypertonic saline infusions upon the fractional reabsorption of urate and other ions in normal and hypertensive man, *Circulation*, 41, 97, 1970.
248. **Ulrych, M., Hofman, J., and Hejl, Z.**, Cardiac and renal hyper-responsiveness to acute plasma volume expansion in hypertension, *Am. Heart J.*, 68, 193, 1964.
249. **Krakoff, L. R., Goodwin, F. J., Baer, L., Torres, M., and Laragh, J. H.**, The role of renin in the exaggerated natriuresis of hypertension, *Circulation*, 42, 335, 1970.
250. **Welner, A. and Groen, J. J.**, Effect of a simple deconditioning procedure on the diuretic and natriuretic response of hypertensive patients to a hypertonic salt load, *Circulation*, 35, 260, 1967.
251. **Thompson, J. E., Silva, T. F., Kinsey, D., and Smithwick, R. H.**, Effect of acute salt loads on the urinary sodium output of normotensive and hypertensive patients before and after surgery, *Circulation*, 10, 912, 1954.
252. **Ulrych, M., Frohlich, E. D., Dustan, H. P., and Page, I. H.**, Cardiac output and distribution of blood volume in central and peripheral circulations in hypertensive and normotensive man, *Br. Heart J.*, 31, 570, 1969.
253. **Luria, M. H., Adelson, E. I., and Lochaya, S.**, Paroxysmal tachycardia with polyuria, *Ann. Intern. Med.*, 65, 461, 1966.
254. **Ledingham, J. M. and Cohen, R. D.**, The role of the heart in the pathogenesis of renal hypertension, *Lancet*, 2, 979, 1963.
255. **Ledingham, J. M. and Cohen, R. D.**, Changes in the extracellular fluid volume and cardiac output during the development of experimental renal hypertension, *Can. Med. Assoc. J.*, 90, 292, 1964.
256. **Ledingham, J. M.**, *Hypotensive Drugs*, Pergamon Press, New York, 1956.
257. **Borst, J. G. J. and Borst-de Gens, A.**, Hypertension explained by Starling's theory of circulatory homeostasis, *Lancet*, 1, 677, 1963.
258. **Trippodo, N. C., Yamamoto, J., and Frohlich, E. D.**, Whole-body venous capacity and effective total tissue compliance in SHR, in *Hypertension*, 3, 104, 1981.
259. **Frohlich, E. D., Kozul, V. J., Tarazi, R. C., and Dustan, H. P.**, Physiological comparison of labile and essential hypertension, *Circ. Res.*, 27 (1), 55, 1970.
260. **Frohlich, E. D. and Pfeffer, M. A.**, Adrenergic mechanisms in human and SHR hypertension, *Clin. Sci. Mol. Med.*, 48, 225s, 1975.
261. **Julius, S., Esler, M. D., and Randall, O. S.**, Role of the autonomic nervous system in mild human hypertension, *Clin. Sci. Mol. Med.*, 48, 243s, 1975.

262. **Kriss, J. P., Futcher, P. H., and Goldman, M. L.**, Unilateral adrenalectomy, unilateral splanchnic nerve resection, and homolateral renal function, *Am. J. Physiol.*, 154, 229, 1948.
263. **Kaplan, S. A., Foman, S. J., and Rapoport, S.**, Effect of splanchnic nerve division on urinary excretion of electrolytes during mannitol loading in the hydropenic dog, *Am. J. Physiol.*, 166, 641, 1951.
264. **Kaplan, S. A. and Rapoport, S.**, Urinary excretion of sodium and chloride after splanchnicotomy, effect on the proximal tubule, *Am. J. Physiol.*, 164, 175, 1951.
265. **Kaurm, D. E. and Levinsky, N. G.** The mechanism of denervation diuresis, *J. Clin. Invest.*, 44, 93, 1965.
266. **Bonjour, J., Churchill, B. D., and Malvin, R. L.**, Change of tubular reabsorption of sodium and water after renal denervation in the dog, *J. Physiol., London*, 204, 571, 1969.
267. **Shear, L.**, Renal function and sodium metabolism in idopathic orthostatic hypotension, *N. Engl. J. Med.*, 268, 347, 1963.
268. **Gill, J. R., Jr., Mason, D. G., and Bartter, F. C.**, Adrenergic nervous system in sodium metabolism: effects of guanethidine and sodium-retaining steroids in normal man, *J. Clin. Invest.*, 43, 177, 1964.
268a. **De Wardener, H. E. and MacGregor, G. A.**, Dahl's hypothesis that a saluretic substance may be responsible for a sustained rise in arterial pressure: its possible role in essential hypertension, *Kidney Int.*, 18, 1, 1980.
269. **Kempner, W.**, Treatment of hypertensive vascular disease with rice diet, *Am. J. Med.*, 4, 545, 1948.
270. **Dole, V. P., Dahl, L. K., Cotzias, G. C., Eder, H. A., and Krebs, M. E.**, Dietary treatment of hypertension. Clinical and metabolic studies of patients on the rice diet, *J. Clin. Invest.*, 29, 1189, 1950.
271. **Watkin, D. M., Froeb, H. F., Hatch, F. T., and Futman, A. B.**, Effects of diet in essential hypertension, *Am. J. Med.* 9, 441, 1950.
272. **Murphy, R. J. F.**, Effect of "rice diet" on plasma volume and extracellular fluid space in hypertensive subjects, *J. Clin. Invest.*, 29, 912, 1950.
273. **Corcoran, A. C., Taylor, R. D., and Page, I. H.**, Controlled observations on the effects of low sodium diethotherapy in essential hypertension, *Circulation*, 3, 1, 1951.
274. **Dustan, H. P., Tarazi, R. C., and Bravo, E. L.**, Diuretic and diet treatment of hypertension, *Arch. Intern. Med.*, 133, 1007, 1974.
275. **Parijs, J., Joosens, J. V., Vander Linden, L., Verstreken, G., and Amery, A. K. P. C.**, Moderate sodium restriction and diuretics in the treatment of hypertension, *Am. Heart J.*, 85, 22, 1973.
276. **Hunt, J. C. and Margie, J. D.**, The influence of diet and hypertension management, in *Hypertension Update: Mechanisms, Epidemiology, Evaluation, Management*, Hunt, J. C., Cooper, T., Frohlich, E. D., Gifford, R. W., Jr., Kaplan, N. M., Laragh, J. H., Maxwell, M. H., and Strong, C. G., Eds., Health Learning Systems, Bloomfield, N.J., 1980.
277. **Londe, S., Bourgoignee, J. J., Robson, A. M., and Goldring, D.**, Hypertension in apparently normal children, *J. Pediatr.*, 78, 659, 1971.
278. **Holland, W. W. and Beresford, S. A. A.**, Factors influencing blood pressure in children, in *Epidemiology and Control of Hypertension*, Paul, O., Ed., Symposia Specialists, Miami, 1975.
279. **Lauer, R. M., Filer, L. J., Jr., Reiter, M. A., and Clarke, W. R.**, Blood pressure, salt preference, salt threshold, and relative weight, *Am. J. Dis. Child.*, 130, 493, 1976.
280. **Aschinberg, L. C., Zeis, P. M., Miller, R. A., John E. G., and Chan, L. L.**, Essential hypertension in childhood, *JAMA*, 237, 322, 1977.
281. Task Force of Blood Pressure in Children, Blood pressure control in children, *Pediatrics*, 59, 797, 1977.
282. **Goldring, D., Londe, S., Sivaloff, M., Hernandez, A., Britton, C., and Choi, S.**, Blood pressure in a high school population. I. Standards for blood pressure and the relation of age, sex, weight, height, and race to blood pressure in children 14 to 18 years of age, *J. Pediatr.*, 91, 884, 1977.
283. **Weil, W. B., Jr.**, Current controversies in childhood obesity, *J. Pediatr.*, 91, 175, 1977.
284. **Julius, S. and Schork, M. A.**, Borderline hypertension — a critical review, *J. Chronic Dis.*, 23, 723, 1971.
285. **Stamler, J., Farinaro, E., Mojonnier, L. M., Hall, Y., Moss, D., and Stamler, R.**, Prevention and control of hypertension by nutritional means. Long-term experience of the Chicago Coronary Preveton Evaluation Program, *JAMA*, 243, 1819, 1980.
286. 1980 Report of the Joint National Committee on Detection, Evaluation, and Treatment of High Blood Pressure, *Arch. Intern. Med.*, 140, 1280, 1980.
287. **Winer, B. M.**, The antihypertensive mechanisms of salt depletion induced by hydrochlorothiazide, *Circulation*, 24, 788, 1961.
288. **Ram, C. V. S., Garrett, B. N., and Kaplan, N. M.**, Moderate sodium restriction and various diuretics in the treatment of hypertension: effects of potassium wastage and blood pressure control, *Arch. Intern. Med.*, in press.
289. **Kaplan, N. M.**, *Clinical Hypertension*, Williams & Wilkins, Baltimore, 1978.

290. **Payne, A. and Callahan, D.,** *The Fat and Sodium Control Cookbook,* 4th ed., Little Brown, Boston, 1975.
291. **Margie, J. D., and Hunt, J. C.,** *Living with High Blood Pressure: The Hypertension Diet Cookbook,* Health Learning Systems, Bloomfield, N. J.,1978.
291. **Margie, J. D. and Hunt, J. C.,** *Living with High Blood Pressure: The Hypertension Diet Cookbook,* Health Learning Systems, Bloomfield, N. J.,1978.
292. **Claiborne, C.,** The Gourmet Diet Cookbook, New York Times Books, New York, 1980.

Diuretics

Chapter 9

DIURETICS

Thomas L. Whitsett and Steven G. Chrysant

TABLE OF CONTENTS

I. HISTORICAL PERSPECTIVE

Excessive accumulation of water in the tissues has been viewed as a manifestation of ill health from the earliest literature. At various times, treatment has consisted of dietary modification, emetics, purging, diaphoretics, leaching, tapping, and scarification. The Egyptians prescribed juniper berries as a diuretic. Galen prescribed preparations consisting of black pepper, wine, and fir-seed for patients with "moist constitutions". Even Withering was in search of, and thought he had found, a diuretic when he experimented with crude forms of digitalis.

Today's diuretics are linked to our distant past through a circuitous route. Several centuries ago calomel, a form of inorganic mercury used for treating syphilis, was known to possess diuretic qualities. Because of its adverse effects a search for less toxic products resulted in the development of organic compounds, found quite by accident to possess considerable diuretic action. A delightful account of this discovery has been reported.[1] The organic mercurials dominated the field of diuretics for many years but were far from ideal since they required parenteral administration, caused metabolic acidosis that limited their activity, and were associated with renal, hepatic, and cutaneous toxicity. However, their minimal kaliuretic activity remains desirable.

A whole new spectrum of diuretics emerged following the introduction in 1953 of the antibacterial agent sulfanilamide. The recognition that it caused the urine to become alkaline was followed by the explanation that it inhibited the enzyme carbonic anhydrase. Indeed, it was suggested that compounds with a free sulfanilamide group would inhibit carbonic anhydrase.[2] Since this resulted in the excretion of sodium, chloride, and water, a sulfanilamide was soon employed in the treatment of congestive heart failure.[3] Numerous compounds were synthesized and acetazolamide emerged as the best sulfanilamide derivative since it was 300 times more potent as a carbonic anhydrase inhibitor. Unfortunately, its activity was limited since tolerance developed rapidly. Convinced that a compound could be developed that specifically increased sodium excretion, Beyer et al.[4] and Novello and Sprague[5] synthesized and defined chlorothiazide. Once its diuretic characteristics were confirmed in humans, widespread interest was generated following a sequence of four reports of clinical success in the *Journal of the American Medical Association* (January 11, 1958).

In pursuit of a more active diuretic substance, chemists began searching for inhibitors of sulfhydryl enzymes that did not contain heavy metals. This resulted in ethacrynic acid, which was considerably more effective than the thiazides and maintained its activity in the presence of renal insufficiency. Interestingly, it was less active than the organic mercurials as an inhibitor of sulfhydryl enzymes, but was considerably more active as a diuretic. Subsequently, there were several new chemicals introduced that possessed similar diuretic activity; however, none had characteristics that constituted a major discovery.

II. CHEMISTRY AND STRUCTURE – ACTIVITY RELATIONSHIPS

The chemical structures of some commonly used diuretics are presented in Figure 1. The presence of a free primary amine group in the sulfonamide moiety is associated with carbonic anhydrase inhibition and is present in all of the thiazides and the loop diuretics, except ethacrynic acid. Curiously, antibacterial activity requires a free amine group that is para to the sulfonamide moiety, which none of the diuretics possess.

A. Thiazides

This category contains a large number of compounds with complex structure-activity relationships.[6] An important feature for virtually all sulfonamides is a halogen atom

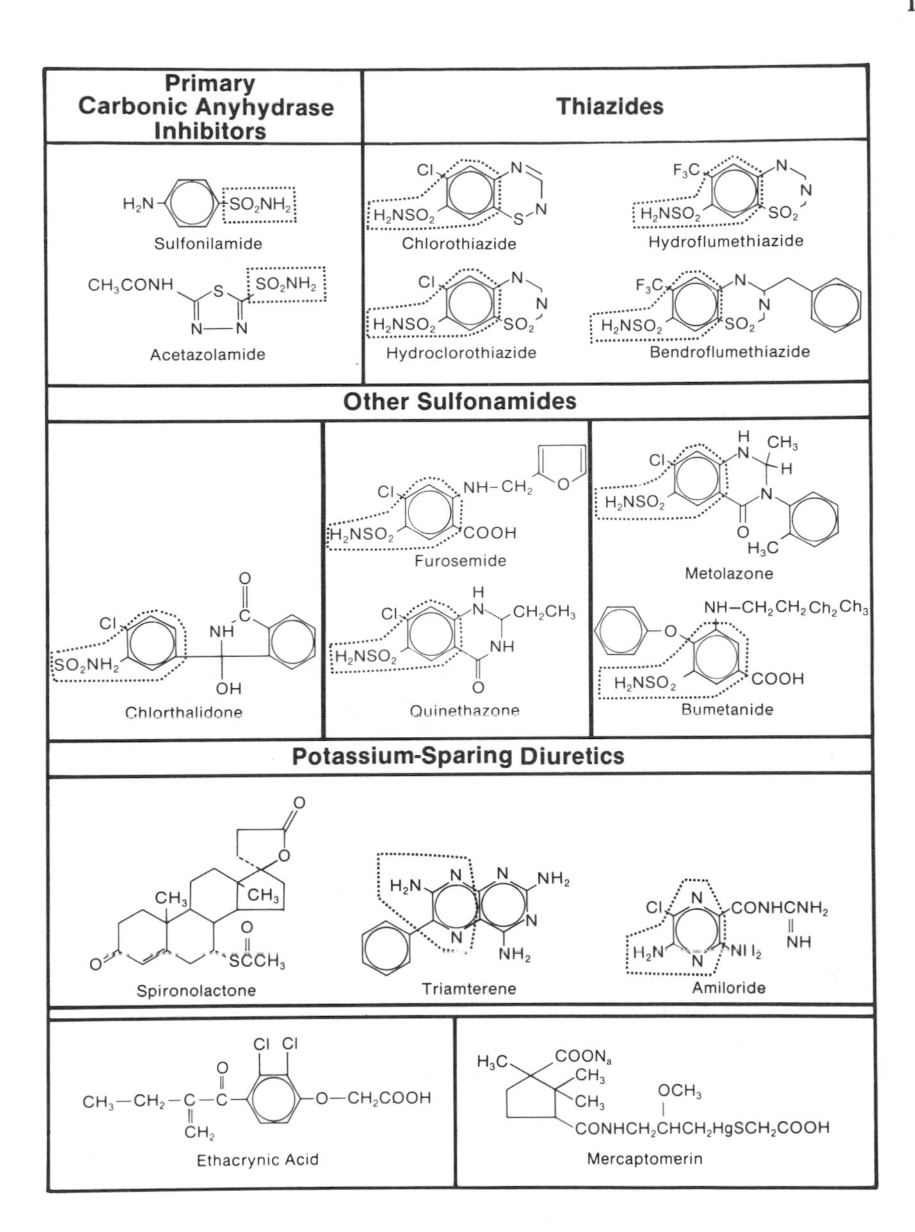

FIGURE 1

adjacent to the sulfonamide moiety. While diuretic potency varies over a 10,000-fold range, all thiazides possess parallel dose response curves; this is typical of compounds sharing the same mechanism of action. Indeed, the largest dose of the least potent member of this group produces diuresis similar to the largest dose of the most potent member. Diuretic activity is not dependent on carbonic anhydrase inhibition, since chlorothiazide is 1/20 as active an inhibitor as acetazolamide, but is a considerably more powerful diuretic.

Highly substituted hydrothiazides, e.g., methyclothiazide, polythiazide, and cyclothiazide, are more lipid-soluble, and have a larger volume of distribution, and consequently a decreased clearance resulting in a 24-hr duration of action. Flumethiazide has a trifluromethane substitution for chloride in the six position resulting in a further decrease in carbonic anhydrase activity. Both hydrochlorothiazide and hydroflumethiazide have less enzyme inhibition than their parent compounds, and are considerably more potent. The addition of a benzene ring in the two position results in benzhydroflumethiazide, which is an extremely potent compound with virtually no carbonic anhydrase inhibitory activity. In general, structure alteration has produced compounds with longer durations of action, and greater potency, but similar efficacy and toxicity.

B. Chlorthalidone, Quinethazone, and Metolazone

These compounds differ chemically from the thiazides by nature of their heterocyclic rings, but are similar regarding sulfamoyl and halogen substituents in the benzene ring. They are thought to act by a mechanism similar to the thiazides. Quinethazone and metolazone are both quinazolinone derivatives.

C. Furosemide, Azosemide, Bumetamide, and Piretanide

These are rather complex monosulfamyl, carboxylic acids that possess considerably greater activity than other sulfonamides. Furosemide is an anthranilic acid derivative. Bumetamide is derived from metanilamide and deviates from the norm in that it has a phenoxy group rather than a halogen atom adjacent to the sulfonamide, thus providing an exception to what has been considered a critical relationship.

D. Ethacrynic Acid

Ethacrynic acid is an unsaturated ketone derivative of aryloxyacetic acid, with activity similar to furosemide. The methylene and adjacent ketone groups react with the sulfhydryl radicals at receptor sites.

E. Aldosterone Antagonists

Spironolactone is structurally similar to the mineralocorticoids, but has a lactone ring as a substitutent at C-17. It competitively inhibits the binding of mineralocorticoids to their intracellular receptors.[7] Its presence in plasma interferes with fluorometric determinations of plasma cortisol.

F. Potassium-Sparing Agents

Triamterene and amiloride are pteridine derivatives, are important for their effects on potassium excretion, and possess minimal natriuretic activity. Triamterene is structurally similar to folic acid, and is a weak folic acid antagonist.[8] Amiloride is an organic base that is more potent than triamterene, but similar otherwise.

G. Organic Mercurials

The organic mercurials are thought to act as a result of intrarenal rupture of their carbon-to-mercury bonds, which release mercuric ions.[9] Those compounds with diuretic activity are acid-labile; while the acid-stabile derivatives lack diuretic effects.

III. MECHANISM AND SITE OF DIURETIC ACTION

The main tubular action of the various diuretics to be discussed here are listed in Table 1.

A. Osmotic Diuretics

Osmotic diuretics are substances that are freely filterable by the glomerulus but not reabsorbed. While urea, sodium sulfate, sodium chloride, and hypertonic glucose have been used as osmotic diuretics, only mannitol has these qualities. Their diuretic action which is attributed to both renal hemodynamic and tubular changes is weak and short-lived and therefore their clinical use is seldom warranted.

Table 1
EFFECTS OF DIURETICS ON SOLUTE AND WATER EXCRETION

Diuretic	Electrolyte excretion						
	Na^+	K^+	Cl^-	HCO_3^-	Ca^{++}	C_{H_2O}	FE_{Na^+} (%)
Osmotic + xanthines	↑	↑	↑	↑	↑	↑	≤4
Acetazolamide	↑	↑	(↑)	↑↑	↑	±	5
Thiazides	↑↑	↑↑	↑	↑	↓	↓→	10
Chlorthalidone	↑↑	↑↑	↑	↑	↓	↓→	10
Metolazone	↑↑	↑	↑	→	↓	↓→	8
Ethacrynic acid	↑↑↑	↑↑	↑	→	↑	↑	25
Furosemide	↑↑↑	↑↑	↑	↑	↑	↑	25
Mercurials	↑↑↑	↑	↑↑	→	→	↓	20
Spironolactone	↑	↓	↑	→	→	→	2
Triamterene	↑	↓	↑	→	→	→	2
Amiloride	↑	↓	↑	→	→	→	2

Note: Number of arrows indicates potency. ↑ Increase, ↓ Decrease, → No change.

Modified from Goldberg, M., *Handbook for Physiology*, Sect. 8, American Physiological Society, Washington, D.C., 1973, 1003. With permission.

1. Renal Hemodynamic Effects

Osmotic diuretics given intravenously cause volume expansion with an increase in cardiac output, glomerular filtration rate (GFR), and renal blood flow (RBF). These changes lead to increased filtration and decreased reabsorption of sodium by the proximal tubule due to increased hydrostatic pressure of proximal peritubular capillaries.[10]

2. Renal Tubular Effects

The proximal tubule is the major site of osmotic diuretic action.[11] The addition of mannitol to the proximal tubular fluid reduces its sodium concentration and decreases the percentage of sodium reabsorption below the 60 to 70% normally absorbed.[12] Thus, increased amounts of sodium and water are delivered to the distal sites of the nephron for reabsorption.

The increased delivery of Na^+ and Cl^- to the loop of Henle and their active reabsorption by this segment of the nephron increases the formation of free water (C_{H_2O}).[10] Despite loop reabsorption, increased amounts of solutes and water are delivered to the distal tubule, overwhelming its reabsorptive capacity and causing increased urinary sodium and water excretion. The increased K^+ losses in the urine by osmotic diuretics are due to large amounts of sodium delivered to the distal tubule that is exchanged for K^+ and not to a direct effect. The xanthines are frequently included with the osmotic diuretics because of their similar mode of action. They increase the cardiac output and renal blood flow which interferes with sodium reabsorption by the proximal tubule.[13] These findings are similar to those observed with the use of various vasodilators.[14] However, the diuretic and natriuretic effect of xanthines, like the osmotic diuretics, are weak and short-lived and, therefore, these agents are seldom used today.

B. Carbonic Anhydrase Inhibitors

These are weak diuretics which act on renal hemodynamics and tubular function. However, the diuresis they cause is self-limiting, because the ensuing metabolic acidosis decreases their effectiveness despite the continued inhibition of the enzyme.

1. Renal Hemodynamic Effects

Carbonic anhydrase inhibitors decrease GFR and RBF by about 30%[15] and redistribute the RBF from the inner to the outer renal cortex.[16] These hemodynamic effects should be considered in the interpretation of micropuncture data.

2. Renal Tubular Effects

The kidney serves an important function in maintaining the acid-base balance by absorbing the filtered bicarbonate (HCO_3) and by secreting H^+. Hydrogen ions are formed from the dissociation of carbonic acid (H_2CO_3), a product of CO_2 hydration, a reaction catalyzed in the tubular cell by the enzyme carbonic anhydrase (CA).[17] The H_2CO_3 is quite unstable and quickly dissociates into H^+ and HCO_3^-, according to the following equation:

$$CO_2 + H_2O \overset{CA}{\rightleftarrows} H_2CO_3 \rightleftarrows H^+ + HCO_3^-$$

The hydrogen ions are secreted into the tubular lumen in exchange for sodium ions, and together with HCO_3^- enter the peritubular blood. Thus, carbonic anhydrase inhibition decreases the supply of H^+, and Na^+ and HCO_3^- remain in the urine, resulting in an alkaline diuresis and a decrease in the excretion of titratable acid and ammonium anions, with little change in Cl^- excretion.[18]

a. Proximal Tubule

Clearance and micropuncture studies show that the proximal tubule is the principal site of action of carbonic anhydrase inhibitors.[19] During administration of carbonic anhydrase inhibitors, the reabsorption of sodium, bicarbonate and chloride is decreased and the fractional excretion of both bicarbonate and phosphorus rise by 30%, another indication of proximal tubular action.[20,21]

b. Loop of Henle

The loop of Henle is responsible for the concentration and dilution of urine, and clearance studies have provided little evidence for any direct effect by carbonic anhydrase inhibitors on these processes.[22]

c. Distal Renal Tubule

While carbonic anhydrase is present in both proximal and distal tubules,[23] it is unclear whether enzyme inhibition affects distal tubular sodium and bicarbonate reabsorption. Acetazolamide caused sodium reabsorption to decrease in the isolated perfused cortical collecting tubules of the rabbit,[24] but the result is difficult to interpret, since high concentrations of the drug were required (10^{-3} to 10^{-4} *M*) and the effect was not readily reversible, in contrast to the proximal tubular effect. Also, the demonstration by free flow micropuncture studies of increased bicarbonate concentration in the distal tubule during administration of carbonic anhydrase inhibitors is compatible but not confirmatory with a distal tubular effect.[25] The rise in distal tubular potassium concentration observed with carbonic anhydrase inhibitors has been attributed not only to increased distal tubular delivery of sodium, but also to increased transepithelial potential caused by large accumulations of bicarbonate, and elevation of the distal tubular fluid pH both of which could increase potassium secretion.[26]

In summary, carbonic anhydrase inhibitors are weak diuretics which exert their maximal effect on the proximal tubule. They lead to increased excretion of bicarbonate, phosphate, and potassium and loss of sodium and chloride. Prolonged administration produces metabolic acidosis that is self-defeating.

C. Mercurial Diuretics

These diuretics are potent saluretic agents with their major site of action the ascending limb of Henle's loop. They seem to promote chloride excretion and to decrease the excretion of bicarbonate and potassium. However, because of their cardiotoxicity, nephrotoxicity, need for parenteral administration and availability of other equipotent and safer diuretic agents, these drugs have become obsolete in the present-day practice of medicine. They have been largely replaced by furosemide and ethacrynic acid which have similar potency and mode of action, but are safer and easier to administer.

D. Ethacrynic Acid and Furosemide

These are very popular drugs because of their strong activity, rapid action, ease of administration, lack of serious toxic side effects, and similar renal hemodynamic and tubular effects.

1. Renal Hemodynamic Effects

Part of the diuretic action of ethacrynic acid and furosemide is due to their renal hemodynamic effects. In usual therapeutic dosages they do not appreciably affect GFR or effective renal plasma flow (ERPF),[27] but in larger amounts they cause renal vasodilation, increase in RBF and redistribution of RBF from the inner cortical (juxtaglomerular) to the outer cortical nephrons.[28,29] Increase in RBF and its diversion from the inner cortical (long-looped) to the outer cortical (short-looped) nephrons increases sodium excretion.[10,14,29] The natriuresis has been attributed to energy (adenosine triphosphate [ATP]) deprivation of the sodium pumps of the inner cortical nephrons and the decreased ability of short-looped outer cortical nephrons to reabsorb sodium.[29] These views, however, have been questioned by studies in the dog which indicated no relationship between renal hemodynamic changes and solute excretion.[16] Despite this controversy, the renal hemodynamic changes induced by ethacrynic acid and furosemide seem to play an important role in the overall diuretic effectiveness of these drugs. Thus, the beneficial effects of furosemide in patients with heart failure have been attributed at least in part to reversal of the abnormal outer to inner cortical RBF distributional changes.[30]

The mechanism for these renal hemodynamic and tubular effects is not well understood and it has been suggested that they are mediated through prostaglandins, since pretreatment of animals with indomethacin prevented the renal hemodynamic changes produced by furosemide.[31-33] It was also found for example, that prostaglandin E concentration increased in renal venous blood after ethacrynic acid or furosemide in untreated, but not in indomethacin-pretreated animals.[31,32] The increased levels of intrarenal prostaglandins could be attributed either to stimulation of their synthesis or more probably to decrease in their degradation. In vitro studies have demonstrated that both diuretics inhibited the enzyme 15-OH PG-dehydrogenase, which is responsible for the catabolism of prostaglandins,[34] and in vivo experiments in our laboratory have shown that indomethacin had no direct effect when infused into the renal artery of the anesthetized dog.[35]

2. Renal Tubular Effects

The molecular mechanism by which both ethacrynic acid and furosemide inhibit solute and water transport in the renal tubule is not clear at present and several hypotheses have been advanced. These are (1) renal hemodynamic changes, (2) altered permeability of the epithelial membrane of the renal tubule affecting passive transfer of solutes and water, and (3) increased levels of renomedullary prostaglandins, since administration of prostaglandin synthesis inhibitors antagonizes the diuretic action of these drugs.[36,37] However, the role of prostaglandins in the RBF distributional changes may be less important, since it has been found that these changes do not contribute significantly to the diuretic action of ethacrynic acid and furosemide.[16] Also, studies in our laboratory have shown that indomethacin does not interfere with the renal hemodynamic effects of furosemide but does with its tubular actions.[38,39] It appears that the most likely explanation for the diuretic effects of both drugs is their action on several cellular metabolic processes. Ethacrynic acid binds to sulfhydryl groups, but whether this property is directly related to its mechanism of action is unknown.[7] Furosemide is a sulfonamide and in this respect is related to the thiazide diuretics and acetazolamide, all carbonic anhydrase inhibitors; but it is unlikely that the strong diuretic potency of furosemide could be attributed solely to carbonic anhydrase inhibition. The demonstration that both drugs inhibit Na^+ — K^+ ATPase in renal cells in vitro suggests that these agents may interfere with active electrolyte transport by the renal tubule.[40] Other possible mechanisms for interference with energy supply include the inhibition of glycolysis by both drugs,[41] respiration of kidney slices,[42] and oxidative phosphorylation by mitochondria.[43] The studies of Cunarro and Weiner,[42] who demonstrated that ethacrynic acid and furosemide inhibited chloride transport by medullary tubules, and Quintanilla et al.,[43] who found that loop diuretics (ethacrynic acid, furosemide, meralluride) but not chlorothiazide, inhibited the incorporation of adenosine diphosphate (ADP) in rabbit cortical and medullary mitochondria, are of particular interest because they provide direct evidence for the action of these diuretics at the loop of Henle. The evidence for other tubular sites of action by these diuretics will be reviewed here.

a. Proximal Tubule

Both ethacrynic acid and furosemide appear in the proximal tubular lumen by glomerular filtration and tubular secretion through the weak acid transport system. Several micropuncture studies have disclosed that both diuretics may have a weak effect on the proximal tubule, since it was found that the (TF:P)IN ratio in the proximal tubule was decreased.[44,45] The evidence for a proximal tubular effect is more conclusive for furosemide than for ethacrynic acid. Furosemide is an inhibitor of carbonic anhydrase and its administration is associated with increased bicarbonate excretion and less propensity to cause metabolic alkalosis than ethacrynic acid.[46,47] In man, furosemide causes phosphaturia but ethacrynic acid does not,[47] whereas in dogs both agents cause phosphaturia.[48] Both agents increase the excretion of calcium and magnesium,[49,50] all markers of proximal tubular reabsorption.

b. Loop of Henle

A diuretic is assumed to act at the ascending limb of Henle's loop when C_{H_2O} or both C_{H_2O} and T^cH_2O are decreased. Studies have shown that ethacrynic acid and furosemide decrease both C_{H_2O} and T^cH_2O.[51,52] The reduction in T^cH_2O indicates that the drugs interfere with chloride and sodium transport in the medullary portion of the ascending limb (Site 2; Figure 2), whereas the reduction of C_{H_2O} indicates that they also act at the cortical diluting segment of the loop (Site 3; Figure 2). Analysis of renal medullary tissue from dogs and rats showed marked reduction of the medullary gradient following administration of ethacrynic acid and furosemide. This suggests inhibition of sodium and chloride transport in the ascending limb of Henle's loop.[52]

c. Distal Tubule and Collecting Ducts

No action on sodium reabsorption in the distal tubule has been observed with either ethacrynic acid or furosemide. However, recent evidence suggests that both agents may have an effect on collecting ducts, decreasing water permeability. Studies in our laboratory have shown that furosemide administration into the renal artery of the anesthe-

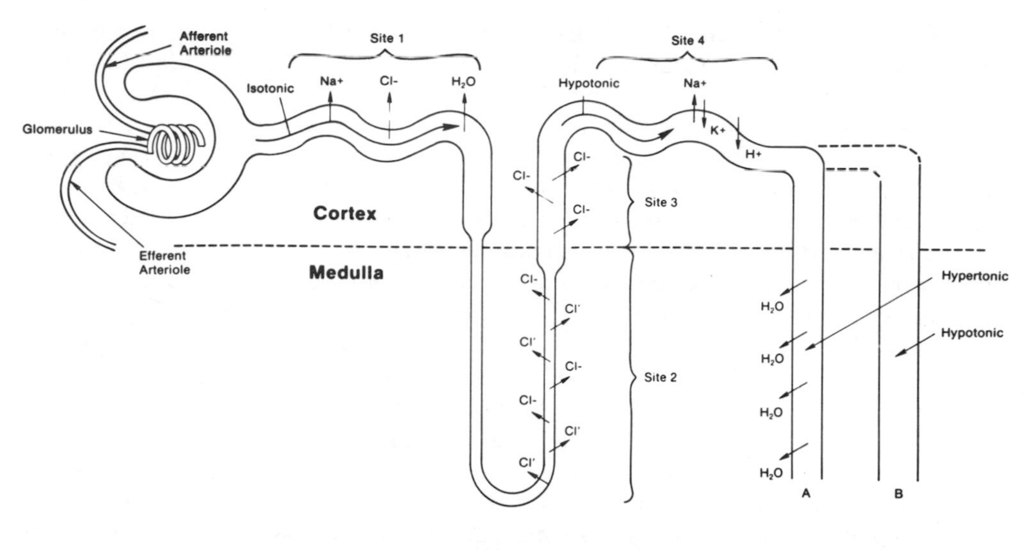

FIGURE 2

tized dog alone and in combination with PGE_1 increased the fractional excretion of water (C_{H_2O}/GFR), whereas indomethacin administration decreased it.[38,39] These data suggest that this effect of furosemide might be mediated through prostaglandins which have been shown to antagonize ADH action.

In summary, ethacrynic acid and furosemide are powerful diuretics with effects on the proximal tubules, loops of Henle, and collecting ducts. Their major site of action, however, is the ascending limb of Henle's loop. They exert their effects through hemodynamic and metabolic changes affecting tubular function.

E. Thiazides, Chlorthalidone, Metolazone

These drugs will be discussed together because they share certain properties, e.g., oral administration, slow onset and prolonged duration of action, similar renal hemodynamic and tubular effects, and common clinical uses. However, certain unique effects exist and these will be brought out in the discussion. Special emphasis will be placed on thiazide diuretics, because they are the most prominent and oldest members of this group.

1. Renal Hemodynamic Effects

GFR is acutely reduced by thiazide diuretics, as a result of volume contraction, and remains relatively low even after prolonged administration.[53] However, their effects as a group on RBF are small and transient, and they do not cause distributional changes as do furosemide and ethacrynic acid.[16,54]

2. Renal Tubular Effects

These agents are relatively weak, with a maximal diuretic effect equal to 10% of filtered load for thiazides and 7% for metolazone.[54] However, certain differences exist regarding potency and electrolyte composition of the final urine. Metolazone is longer acting and is eight to ten times more potent on a weight basis than the thiazide diuretics and does not appreciably increase the excretion of potassium and bicarbonate.[54]

a. Proximal Tubule

The thiazides are weak carbonic anhydrase inhibitors, and with small doses the bicarbonate excretion is relatively small. whereas with larger doses, it increases to approximately 10% of filtered load.[55] Excretion of sodium, chloride, and bicarbonate also increases with thiazide diuretics. These changes have been attributed to a proximal tubular effect by both stop-flow[19] and micropuncture studies.[56] Also, the excretion of phosphorus increases with thiazide diuretics,[54] excretion of calcium and uric acid increases acutely, and decreases after prolonged administration.[57] After several days of thiazide administration the negative sodium balance is corrected and the excretion of sodium and water decrease,[58] and for this reason, these agents have been used successfully to treat patients with diabetes insipidus and idiopathic hypercalciuria.[58,59] Unlike pure carbonic anhydrase inhibitors, thiazides are more potent diuretics, their action on chloride excretion is not inhibited by metabolic acidosis, and prolonged administration leads to hypokalemic, hypochloremic metabolic alkalosis instead of the metabolic acidosis produced by pure carbonic anhydrase inhibitors.[19] Chlorthalidone, like the thiazide diuretics, has a strong proximal tubular effect and micropuncture studies in the dog have shown that this drug decreases proximal tubular chloride and sodium reabsorption.[60] Also metolazone has been found to have a weak proximal tubular effect, probably unrelated to carbonic anhydrase inhibition, since its administration increases phosphate but not bicarbonate excretion.[54] All these agents exert their effect by active secretion in the proxima tubule through the organic acid transport system[19] (Site 1; Figure 2).

b. Loop of Henle

There is experimental evidence indicating that the thiazides, chlorthalidone and metolazone may have an effect on the cortical segment of the loop of Henle (Site 3; Figure 2), because they decrease C_{H_2O} but not T^cH_2O. Agents which inhibit both C_{H_2O} and T^cH_2O work at the medullary portion of the loop (Site 2; Figure 2), whereas agents affecting only C_{H_2O}, work at the cortical segment of the loop.[19]

c. Distal Tubule

Rat micropuncture studies suggested that the thiazide diuretics exert their major saluretic effect at the distal convoluted tubule (Site 4; Figure 2). It was found that the distal tubule (TF:P)IN ratio decreased, whereas the (TF:P)Cl^- ratio increased. The fractional excretion of chloride increased with chlorothiazide from 0.29 to 3.44%, in contrast to benzolamide which did not increase chloride excretion. Similar findings have also been reported by other investigators using both clearance,[61] stop-flow,[62] and micropuncture[63] studies. The mechanism for the inhibition of chloride reabsorption at the distal tubule is not well understood, but it could represent an interference with active chloride transport.[64] The kaliuretic effect of the thiazide diuretics is the result of increased distal exchange of Na^+ for K^+ due to increased solute concentration at this site, and does not imply a direct action by these agents on potassium transport. Other factors leading to increased potassium excretion include: increased levels of poorly reabsorbable anions (bicarbonate and phosphate) which enhance intratubular negativity and increase potassium secretion, and increased aldosterone levels which facilitate potassium loss. Administration of metolazone also has a distal tubular effect but lacks the strong kaliuresis seen with thiazide diuretics and chlorthalidone.[54,63]

In summary, the thiazides, chlorthalidone, and metolazone are weak diuretics, but with wide clinical application because of their prolonged effects. They all have actions in the proximal and distal tubule and the cortical segment of the loop of Henle. Their administration is associated with an increased elimination of sodium, chloride, bicarbonate, and potassium, except for metolazone, which does not appreciably increase the excretion for potassium and bicarbonate.

F. Potassium-Sparing Diuretics

Spironolactone, triamterene, and amiloride are the three diuretics currently used to spare potassium losses through the kidney. Spironolactone is a hormonal substance that acts by antagonizing the effects of aldosterone and deoxycorticosterone at receptor sites,[65] whereas triamterene and amiloride work independently of the presence or absence of aldosterone.[66,67] Their natriuretic effect is very modest accounting for only 2% of the filtered load of sodium.

1. Renal Hemodynamic Effects

In therapeutic doses, spironolactone does not appreciably affect renal hemodynamics, whereas triamterene reduces GFR and ERPF and increases the level of BUN and creatinine.[68] Administration of amiloride did not cause any adverse effects on GFR and RBF in patients with mild essential hypertension and normal renal function.[69]

a. Proximal Tubule and Loop of Henle

Studies in various animal species have not disclosed any effect of spironolactone, triamterene, and amiloride on proximal tubule or loop of Henle.[65-68] No effect on C_{H_2O} was shown with triamterene[66] and studies with amiloride both in vivo[70] and in vitro[71] did not show any changes on chloride and sodium reabsorption at the ascending limb of Henle's loop.

b. Distal Tubule

This segment of the nephron (Site 4; Figure 2) appears to be the main site of action of potassium-sparing diuretics.[65-67] Spironolactone and its congeners block the action of mineralocorticoids on sodium reabsorption in the rat,[72] dog,[65] and man.[73] Aldosterone and other mineralocorticoids penetrate the epithelial cells and stimulate the nucleus to generate new DNA and RNA which control the production of new proteins involved in the transport of sodium and potassium.[74] Spironolactone appears to exert its effects by interfering with the nuclear binding of aldosterone and the suppression of protein synthesis for sodium and potassium transport. This explains why spironolactone is active only in the presence of mineralocorticoids, and is ineffective in adrenalectomized animals or humans.[65,72,73] The mechanism by which amiloride and triamterene interfere with sodium transport has been studied in various tissues. In the renal tubule, the drugs exert their effects on the outside (luminal), instead of the inside (contraluminal) surface of the epithelial cells.[75] The kinetics of transepithelial sodium transport involve two different steps, sodium entry into the cell from the outside (luminal) surface and movement of sodium across the inside (contraluminal) surface into the interstitium and bloodstream. Amiloride and triamterene probably affect the first step by inhibiting sodium transport into the cell,[76] and also affect potassium transport and urinary acidification in the renal tubule. These effects have been attributed to inhibition of sodium transport, since conditions which inhibit sodium reabsorption, such as addition of ouabain, and removal of sodium from the lumen, also inhibits potassium transport.[70]

In summary, spironolactone, amiloride, and triamterene given in therapeutic doses, produce a modest natriuresis and effectively prevent potassium and hydrogen excretion by the distal tubule. Spironolactone competitively inhibits mineralocorticoids at receptor sites and is only effective in the presence of mineralocorticoids, while amiloride and triamterene operate independently of mineralocorticoids and aldosterone, through an unknown mechanism.

IV. CLINICAL PHARMACOLOGY

A. Thiazides

Hydrochlorothiazide is the most commonly prescribed thiazide diuretic. It is more reliably absorbed than chlorothiazide, and consequently preferred for clinical use. Other thiazides offer no distinct advantages.

1. Pharmacokinetics

In healthy volunteers hydrochlorothiazide absorption varies from 60 to 80% compared to 33 to 58% for chlorothiazide.[77] The peak plasma level occurs between 2 to 4 hr and the beta phase half-life varies between 6 and 15 hr. Since hydrochlorothiazide (like the other thiazides) is primarily excreted unchanged by proximal renal tubular secretion, clearance is usually greater than glomerular filtration rate. The elimination rate is prolonged by impaired renal function, while it is unchanged by liver disease.

2. Side Effects and Adverse Reactions

These have been extensively reviewed,[78] and rarely require discontinuation of therapy. Pharmacologic type reactions include:

1. *Hypokalemia* — generally plateaus by one month and is positively correlated with sodium intake and mineralocorticoid activity.
2. *Hyperuricemia* — related to decreased uric acid excretion; rarely causes problems with gout.
3. *Azotemia* — thought to be prerenal in origin, i.e., the decreased plasma volume and renal blood flow reduce glomerular filtration rate.
4. *Hyperglycemia* — rare, and may be related to hypokalemia, pancreatic disturbance, or preexisting impaired glucose tolerance.
5. *Hyponatremia* — results from diminished free water clearance, excessive water ingestion, and restricted sodium intake.
6. *Hypercalcemia* — secondary to decreased urinary calcium excretion, increased plasma protein binding (due to hemoconcentration), and potentiation of the renal actions of parathyroid hormone.

Other less common and unpredictable adverse reactions have included: (1) a wide variety of allergic type reactions (Note: ethacrynic acid is the only standard nonsulfonamide diuretic); (2) pancreatitis; (3) blood dyscrasias; (4) hypertriglyceridemia; and (5) a variety of other phenomena and nonspecific constitutional symptoms.

3. Drug Interactions

1. Diuretics decrease lithium clearance allowing accumulation.
2. Thiazides potentiate skeletal muscle relaxants, e.g., tubocurarine.
3. Thiazides may decrease the hypoprothrombinemic effect of warfarin.
4. The diuretic effect of thiazides are prolonged by probenecid.

B. Loop diuretics

Ethacrynic acid is used less than furosemide because of an occasional instance of irreversible hearing loss, and a narrow dose-response curve. Other compounds that are under clinical investigation are bumetanide, azosemide, piretanide, and indepamide.

1. Pharmacokinetics

The pharmacokinetics of furosemide were recently reviewed.[79] Absorption ranges from 60 to 69%, and fasting peak plasma levels occur between 60 to 90 min. When administered with food, the peak levels are decreased by 50% and delayed to 4 hr, but by 12 hr, the natriuretic effect is similar to that seen in the fasting state.[80] Decreased absorption has been described in patients with congestive heart failure,[81] cirrhosis and ascites,[82] and end-stage renal disease.[83] Thus, it is likely that such patients will require larger oral doses.

Furosemide is 91 to 99% bound to plasma protein. The free fraction is increased in patients with acute and chronic renal failure and in individuals with a serum albumin concentration < 2 g/dℓ.[84]

In individuals with normal renal function the majority of furosemide is excreted unchanged in the urine, and smaller amounts are excreted as the glucuronide and free amine metabolite. The beta phase half-life is 20 to 120 min. Like the thiazides, the major route of elimination is approximal renal tubular secretion, and liver failure does not alter the furosemide elimination rate. In general, furosemide clearance is greater than creatinine clearance except in end-stage renal disease when it is considerably less. Under such circumstances, the half-life may be as long as 15 hr. Furosemide clearance is also decreased in neonates and the elderly and during probenicid administration.

2. Side-Effects and Adverse Reactions

a. Pharmacological

Furosemide is an extremely active diuretic capable of producing serious hyponatremia, hypokalemia, hypochloremia, metabolic alkalosis, and volume depletion. In patients who have complicating circumstances these reactions may be life threatening. Hyperuricemia, hyperglycemia, and deterioration of the oral glucose tolerance test is comparable to that discussed previously for the thiazides. Hypercalcinuria and lowering of serum calcium is an effect opposite to that of the thiazides.

b. Unpredictable Reactions

1. Ototoxicity seems to be more common following rapid i.v. injection and may vary from a transient mild hearing loss to permanent deafness. The presence of other ototoxic drugs and a preexisting hearing deficit predisposes to toxicity.
2. Allergic reactions are uncommon and have consisted of various cutaneous lesions, anaphylactic shock, interstitial nephritis, and vasculitis.
3. Blood dyscrasias, mesenteric infarction, acute pancreatitis, and CNS reactions have been reported, but are rare.

3. Drug Interactions

1. Indomethacin, and probably other nonsteroidal antiinflammatory agents blunt the diuretic and natriuretic effect of furosemide.
2. Furosemide potentiates the nephrotoxicity of the cephalosporins.
3. Furosemide enhances skeletal muscle relaxants, e.g., tubocurarine.
4. Diuretics decrease the clearance of lithium allowing accumulation.
5. Furosemide potentiates ototoxicity of the aminoglycoside antibiotics.
6. Anticonvulsants may attenuate the effects of furosemide.

C. Triamterene

Triamterene has a weak natriuretic effect, augments that of other diuretics, and promotes potassium retention. It is primarily used with other diuretics to prevent or correct diuretic-induced hypokalemia.

1. Pharmacokinetics

Triameterene absorption varies from 30 to 70%, and is rapidly metabolized by the liver.[85] The beta phase half-life is 6 to 12 hr. In the presence of liver disease or congestive heart failure the half-life is prolonged.

2. Side-Effects and Adverse Reactions

Hyperkalemia may become severe in patients with impaired renal function, individuals receiving potassium supplementation, and in insulin-dependent diabetics, during hyperglycemia, and in individuals with nephropathy. Hyperuricemia and hyperglycemia are mild; megaloblastic bone marrow changes have been reported in patients with dietary deficiencies and in individuals with excessive folic acid utilization; nausea, weakness, and dermatological reactions infrequently occur. An increased incidence of nephrolithiasis has been associated with triamterene administration, particularly in individuals with a history of renal stone formation.

D. Spironolactone

Spironolactone has been extensively reviewed.[86,87] It is used to prevent hypokalemia associated with elevated aldosterone levels and in treating the hypokalemia associated with diuretic therapy.

1. Pharmacokinetics

Spironolactone is >90% absorbed. It is rapidly metabolized to canrenone, the active form, which develops a peak plasma level by 2 to 4 hr. Both substances are greater than 90% bound to plasma protein. Canrenone is primarily excreted in the urine, with a beta phase half-life between 12 to 36 hr. Plasma canrenoate levels vary 15-fold among individuals receiving similar doses. Curiously, absorption is improved when it is administered with food.

2. Side-Effects and Adverse Reactions

These have been reported in 20% of patients receiving spironolactone. Approximately 9% develop hyperkalemia, and 2 to 4% develop dehydration, hyponatremia, gastrointestinal symptoms, and neurological disturbances. Gynecomastia is a common complaint. While tumorigenesis has been seen in animals, no cause and effect relationships with tumors in humans has been observed.

3. Drug Interactions

Aspirin and nonsteroidal antiinflammatory agents reduce the excretion of canrenoate, and also its effects.

E. Amiloride

This is a mild natriuretic and antihypertensive agent with potassium-sparing properties. It is well absorbed, and 20 to 50% is excreted unchanged in the urine. It is not metabolized by the liver; thus, dosage modification is not required in liver disease.

Hyperkalemia is the most frequent adverse reaction. Gastrointestinal symptoms and dermatological reactions occasionally occur.

V. HYPERTENSION

The effect of diuretics in hypertension has recently been reviewed.[88] Diuretics were first shown to possess antihypertensive action in patients in 1948 when Begilow used repeated i.m. injections of a mercurial diuretic.[89] However, the impracticality of chronic use, coupled with uncertainty at that time regarding the importance of treating hypertension, precluded this observation from becoming important therapeutically. However, the availability of chlorothiazide resulted in a practical method for administering diuretics to patients with essential hypertension.[90,91] Subsequently, this opened the door to a myriad of studies relating to diuretic use in hypertension.

A. Mechanism of Action

The current understanding of the antihypertensive mechanisms of diuretic action has evolved along with technological advances. General antihypertensive mechanisms for diuretics have centered around contracted extracellular fluid volume (ECFV), decreased plasma volume and cardiac output, diminished vasopressor receptor sensitivity, and direct vasodilator activity.

1. Body Fluids and Hemodynamic Effects

Contraction of ECFV and plasma volume following the use of diuretics is undisputed. By 48 to 72 hr there is in general a 2 ℓ decrease in ECFV. In hemodynamic studies the early decrease in plasma volume has been associated with a decrease in cardiac output and a compensatory increase in total peripheral resistance.[92] While there is generally thought to be a continued effect of diuretics on ECFV, there have been conflicting reports. The majority of studies suggest a continued effect, and indeed, studies following discontinuation of diuretics after long-term use have shown that plasma volume increases towards normal.[93] The early reduction in blood pressure seems to be correlated with a decrease in cardiac output. After 2 to 3 months there is a return of cardiac output to normal levels followed by a decrease in peripheral resistance.

The precise mechanism of the delayed decrease in peripheral resistance remains unexplained. A current theory suggests that a decrease in tissue perfusion secondary to depressed cardiac output eventually causes autoregulatory changes that reduce peripheral resistance.[94] This concept was originally suggested by Tobian,[95] and was referred to as "reverse autoregulation". However, it seems unnecessary to use the term "reverse", since autoregulatory changes normally occur with either a decrease or increase in tissue perfusion. It was suggested that the subsequent decrease in peripheral resistance may account for the return of cardiac output to normal as afterload lessens.[94] However, such changes in afterload and cardiac output are seen in states of myocardial depression. It may be that a gradual re-setting of the baroreceptors secondary to a decrease in cardiac output accounts for the lowering of total peripheral resistance. Another very appealing mechanism that was recently suggested raises the possibility that thiazide may reduce peripheral vascular resistance by increasing prostacyclin biosynthesis.[96]

Another important diuretic effect relates to a secondary increase in the renin-angiotensin-aldosterone activity following diuretic administration.[97] This was suggested as a mechanism whereby diuretics thwarted their own effectiveness.[98] More recently, it was demonstrated that nonresponders to diuretics had a greater increase in renin and aldosterone activity than did responders.[99]

2. Arterial Effects

There has been considerable interest in the possibility that diuretics have either a direct or indirect effect on arteriolar smooth muscles. It was postulated that diuretics decrease the sodium and water content of the arterial wall.[100] While this theory is intellectually appealing, there is no compelling evidence for its support.[88,101]

There have been a number of studies demonstrating a decrease in vascular receptor sensitivity to vasopressor substances following a diuretic.[102,103] This is probably explained by the reduced plasma volume since normal pressor responsiveness was restored following blood volume replacement with dextran.[104] Also, there was no change in blood pressure with a thiazide diuretic in dialysis patients, unless a diuresis was established.[105]

B. Clinical Aspects

Diuretics are the most common drugs used in the management of hypertension, either as a single therapeutic agent or in combination with other antihypertensive compounds. There are a variety of factors that influence responsiveness, and in certain situations there is reason to use one preparation in preference to another. Side-effects and drug interactions also influence use and may determine the successfulness of the therapeutic regimen.

1. Blood Pressure Response

a. Thiazides

McMahon summarized the effects of thiazides or thiazide-like drugs as single modes of therapy in 16 studies in patients with essential hypertension.[106] There was an average decrease in blood pressure of 21/10 mmHg. In situations where a diuretic was added to another antihypertensive agent there was an additional decrease of 14/8 mmHg. The potentiation depends on a reduction of ECF volume which frequently increases following the use of nondiuretic antihypertensive agents. While the onset of diuresis is rather rapid, the decrease in blood pressure may take several days to become maximum. After discontinuation, blood pressure returns to control values over a period of 1 to 2 weeks. With chronic therapy thiazides seem to work indefinitely. Over a 12-year period 80% of 227 patients had a sustained drop in blood pressure.[107]

The dose-response curve for antihypertensive effect is considered flat and is similar to the natriuretic action. In one study hydrochlorothiazide had minimal antihypertensive effects at 50 mg/day, and it was suggested that 50 mg twice daily should be used.[108] However, in another investigation a much flatter dose-response curve was demonstrated, and it was suggested that there was little to be gained by increasing the dose above 50 mg/day.[77,108] In general, there is no additional blood pressure lowering beyond 100 mg/day.

Chlorthalidone is also a thiazide-like diuretic that is administered once daily. It was demonstrated that 25 mg/day was as effective as the usual, higher doses, but the potassium loss was considerably less.[110]

Metolazone is a thiazide-like diuretic, yet its activity is not depressed as much in patients with impaired renal function. It has a long duration of action and may be administered once daily. It is thought to cause less kaliuresis with equivalent natriuresis.

2. Furosemide

While furosemide has a larger diuretic effect than the thiazides, its antihypertensive action is less.[6,111,112] The effect of furosemide in doses of 75 to 160 mg/day as a sole agent in five studies was reviewed, and there was an average decrease in blood pressure of 10/4 mmHg.[112] As noted previously thiazides may lower blood pressure by 20/10 mmHg. The reason why furosemide, which has greater natriuretic properties, affects the venous capacitance bed and dilates the renal vasculature, has less antihypertensive effect than thiazides remains unexplained. This further supports the hypothesis that thiazides have a vascular effect separate from their diuretic action.

Furosemide use has a special place in four particular circumstances: (1) during a hypertensive crisis when a rapid acting i.v. preparation is necessary;[113] (2) in patients with impaired renal function and volume-dependent hypertension;[114] (3) as an alternate to a thiazide diuretic in severe resistance hypertension;[115] and (4) to reverse the diazoxide and minoxidil-induced sodium retention.

3. Spironolactone

Spironolactone may be used both diagnostically and as specific therapy in the occasional patient with primary aldosteronism.[116] However, it is more commonly used in conjunction with a diuretic in patients with essential hypertension to inhibit the aldosterone-mediated potassium excretion. To prevent or correct hypokalemia, it is generally prescribed on a milligram for milligram basis with hydrochlorothiazide.

4. Triamterene

Triamterene has less intrinsic natriuretic activity than spironolactone, and alone is not recommended for the treatment of hypertension.[117] It is primarily used to prevent or correct the hypokalemia associated with chronic diuretic therapy. The usual dose is 100 to 300 mg daily.

5. Amiloride

Amiloride by itself has only mild natriuretic and antihypertensive properties.[118] It is primarily used to prevent or correct the hypokalemia associated with chronic diuretic therapy. The usual daily dose is 5 to 10 mg.

VI. RENAL DISEASE

Diuretics are often employed in patients with renal disease for accumulation of salt and water. While renal mechanisms may be responsible, this is often complicated by hypertension and congestive heart failure. Therapeutic efforts directed towards the latter conditions could minimize the necessity of excessive reliance on diuretics.

A. Acute Renal Failure

While diuretics are used for both prevention and modification of acute renal failure, questions still remain regarding efficacy.

1. Prevention

Patients with aortic surgery, transfusion reactions, rhabdomyolysis, and shock syndrome have an increased risk of developing acute tubular necrosis. A considerable body of information in animals has been generated regarding prevention of ischemic acute renal failure by mannitol infusion, and this was recently reviewed.[119] Animal studies have also demonstrated protection against acute renal failure in response to a variety of pigments and toxins. The clinical studies, which are obviously not as well controlled and contain considerable amounts of heterogeneity, are less convincing. In a study of 37 patients with nonhypovolemic ischemic renal disease, a response to mannitol infusion (urine flow of 50 mℓ/hr) was associated with a shorter period of oliguria and an absence of renal failure.[120] This study suggested that a mannitol infusion identified the nonresponders that could be expected to progress to acute renal failure, or in some instances, prevent the full development of acute renal failure.

Loop diuretics have been advocated as potentially beneficial in protecting against ischemic and other types of acute renal insufficiency because of their ability to increase cortical renal blood flow, decrease renal vascular resistance, and promote a diuresis. In a study on the effects of furosemide on renal function and hemodynamics in 54 critically ill surgical patients, furosemide did not increase renal blood flow nor protect against renal failure, and further, the hypovolemia produced by furosemide may have contributed to the development of acute renal failure.[121] While disagreement regarding the role of diuretics in this situation remains, there is general agreement that maintenance of an adequate circulatory blood volume is important, and that if diuretics are to be used, it should be early.

The osmotic diuretic mannitol is generally preferred in this situation as it lacks the toxicity associated with loop diuretics. It is given by an i.v. injection of 12.5 to 25 g in a 20% solution. If the urine output in the next hour is less than 50 mℓ, the dose is repeated. If the patient remains unresponsive, it can be assumed that the patient has acute renal failure, and no further mannitol should be given. Some individuals have suggested the addition of furosemide, 2 to 4 mg/kg i.v., with the second dose of mannitol. However, no reported evidence supports this therapy.

2. Diagnosis of Acute Renal Failure

Response to diuretics has been advocated as a diagnostic test in differentiating acute tubular necrosis from prerenal azotemia. While the clinical problem can generally be resolved by urine studies or a saline challenge, an increase in urine flow to 40 mℓ/hr following mannitol, 12.5 g i.v., or furosemide, 1 to 2 mg/kg, suggests a prerenal cause. Although this is generally safe, it is possible that adverse effects could complicate the status of patients with established acute tubular necrosis. Further, furosemide use may obscure the diagnosis of acute renal failure by altering urine electrolyte composition.

B. Chronic Renal Disease

Edema in the presence of chronic renal insufficiency needs careful evaluation prior to diuretic use. If it is complicated by congestive heart failure or pulmonary edema, then intervention is more urgent. If it is the result of nephrotic syndrome, urgent measures may cause hypovolemia, impaired cardiac output and a decrease in renal blood flow which predispose to development of the shock syndrome.

1. Nephrotic Syndrome

Edema secondary to nephrotic syndrome is primarily the result of deficient plasma colloidal osmotic pressure and does not represent consequences of hypervolemia. Thus, diuretics do not play an important part in management. They may be used if conservative measures such as salt and water restriction fail to reduce edema sufficient to allow comfortable ambulation. When initiating diuretic therapy, it is important to begin with a small dose and gradually increase it as required. Also, intermittent or pulsed diuretic administration may avoid the potentially serious problems of hypovolemia, postural hypotension, decreased renal blood flow, and hypokalemia.

2. Chronic Renal Insufficiency

Patients with chronic renal disease are ideally managed by conservative means. However, this condition is often complicated by hypertension and congestive heart failure where diuresis assumes greater importance. Occasionally, patients with end-stage renal disease are relatively free of uremic symptoms other than fluid retention and hypertension. In these instances salt restriction may be inadequate to control the edema and furosemide is beneficial and might otherwise postpone the necessity of initiating chronic dialysis. In these instances it may be necessary to employ doses up to and occasionally above 1000 mg.[122] While occasional patients may increase glomerular filtration rates following furosemide, this is primarily seen in patients with congestive heart failure and is unlikely to occur in the absence of heart failure, although this appears unsettled.[123]

3. Dialysis

Diuretics have been used in patients with residual renal function who require hemo- or peritoneal dialysis. Although not routinely recommended, it may increase the interdialysis interval.[124] In patients requiring hemodialysis, furosemide may allow an increasing daily fluid intake up to 1300 mℓ without interdialysis weight gain.[125]

VII. CONGESTIVE HEART FAILURE

Sodium retention and hypervolemia are normal compensatory mechanisms for increasing cardiac output in patients with impaired cardiac function. Excessive expansion of ECFV and edema formation account for the majority of signs and symptoms of congestive heart failure. With the emergence of thiazide diuretics, it became possible to chronically remove edema fluid. With the subsequent availability of the loop diuretics, it became possible to promote excessive diuresis and interfere with cardiac output. Thus, it is important to appreciate the limits of benefit and the possibility of hazards associated with diuretic use.

A. Pulmonary Edema

With the advent of parenteral ethacrynic acid and furosemide another dimension evolved in the emergency management of acute pulmonary edema. Excessive capillary hydrostatic pressure, a relative deficiency of plasma collodial osmotic pressure, and increased permeability of pulmonary capillaries may produce pulmonary edema. Acute pulmonary edema, especially following a myocardial infarction, may occur in the presence of normovolemia, as well as in patients with chronic circulatory congestion. It is important to recognize and differentiate between these situations, since excessive diuresis may cause hypovolemia and depressed cardiac output in normovolemic individuals.

The hemodynamic changes secondary to furosemide in patients with acute myocardial infarction have been studied.[126] At 1 hr there was a decrease in cardiac output and pulmonary arterial pressure and a rise in brachial artery pressure. By 6 hr the cardiac output returned to pretreatment levels and the pulmonary and systemic arterial pressures were reduced. While most studies have not differentiated between the immediate and short-term hemodynamic effects, it has been recognized that some patients with pulmonary edema have a sudden and dramatic improvement with furosemide even in the absence of a diuresis.[127] This phenomenon was considered to be a biphasic effect of furosemide.[128] By 15 min a fall in left ventricular filling pressure was secondary to an increase in venous capacitance, not a diuresis. The peak urine flow occurred by 30 min and the peak natriuretic effect was at 60 min. Another study has suggested that furosemide decreases pulmonary transvascular fluid filtration rate by a nondiuretic effect.[129] Thus, it would appear that both vascular and diuretic effects with furosemide contribute to its beneficial effect in acute pulmonary edema. Both of these actions would be expected to decrease myocardial oxygen consumption. However, if excessive effects occur, hypovolemia and a reflex increase in cardiac activity may compound coronary blood flow problems and enhance myocardial oxygen demand.

B. Chronic Congestive Heart Failure

It has been known for some time that the removal of signs and symptoms of circulatory congestion by a diuretic does not indicate an improvement in cardiac function.[130] The hemodynamic effects of diuresis at rest and during exercise in patients with heart failure has been evaluated.[131] There was a decrease in the pulmonary arterial wedge pressure and cardiac output at a time when substantial clinical improvement had occurred. As discussed previously, excessive diuresis may accentuate the hypovolemia and further reduce cardiac output, since the beneficial Starling effect secondary to volume expansion is lost. Thus, continuous observation is essential to obtain optimal therapeutic results.

A thiazide diuretic along with sodium restriction may be satisfactory to control the signs and symptoms of congestive heart failure. However, patients who have impaired renal function, or who are refractory to the thiazides should be managed with furosemide, and likely require a dose of 80 mg or higher to establish satisfactory diuresis.

VIII. LIVER DISEASE

The use of diuretics in patients with edema secondary to liver disease has often been very tempting. However, it was not until the advent of the more potent loop diuretics that associated hazards became manifest. Indeed, numerous situations exist where the risk is clearly greater than the benefit.

There are several factors thought to account for the sodium retention and edema formation in patients with liver disease.[132-134] Decreased albumin synthesis may cause edema by diminishing colloidal osmotic pressure. Portal hypertension develops secondary to obstruction of intrahepatic blood flow with an accompanying increase in hydrostatic pressure and overproduction of hepatic lymph. With these changes in the Starling's forces, there is a transudation of fluid into the intra-abdominal cavity. This loss of fluid to the extravascular space causes a diminished central, or effective, blood volume. Another consideration suggests primary renal sodium retention with "overflow" into the peritoneal cavity as a major cause of ascites formation.[135] In situations where this seems to be the major cause of ascites formation, the skillful use of a diuretic is appropriate. Renal vasoconstriction with diminished renal blood flow may also contribute to sodium retention. Further, there is evidence of elevated plasma renin activity and aldosterone levels in patients with cirrhosis and ascites. Excessive aldosterone production and possibly a decreased clearance may explain the increased plasma levels. Thus, with a decrease in colloidal osmotic pressure and trapping of fluid in the abdominal cavity secondary to elevated hepatic hydrostatic pressure, the use of a diuretic that removes volume from the intravascular space is not an entirely rational form of therapy. However, the cautious use of loop diuretics with compounds designed to interfere with the distal tubular sodium retention and potassium excretion occasionally may be helpful.

A. Standard Diuretics

Prior to the use of diuretics, patients should be tried on a low sodium diet, placed at bed rest, and given adequate nutritional support. If this fails to promote diuresis, then careful diuretic therapy should be considered to achieve a 1-lb weight loss per day.

As previously discussed, there is considerable sodium retention by both proximal and distal tubular mechanisms in patients with liver failure. Agents that act primarily by inhibiting proximal tubular reabsorption, e.g., carbonic anhydrase inhibitors, organomercurials, and osmotic diuretics, generally do not promote vigorous natriuresis, since there is enhanced sodium retention at distal tubular sites.

The loop diuretics are capable of delivering larger amounts of sodium to the distal tubule than can be reabsorbed, thus stimulating vigorous natriuresis. Also, they can produce an increase in free-water clearance.[136] This could result in intravascular volume depletion with vasopressin stimulation that leads to hyponatremia.

The thiazides, like the loop diuretics, are associated with enhanced potassium excretion. However, they differ in their inhibition of urinary diluting mechanisms, which interferes with free-water excretion, especially in the hyponatremic cirrhotic patient.[137]

Adverse reactions are fairly common with furosemide in patients with cirrhosis of the liver.[138] In 172 patients with cirrhosis of the liver, approximately 50% had some event related to furosemide and 24% of these were judged severe. Electrolyte disturbance and volume depletion were the most common. Hepatic encephalopathy occurred in 11.6% of the patients and was generally associated with hypokalemia, alkalosis, and a rising BUN. Increased ammonia secondary to hypokalemia may contribute to the encephalopathy.[139] Azotemia is often prerenal, but may be related to hypokalemia. In patients who are refractory to furosemide and require high doses, there is an increased likelihood of developing hepatorenal failure.[140]

B. Potassium-Sparing Diuretics

After conservative attempts to manage patients with excessive salt and water, a potassium-sparing agent should be employed in preference to a standard diuretic. A diuretic that acts on the distal tubule is generally required to reverse the pronounced sodium retention associated with cirrhosis. While spironolactone specifically antagonizes aldosterone and is generally preferred, triamterene and ameloride are also effective. In order to establish a diuresis, a simple clinical method using urinary sodium/potassium ratio as a guide has been devised.[141] Patients with a ratio greater than one responded to 100 mg/day of spironolactone; patients with a ratio less than one required doses of 200 mg or more per day. While this is likely to be effective in a larger percentage of patients, the possibility of adverse reactions remains.

Major problems with use of this class of diuretics include the development of hyperchloremic acidosis with bicarbonate levels in the range of ten. Hyperkalemia may occur particularly in patients receiving higher doses. It is necessary to monitor serum potassium in this situation as there have been fatalities reported secondary to hyperkalemia. If patients fail to respond to a potassium-sparing diuretic, then a compound such as furosemide or a thiazide diuretic could be administered. It is best to prescribe single doses and repeat them as necessary rather than begin a daily regimen that may be continued beyond a time when it is safe.

IX. PREGNANCY

Considerable controversy has existed regarding the use of diuretics in preeclampsia, hypertension, and pregnancy, and for the physiological and mechanical consequences of pregnancy. A discussion on sodium balance and diuretics in pregnancy has been reviewed.[142-145] Diuretic administration is associated with a decrease in placental perfusion;[146] however, except for rare occurrences of maternal pancreatitis and maternal and fetal thrombocytopenia, there is no convincing evidence that the judicious use of thiazides is dangerous. Caution has been advised regarding the use of sodium restriction and diuretic use in pregnancy,[147] and the American College of Obstetrics and Gynecology has concurred. Since animal reproductive studies with furosemide have demonstrated fetal developmental abnormalities, furosemide is contraindicated during pregnancy except in life-threatening situations.

A. Hypertension and Pregnancy

Hypertension during pregnancy is associated with fetal as well as maternal risk.[148] The use of diuretics has been fundamental, just as it is in the nonpregnant hypertensive. However, special care should be taken to avoid excessive hypovolemia. Beneficial effects have been reported on fetal outcome with methyldopa as a single antihypertensive agent.[149] Thus, in selected situations, diuretics may not be necessary. Thiazides are excreted in breast milk, and if their use is essential, the patient should discontinue nursing.

B. Eclampsia

1. Prevention

Numerous studies have assessed the role of diuretics in preeclampsia. Originally, it was felt that sodium retention was responsible for the vasoconstriction, and further, that it was amenable to diuretic therapy.[150,151] However, other attempts to demonstrate a benefit in the incidence and severity of toxemia have failed.[152] In a randomized double-blind fashion, the prophylactic use of hydrochlorothiazide in 1030 obstetrical patients was investigated, and it was found that the incidence of preeclampsia, hypertension, prematurity, and congenital abnormalities or perinatal mortality were unaffected by hydrochlorothiazide.[153] Further, evidence has been presented that intravenous saline and albumin returned central venous pressure to normal and restored renal function.[154] Thus, the bulk of the evidence does not support a beneficial effect of the prophylactic use of diuretics in pregnancy, and further, their use may alter the clinical findings important in the diagnosis of preeclampsia.[146]

2. Diuretic Treatment

The use of diuretics in eclampsia is not advisable, unless it is complicated by pulmonary edema, anasarca, chronic renal failure, or unresponsive hypertension. Pritchard and Pritchard reported on the successful treatment of 154 cases of eclampsia and did not feel there was a routine indication for the use of a diuretic.[155]

REFERENCES

1. **Vogl, A.,** The discovery of the organic mercurial diuretics, *Am. Heart J.,* 39, 881, 1950.
2. **Mann, T. and Keilin, D.,** Sulfanilamide as a specific inhibitor of cardiac anhydrase, *Nature,* 146, 164, 1940.
3. **Schwartz, W. B.,** The effect of sulfanilamide on salt and water excretion in congestive heart failure, *N. Engl. J. Med.,* 240, 173, 1949.
4. **Beyer, K. H., Baer, J. E., Russo, H. F., and Haimbach, A. S.,** Chlorothiazide (6-chloro-7-sulfamyl-1,2,4-benzothiadiazine-1-1-dioxide). The enhancement of sodium chloride excretion, *Fed. Proc.,* 16, 333, 1957.
5. **Novello, F. C. and Sprague, J. M.,** Benzothiadiazine dioxides as novel diuretics, *J. Am. Chem. Soc.,* 79, 2028, 1957.
6. **Finnerty, F. A., Jr., Davidov, M., Mroczek, W. J., and Gavrilovich, L.,** Influence of extracellular fluid volume on response to antihypertensive drugs, *Circ. Res.,* 26, 71, 1970.
7. **Baer, J. E. and Beyer, K. H.,** Subcellular pharmacology of natriuretic and potassium-sparing drugs, *Prog. Biochem. Pharmacol.,* 7, 59, 1972.
8. **Mass, A. R., Wiebelhaus, V. D., Sosnowski, G., Jenkins, B., and Gessner, G.,** Effect of triamterene on folic reductase activity and reproduction in the rat, *Toxicol. Appl. Pharmacol.,* 10, 413, 1967.
9. **Weiner, I. M., Levy, R. I., and Mudge, G. H.,** Studies on mercurial diuresis: renal excretion, acid stability and structure-activity relationships of organic mercurials, *J. Pharmacol. Exp. Ther.,* 138, 96, 1962.
10. **Earley, L. E. and Friedler, R. M.,** Observations on the mechanism of decreased tubular reabsorption of sodium and water during saline loading, *J. Clin. Invest.,* 43, 1928, 1964.
11. **Rapoport, S., Brodsky, W. A., West, C. D., and Mackler, B.,** Urinary flow and excretion of solutes during osmotic diuresis in hydropenic man, *Am. J. Physiol.,* 156, 433, 1949.
12. **Wesson, L. G., Jr. and Anslow, W. P.,** Excretion of sodium and water during osmotic diuresis in the dog, *Am. J. Physiol.,* 153, 465, 1948.
13. **Earley, L. E. and Friedler, R. M.,** Studies on the mechanism of natriuresis accompanying increased renal blood flow and its role in the renal response to extracellular volume expansion, *J. Clin. Invest.,* 44, 1857, 1965.
14. **Chrysant, S. G. and Lavender, A. R.,** Direct renal hemodynamic effects of two vasodilators: diazoxide and acetylcholine, *Arch. Int. Pharmacodyn. Ther.,* 217, 44, 1975.
15. **Rosin, J. M., Katz, M. A., Rector, F. C., Jr., and Seldin, D. W.,** Acetazolamide in studying sodium reabsorption in diluting segment, *Am. J. Physiol.,* 219, 1731, 1970.
16. **Puschett, J. B. and Kuhrman, M. A.,** Differential effects of diuretic agents on electrolyte excretion in the dog. Role of renal hemodynamics, *Nephron,* 23, 38, 1979.
17. **Schwartz, W. B., Falbriard, A., and Relman, A. S.,** An analysis of bicarbonate reabsorption during partial inhibition of carbonic anhydrase, *J. Clin. Invest.,* 37, 744, 1958.
18. **Pitts, R. F. and Alexander, R. S.,** The nature of the renal tubular mechanism for acidifying the urine, *Am. J. Physiol.,* 144, 239, 1945.
19. **Goldberg, M.,** The renal physiology of diuretics, in *Handbook for Physiology,* Sect. 8, American Physiological Society, Washington, D.C., 1973, 1003.
20. **Kunau, R. T., Jr.,** The influence of the carbonic anhydrase inhibitor, benzolamide (C1-11,366) on the reabsorption of chloride, sodium and bicarbonate in the proximal tubule of the rat, *J. Clin. Invest.,* 51, 294, 1972.
21. **Malvin, R. L. and Lottspeich, W. D.,** Relation between tubular transport and inorganic phosphate and bicarbonate in the dog, *Am. J. Physiol.,* 187, 51, 1956.
22. **Clapp, J. P. and Robinson, R. R.,** Distal sites of action of diuretic drugs in the dog nephron, *Am. J. Physiol.,* 215, 225, 1968.

23. **Lonnerholm, G. and Ridderstrale, Y.**, Distribution of carbonic anhydrase in the frog nephron, *Acta Physiol. Scand.*, 90, 764, 1974.
24. **Burg, M.**, Mechanism of action of diuretic drugs, in *The Kidney*, Vol. 1, Brenner, B. M. and Rector, F. C., Eds., W. B. Saunders, Philadelphia, 1976, chap. 19.
25. **Bernstein, B. A. and Clapp, J. R.**, Micropuncture study of bicarbonate reabsorption by the dog nephron, *Am. J. Physiol.*, 244, 251, 1968.
26. **Boudry, J., Stoner, L., and Burg, M.**, Effect of acid lumen pH on potassium transport in renal cortical collecting tubules, *Am. J. Physiol.*, 230, 239, 1976.
27. **Laragh, J. H., Cannon, P. J., Stason, W. B., and Heinemann, H. O.**, Physiological and clinical observations on furosemide and ethacrynic acid, *Ann. N.Y. Acad. Sci.*, 139, 453, 1966.
28. **Ludens, J. H., Hook, J. B., Brody, M. J. , and Williamson, H. E.**, Enhancement of renal blood flow by furosemide, *J. Pharmacol. Exp. Ther.*, 163, 456, 1968.
29. **Birtch, A. G., Zakheim, R. M., Jones, L. G., and Barger, A. C.**, Redistribution of renal blood flow produced by furosemide and ethacrynic acid, *Circ. Res.*, 21, 869, 1967.
30. **Barger, A. C.**, Renal hemodynamic factors in congestive heart failure, *Ann. N.Y. Acad. Sci.*, 139, 276, 1966.
31. **Bailie, M. O., Barbour, J. A., and Hook, J. G.**, Effects of indomethacin on furosemide-induced changes in renal blood flow, *Proc. Soc. Exp. Biol. Med.*, 148, 1173, 1975.
32. **Williamson, H. E., Bourland, W. A., and Marchand, G. R.**, Inhibition of ethacrynic acid-induced increase in renal blood flow by indomethacin, *Prostaglandins*, 8, 53, 1973.
33. **Williamson, H. E., Bourland, W. A., and Marchand, G. R.**, Inhibition of furosemide-induced increase in renal blood flow by indomethacin, *Proc. Soc. Exp. Biol. Med.*, 148, 164, 1975.
34. **Paulsrud, J. R. and Miller, O. N.**, Inhibition of 15-OH prostaglandin dehydrogenase by several diuretic drugs, *Fed. Proc.*, 33, 590, 1974.
35. **Chrysant, S. G.**, Renal functional changes induced by prostaglandin E_1 and indomethacin in the anesthetized dog, *Arch. Int. Pharmacodyn. Ther.*, 234, 156, 1978.
36. **Oliw, E., Kövér, G., Larsson, C., and Änggård, E.**, Reduction by indomethacin of furosemide effects in the rabbit, *Eur. J. Pharmacol.*, 38, 95, 1976.
37. **Miller, R. F., Amonette, R. L., Baxter, P. R., and Chrysant, S. G.**, The interaction of diuretics and prostaglandin synthesis inhibitors in the rat, *Fed. Proc.*, 38, 1439, 1979.
38. **Chrysant, S. G., Amonette, R. L., and Baxter, P. R.**, Renal effects of furosemide and indomethacin in volume expanded and non-volume expanded dogs, *Kidney Int.*, 16, 792, 1979.
39. **Chrysant, S. G., Amonette, R. L., and Baxter, P. R.** Direct renal hemodynamic effects of indomethacin, furosemide and prostaglandin E_1 in the anesthetized dog, *Fed. Proc.*, 39, 950, 1980.
40. **Ebel, H., Ehrich, J., Desanto, N. G., and Doerken, J.**, Plasma membranes of the kidney. III. Influence of diuretics on ATPase activity, *Pflügers Arch. Eur. J. Physiol.*, 335, 224, 1972.
41. **Klahr, S., Yates, J., and Bourgoignie, J.**, Inhibition of glycolysis by ethacrynic acid and furosemide, *Am. J. Physiol.*, 221, 1038, 1971.
42. **Cunarro, J. A. and Weiner, M. W.**, Effects of ethacrynic acid and furosemide on respiration of isolated kidney tubules: the role of ion transport and the source of metabolic energy, *J. Pharmacol. Exp. Ther.*, 206, 198, 1978.
43. **Quintanilla, A. P., Levin, M. L., Lastre, C. C., Yokoo, H., and Levin, N. W.**, Effect of diuretics on ADP incorporation in kidney mitochondria, *J. Pharmacol. Exp. Ther.*, 211, 456, 1979.
44. **Brenner, B. M., Keimowitz, R. I., Wright, F. S., and Berliner, R. W.**, An inhibitory effect of furosemide on sodium reabsorption in the proximal tubule of the rat nephron, *J. Clin. Invest.*, 48, 290, 1969.
45. **Clapp, J. R., Nottebohm, G. A., and Robinson, R. R.**, Proximal site of action of ethacrynic acid: importance of filtration rate, *Am. J. Physiol.*, 220, 1355, 1971.
46. **Alguire, P. C., Bailie, M. D., Weaver, W. J., Taylor, D. G., and Hook, J. B.**, Differential effects of furosemide and ethacrynic acid on electrolyte excretion in anesthetized dogs, *J. Pharmacol. Exp. Ther.*, 190, 515, 1974.
47. **Stein, J. H., Wilson, C. B., and Kirkendall, W. M.**, Differences in the acute effects of furosemide and ethacrynic acid in man, *J. Lab. Clin. Med.*, 71, 654, 1968.
48. **Eknoyan, G., Suki, W. N., Rector, F. C., Jr., and Seldin, D. W.**, Functional characteristics of the diluting segment of the dog nephron and the effect of extracellular volume expansion on its reabsorptive capacity, *J. Clin. Invest.*, 46, 1178, 1967.
49. **Puschett, J. B. and Goldberg, M.**, The acute effects of furosemide on acid and electrolyte excretion in man, *J. Lab. Clin. Med.*, 71, 666, 1968.
50. **Eknoyan, G., Suki, W. N., and Martinez-Maldonado, M.**, Effect of diuretics on urinary excretion of phosphate, calcium, and magnesium in thyroparathyroidectomized dogs, *J. Lab. Clin. Med.*, 76, 257, 1970.

51. **Goldberg, M., McCurdy, D. K., Foltz, E. L., and Bluemle, L. W., Jr.,** Effects of ethacrynic acid (a new saluretic agent) on renal diluting and concentrating mechanisms: evidence for site of action in the loop of Henle, *J. Clin. Invest.,* 43, 201, 1964.
52. **Fraser, A. G., Cowie, J. F., Lambie, A. T., and Robson, J. S.,** The effects of furosemide on the osmolality of the urine and the composition of renal tissue, *J. Pharmacol. Exp. Ther.,* 158, 475, 1967.
53. **Villareal, H., Revollo, A., Exaire, J. E., and Larrondo, F.,** Effects of chlorothiazide in renal hemodynamics, *Circulation,* 26, 409, 1962.
54. **Steinmuller, S. R. and Puschett, J. B.,** Effects of metolazone in man: comparison with chlorothiazide, *Kidney Int.,* 1, 169, 1972.
55. **Beyer, K. H.,** The mechanism of action of chlorothiazide, *Ann. N.Y. Acad. Sci.,* 71, 363, 1958.
56. **Kunau, R. T., Jr., Weller, D. R., and Webb, H. L.,** Clarification of the site of action of chlorothiazide in the rat nephron, *J. Clin. Invest.,* 56, 401, 1975.
57. **Suki, W. N., Hull, A. R., Rector, F. C., Jr., and Seldin, D. W.,** Mechanism of the effect of thiazide diuretics on calcium and uric acid, *J. Clin. Invest.,* 46, 1121, 1967.
58. **Earley, L. E. and Orloff, J.,** The mechanism of antidiuresis associated with the administration of hydrochlorothiazide to patients with vasopressin-resistant diabetes insipidus, *J. Clin. Invest.,* 41, 1988, 1962.
59. **Yendt, E. R., Gagné, R. J. A., and Cohanim, M.,** The effect of thiazides in idiopathic hypercalciuria, *Am. J. Med. Sci.,* 251, 449, 1966.
60. **Ullrich, K. J., Baumann, K., Loeschke, K., Rumrich, G., and Stolte, H.,** Micropuncture experiments with saluretic sulfonamides, *Ann. N.Y. Acad. Sci.,* 139, 416, 1966.
61. **Earley, L. E., Kahn, M., and Orloff, J.,** The effects of infusion of chlorothiazide on urinary dilution and concentration in the dog, *J. Clin. Invest.,* 40, 857, 1961.
62. **Cafruny, E. J. and Ross, C.,** Involvement of the distal tubule in diuresis produced by benzothiadiazines, *J. Pharmacol. Exp. Ther.,* 137, 324, 1962.
63. **Fernandez, P. C. and Puschett, J. B.,** Proximal tubular actions of metolazone and chlorothiazide *Am. J. Physiol.,* 225, 954, 1973.
64. **Rector, F. C., Jr. and Clapp, J. R.,** Evidence for active chloride reabsorption in the distal renal tubule of the rat, *J. Clin. Invest.,* 41, 101, 1962.
65. **Liddle, G. W.,** Sodium diuresis induced by steroidal antagonists of aldosterone, *Science,* 126, 1016, 1957.
66. **Baba, W. I., Tudhope, G. R., and Wilson, G. M.,** Site and mechanism of action of the diuretic triamterene, *Clin. Sci.,* 27, 181, 1964.
67. **Bull, M. and Laragh, J. H.,** Amiloride, a potassium-sparing natriuretic agent, *Circulation,* 37, 45, 1968.
68. **Ginsberg, D. J., Saad, A., and Gubuzda, G. J.,** Metabolic studies with the diuretic triamterene in patients with cirrhosis and ascites, *N. Engl. J. Med.* 271, 1229, 1964.
69. **Chrysant, S. G. and Luu, T. M.,** Effects of amiloride on arterial pressure and renal function, *J. Clin. Pharmacol.,* 20, 332, 1980.
70. **Duarte, C. G., Chomety, F., and Giebisch, G.,** Effect of amiloride, ouabain and furosemide on distal tubular function in the rat, *Am. J. Physiol.,* 221, 632, 1971.
71. **Vander, A. J., Wilde, W. S., and Malvin, R. L.,** Stop-flow analysis of aldosterone and steroidal antagonist SC 8109 on renal tubular transport kinetics, *Proc. Soc. Exp. Biol. Med.,* 103, 525, 1960.
72. **Kagawa, C. M., Sturtevant, F. M., and van Arman, C. G.,** Pharmacology of a new steroid that blocks salt activity of aldosterone and deoxycorticosterone, *J. Pharmacol. Exp. Ther.,* 126, 123, 1959.
73. **Liddle, G. W.,** Aldosterone antagonists, *Arch. Int. Med.,* 102, 998, 1958.
74. **Edelman, I. S., Bogoroch, R., and Porter, G. A.,** On the mechanism of action of aldosterone on sodium transport: the role of protein synthesis, *Proc. Natl. Acad. Sci. U.S.A.,* 50, 1169, 1963.
75. **Stoner, L., Burg, M., and Orloff, J.,** Ion transport in cortical collecting tubule; effect of amiloride, *Am. J. Physiol.,* 227, 453, 1974.
76. **Crabbe, J.,** A hypothesis concerning the mode of action of amiloride and triamterene, *Arch. Int. Pharmacodyn. Ther.,* 173, 474, 1968.
77. **Beerman, B. and Groschinsky-Grind, M.,** Pharmacokinetics of hydrochlorothiazide in man, *Eur. J. Clin. Pharmacol.,* 12, 297, 1977.
78. **McMahon, F. G.,** Furosemide, in *Management of Essential Hypertension,* Futura Publishing, New York, 1978, chap. 2.
79. **Cutler, R. E. and Blair, A. D.,** Clinical pharmacokinetics of frusemide, *Clin. Pharmacokinetics,* 4, 279, 1979.
80. **Kelly, M. R., Cutler, R. E., Forrey, A. W., and Kimpel, B. M.,** Pharmacokinetics of orally administered furosemide, *Clin. Pharmacol. Ther.,* 15, 178, 1974.

81. **Greither, A., Goldman, S., Edelen, J. S., Cohn, K., and Benet, L. Z.,** Erratic and incomplete absorption of furosemide in congestive heart failure, *Am. J. Cardiol.*, 37, 139, 1976.
82. **Kelly, M. R., Blair, A. D., Forrey, A. W., Smidt, N. A., and Cutler, R. E.,** A comparison of the diuretic response to oral and intravenous furosemide in 'diuretic-resistant' patients, *Current Therapeutic Research: Clinical and Experimental*, 21, 1, 1977.
83. **Tilstone, W. J. and Fine, A.,** Furosemide kinetics in renal failure, *Clin. Pharmacol. Ther.*, 23, 644, 1978.
84. **Prandota, J. and Pruitt, A. W.,** Furosemide binding to human albumin and plasma of nephrotic children, *Clin. Pharmacol. Ther.*, 17, 159, 1975.
85. **Gundert-Remy, U., von Kenne, D., Weber, E., Geissler, H. E., and Mutschler, E.,** Plasma and urinary levels of triamterene and certain metabolites after oral administration to man, *Eur. J. Clin. Pharmacol.*, 16, 39, 1979.
86. **Ochs, H. R., Greenblatt, D. J., Bodem, G., and Smith, T. W.,** Spironolactone, *Am. Heart J.*, 96, 389, 1978.
87. **Karim, A.,** Spironolactone: disposition, metabolism, pharmacodynamics, and bioavailability, *Drug Metab. Rev.*, 8, 151, 1978.
88. **Freis, E. D.,** Salt in hypertension and the effects of diuretics, in *Annual Review of Pharmacology and Toxicology*, Vol. 19, 1st ed., George, R., Okun, R., and Cho, A. K., Eds., Annual Reviews, Palo Alto, Calif., 1979, 13.
89. **Begilow, R. S., Pollack, H., Stollerman, G. H., Roston, E. H., and Bookman, J. J.,** The therapy of hypertension by accelerated sodium depletion, *J. Mount Sinai Hosp. (N.Y.)*, 15, 233, 1948.
90. **Hollander, W. and Wilkins, R. W.,** Chlorothiazide: a new type of drug for the therapy of arterial hypertension, *Boston Med. Q.*, 8, 69, 1957.
91. **Freis, E. D., Wanko, A., Wilson, I. M., and Parrish, A. E.,** Treatment of essential hypertension with chlorothiazide (Diuril), *JAMA*, 166, 137, 1957.
92. **Frohlich, E., Schnaper, H. W., Wilson, I. M., and Freis, E. D.,** Hemodynamic alterations in hypertensive patients due to chlorothiazide, *N. Engl. J. Med.*, 262, 1261, 1960.
93. **Tarazi, R. C., Dustan, H. P., and Frohlich, E. D.,** Long-term thiazide therapy in essential hypertension: evidence for persistent alteration in plasma volume and renin activity, *Circulation*, 41, 709, 1970.
94. **Shah, S., Khatri, I., and Freis, E. D.,** Mechanism of antihypertensive effect of thiazide diuretics, *Am. Heart J.*, 95, 611, 1978.
95. **Tobian, L., Jr.,** How sodium and the kidney relate to the hypertensive arteriole, *Fed. Proc.*, 33, 138, 1975.
96. **Webster, J., Dollery, C. T., Hensby, C. H., and Friedman, L. A.,** Anti-hypertensive action of bendroflumethiazide: increased prostacyclin production? *Clin. Pharmacol. Ther.*, 28, 751, 1980.
97. **Laragh, J. H., Baer, L., Brunner, H. R., Buhler, F. R., Sealey, J. E., and Vaughn, E. D., Jr.,** Renin, angiotensin and aldosterone system in pathogenesis and management of hypertensive vascular disease, *Am. J. Med.*, 52, 633, 1972.
98. **Dustin, H. P.,** Evaluation and therapy of hypertension, *Mod. Concepts Cardiovasc. Dis.*, 45, 97, 1976.
99. **van Brummelen, P., Man in't Veld, A. J., and Schalekamp, M. A. D. H.,** Hemodynamic changes during long-term thiazide treatment of essential hypertension in responders and nonresponders, *Clin. Pharmacol. Ther.*, 27, 328, 1980.
100. **Aleksandrow, D., Wyszacka, W., and Gajewski, J.,** Influence of chlorothiazide upon arterial responsiveness to nor-epinephrine in hypertensive subjects, *N. Engl. J. Med.*, 261, 1052, 1959.
101. **Tobian, L., Janecek, J., Foker, J., and Ferreira, D.,** Effect of chlorothiazide on renal juxtaglomerular cells and tissue electrolytes, *Am. J. Physiol.*, 202, 905, 1962.
102. **Freis, E. D., Wanko, A., Schnaper, H. W., and Frohlich, E. D.,** Mechanism of the altered blood pressure responsiveness produced by chlorothiazide, *J. Clin. Invest.*, 39, 1277, 1960.
103. **Weinberger, M. H., Ramsdell, J. W., Rosner, D. R., and Geddes, J. J. L.,** Effect of chlorothiazide and sodium on vascular responsiveness to angiotensin II, *Am. J. Physiol.*, 223, 1049, 1972.
104. **Wilson, I. M. and Freis, E. D.,** Relationship between plasma and extracellular fluid volume depletion and antihypertension effect of chlorothiazide, *Circulation*, 20, 1028, 1959.
105. **Bennett, W. M., McDonald, W. J., Kuehnel, E., Hartnett, M. N., and Porter, G. A.,** Do diuretics have antihypertensive properties independent of natriuresis?, *Clin. Pharmacol. Ther.*, 22, 499, 1977.
106. **McMahon, F. G.,** Thiazides, in *Management of Essential Hypertension*, Futura Publishing, New York, 1978, chap. 2.
107. **Beevers, D. G., Hamilton, M., and Harpur, J. E.,** The long-term treatment of hypertension with thiazide diuretics, *Postgrad. Med. J.*, 47, 639, 1971.
108. **Berglund, G., Andersson, O., Larsson, O., and Wilhelmsen, L.,** Antihypertensive effect and side-effects of bendroflumethiazide and propranolol, *Acta Med. Scand.*, 199, 499, 1976.

109. **Beerman, N. B. and Groschinsky-Grind, M.,** Antihypertensive effect of various doses of hydrochlorothiazide and its relation to the plasma level of the drug, *Eur. J. Clin. Pharmacol.,* 13, 195, 1978.
110. **Materson, B. J., Oster, J. R., Michael, U. F., Bolton, S. M., Burton, Z. C., Stambaugh, J. E., and Morledge, J.,** Dose response to chlorthalidone in patients with mild hypertension, *Clin. Pharmacol. Ther.,* 24, 192, 1978.
111. **Anderson, J., Godfrey, B. E., Hill, D. M., Munro-Faure, A. D., and Sheldon, J.,** A comparison of the effects of hydrochlorothiazide and of furosemide in the treatment of hypertensive patients, *Q. J. Med.,* 40, 541, 1971.
112. **McMahon, F. G.,** Furosemide, in *Management of Essential Hypertension,* Futura Publishing, New York, 1978, chap. 3.
113. **Gifford, R. W. and Westbrook, E.,** Hypertensive encephalopathy: mechanisms, clinical features, and treatment, *Prog. Cardiovasc. Dis.,* 17, 115, 1974.
114. **Muth, R. G.,** Diuretic properties of furosemide in renal disease, *Ann. Intern. Med.,* 69, 249, 1968.
115. **Wilson, M., Morgan, T., and Gillies, A.,** A role of furosemide in resistant hypertension, *Med. J. Aust.,* 1, 213, 1977.
116. **Bravo, E. L., Tarazi, R. C., and Dustan, H. P.,** Spironolactone in primary aldosteronism, *Clin. Pharmacol. Ther.,* 15, 201, 1974.
117. **Simpson, F. O. and Waal, H. J.,** Clinical trial of triamterene in hypertensive patients, *N. Z. Med. J.,* 63, 199, 1964.
118. **Gombos, E. A., Freis, E. D., and Moghadam, A.,** Effects of MK-870 in normal subjects and hypertensive patients, *N. Engl. J. Med.,* 275, 1215, 1966.
119. **Nissenson, A. R., Weston, R. E., and Kleeman, C. R.,** Mannitol, *West J. Med.,* 131, 277, 1979.
120. **Sherwood, T., Lavender, J. P., and Russell, S. B.,** Mercury-induced renal vascular shut-down: observations in experimental acute renal failure, *Eur. J. Clin. Invest.,* 4, 1, 1974.
121. **Lucas, C. E., Zito, J. G., Carter, K. M., Cortez, A., and Stebner, F. C.,** Questionable value of furosemide in preventing renal failure, *Surgery,* 82, 134, 1977.
122. **Silverberg, D. S., Ulan, R. A., Baltzan, M. A., and Baltzan, R. B.,** Experiences with high doses of furosemide in renal disease and resistant edematous states, *Can. Med. Assoc. J.,* 103, 129, 1970.
123. **Muth, R. G.,** Diuretics in chronic renal insufficiency, in *Modern Diuretic Therapy in the Treatment of Cardiovascular and Renal Disease,* Excerpta Medica, Amsterdam, 1973, 294.
124. **Spilkin, E. S. and Weller, J. M.,** Effect of furosemide in patients undergoing chronic intermittent peritoneal dialysis, *Postgrad. Med. J.,* 47 (Suppl. 3), 36, 1971.
125. **Sharma, B. K., Schuster, G., Quintanilla, A., and Levin, N. W.,** Improvement of renal function in uraemic patients by furosemide and salt, *Ind. J. Med. Res.,* 67, 406, 1978.
126. **Tattersfield, A. E. and McNicol, M. W.,** Diuretics in acute myocardial infarction, *Clin. Sci.,* 38, 32, 1970.
127. **Biagi, R. W. and Bapat, B. N.,** Furosemide in acute pulmonary edema, *Lancet,* 1, 849, 1967.
128. **Dikshit, K., Vyden, J. K., Forrester, J. S., Chatterjee, K., Prakash, R., and Swan, H. J. C.,** Renal and extrarenal hemodynamic effects of furosemide in congestive heart failure after acute myocardial infarction, *N. Engl. J. Med.,* 288, 1087, 1973.
129. **Demling, R. H. and Will, J. A.,** The effect of furosemide on the pulmonary transvascular fluid filtration rate, *Crit. Care Med.,* 6, 317, 1978.
130. **Rader, B., Smith, W. W., Berger, A. R., and Eichna, L. W.,** Comparison of the hemodynamic effects of mercurial diuretics and digitalis in congestive heart failure, *Circulation,* 29, 328, 1964.
131. **Stampfer, M., Epstein, S. E., Beiser, G. D., and Braunwald, E.,** Hemodynamic effects of diuresis at rest and during intense upright exercise in patients with impaired cardiac function, *Circulation,* 37, 900, 1968.
132. **Klingler, E. L., Jr., Vaamonde, C. A., Vaamonde, L. S., Lancestremere, R., Morosi, H. J., Frisch, E., and Papper, S.,** Renal function changes in cirrhosis of the liver, *Arch. Intern. Med.,* 125, 1010, 1970.
133. **Levinsky, N. G.,** Refractory ascites in cirrhosis, *Kidney Int.,* 14, 93, 1978.
134. **Rosoff, L., Jr., Zia, P., Reynolds, T., and Horton, R.,** Studies of renin and aldosterone in cirrhotic patients with ascites, *Gastroenterology,* 69, 698, 1975.
135. **Lieberman, F. L., Denison, E. K., and Reynolds, T. B.,** The relationship of plasma volume, portal hypertension, ascites, and renal sodium retention in cirrhosis: the overflow theory of ascites formation, *Ann. N.Y. Acad. Sci.,* 170, 202, 1970.
136. **Schrier, R. W., Lehman, D., Zacherle, B., and Earley, L. E.,** Effect of furosemide on free water excretion in edematous patients with hyponatremia, *Kidney Int.,* 3, 30, 1973.
137. **Linas, S. L., Anderson, R. J., Miller, P. D., and Schrier, R. W.,** Rational use of diuretics in cirrhosis, in *The Kidney in Liver Disease,* 1st ed., Epstein, M., Ed., Elsevier North-Holland, New York, 1978, 313.
138. **Naranjo, C. A., Pontigo, E., Valdenegro, C., Gonzalez, G., Ruiz, I., and Busto, Y.,** Furosemide-induced adverse reactions in cirrhosis of the liver, *Clin. Pharmacol. Ther.,* 25, 154, 1979.

139. **Kew, M.**, Progress report on renal changes in cirrhosis, *Gut*, 13, 748, 1972.
140. **Gabuzda, G. J. and Hall, P. W., III.** Relation of potassium depletion to renal ammonium metabolism and hepatic coma, *Medicine*, 45, 481, 1966.
141. **Eggert, R. C.**, Spironolactone diuresis in patients with cirrhosis and ascites, *Br. Med J.*, 4, 401, 1970.
142. **Lindheimer, M. D. and Katz, A. I.**, Sodium and diuretics in pregnancy, *N. Engl. J. Med.*, 288, 891, 1973.
143. **Kincaid-Smith, P.**, The use of diuretics in pregnancy, in *Modern Diuretic Therapy in the Treatment of Cardiovascular and Renal Disease*, Excerpta Medica, Amsterdam, 1973, 281.
144. **Pritchard, J. A.**, Management of preeclampsia and eclampsia, *Kidney Int.*, 18, 259, 1980.
145. **Redman, C. W. G.**, Treatment of hypertension in pregnancy, *Kidney Int.*, 18, 267, 1980.
146. **Gant, N. F., Madden, J. D., Siiteri, P. K., and MacDonald, P. C.**, The metabolic clearance rate of dehydroisoandrosterone sulfate, *Am. J. Obstet. Gynecol.*, 123, 159, 1975.
147. **Pike, R. L. and Smiciklas, H. A.**, A reappraisal of sodium restriction during pregnancy, *Int. J. Gynecol. Obstet.*, 10, 1, 1972.
148. **Roberts, J. M. and Perloff, D. L.**, Hypertension and the obstetrician-gynecologist, *Am. J. Obstet. Gynecol.*, 127, 316, 1977.
149. **Redman, C. W. G., Beilin, L. J., Bonnar, J., and Ounsted, M. K.**, Fetal outcome in trial of antihypertensive treatment in pregnancy, *Lancet*, 2, 753, 1976.
150. **Finnerty, F. A., Jr. and Bepko, F. J., Jr.**, Lowering the perinatal mortality and the prematurity rate: the value of prophylactic thiazides in juveniles, *JAMA*, 195, 429, 1966.
151. **Finnerty, F. A., Jr.**, How to treat toxemia of pregnancy, *GP*, 27, 116, 1963.
152. **Flowers, C. E., Grizzle, J. E., Easterling, W. E., and Bonner, O. B.**, Chlorothiazide as a prophylaxis against toxemia of pregnancy, *Am. J. Obstet. Gynecol.*, 84, 919, 1962.
153. **Kraus, G. W., Marchese, J. R., and Yen, S. S. C.**, Prophylactic use of hydrochlorothiazide in pregnancy, *JAMA*, 198, 1150, 1966.
154. **Maclean, A. B., Doig, J. R., and Aickin, D. R.**, Hypovolaemia, preeclampsia and diuretics, *Br. J. Obstet. Gynecol.*, 85, 597, 1978.
155. **Pritchard, J. A. and Pritchard, S. A.**, Standardized treatment of 154 consecutive cases of eclampsia, *Am. J. Obstet. Gynecol.*, 123, 543, 1975.

Sodium Depletion

Chapter 10

SODIUM DEPLETION

Carlos A. Vaamonde

TABLE OF CONTENTS

I. INTRODUCTION

The maintenance of extracellular fluid (ECF) and plasma volumes (PV) is essential for the support of the circulation and survival of the organism. Many regulatory systems are responsible for the homeostasis of the ECF volume. These mechanisms allow the body's cells to have an environment of remarkable constancy, which in the words of Claude Bernard, is the fundamental "condition for free life".

Under normal circumstances, the kidney compensates for extrarenal salt and water losses (or gains) with great precision, maintaining the ECF volume within narrow limits. When abnormal conditions prevail, the augmented or markedly aberrant extrarenal losses of salt and water will set in motion an array of systemic and renal compensatory responses leading to the restoration of the circulation and ECF volume. On many occasions, however, the kidney itself is the responsible organ as in chronic renal failure, or renal salt-wasting states. In others (abuse of diuretics), the kidney is an important causative factor in the genesis of the perturbation of volume homeostasis.

Extracellular fluid sodium and the accompanying anions chloride and bicarbonate constitute over 90% of the ECF solutes. Thus, the ECF osmolality depends on the concentration of sodium and its attendant anions. Perturbations of sodium ion homeostasis result in alterations of ECF osmolality and water shifts to or from cells, finally leading to ECF volume alterations.

Sodium depletion states are frequently encountered clinical problems and require a clear understanding of their causative factors, of the pathogenesis of the body's homeostatic responses, and of their diagnosis, if treatment and restoration of volume homeostasis are to be accomplished with rapidity and efficiency minimizing morbidity and mortality.

This chapter describes the following: the normal response to sodium depletion; clinical characteristics and laboratory findings; classification; and the sodium depletion states and their treatment.

The definition of certain terms is perhaps in order.

Sodium depletion refers to an actual loss of sodium from the body. This can be directly quantified (diarrhea, drainages, diuresis, etc.) or the deficit estimated by the measurement of exchangeable total body sodium (approximately 45 to 50 mEq/kg body weight (BW) in adult men and 35 to 40 mEq/kg BW in adult women). The first method is frequently used in hospitalized patients, while the latter is seldom employed in clinical medicine except for research purposes. In practice, sodium depletion is estimated by clinical evaluation of the consequences of volume loss: circulatory status, orthostatic hypotension and tachycardia, signs of dehydration, loss of body weight, etc. The terms *depletion, deficiency,* and *deficit* will be used interchangeably in this chapter.

Dehydration is perhaps an unfortunate term.[1] Literally, dehydration refers to the loss of water from the body and for this circumstance "water deficit" is perhaps a better term. Dehydration, however, as it is clinically used refers to the clinical consequences of deficits of water or salt or more commonly both. The literal interpretation of the term dehydration thus can lead to an erroneous clinical assessment of a patient's body fluid alteration and may result in improper and inadequate therapy. Therefore precise terminology should be used to characterize body fluid abnormalities.

Sodium depletion frequently results in hyponatremia. Normal serum sodium concentration and even hypernatremia can be associated with sodium depletion as well. The relationship of the body stores of sodium to its concentration in the ECF volume depends not only on the actual losses of sodium, but also on the concomitant losses of water and on the type of fluid replacement of the volume deficit incurred. Thus, sodium depletion and hyponatremia are not to be considered as synonymous terms. In fact, the great majority of the hyponatremic states (to be discussed in a subsequent chapter) observed in medicine are of the dilutional type and are associated with normal or excessive body sodium stores.

II. THE NORMAL RESPONSE TO SODIUM DEPLETION

Sodium depletion results in cardiovascular, renal, and to some limited extent in other extrarenal homeostatic efforts to sustain the circulation, minimize additional losses, and restore volume. The cardiovascular responses are initial and rapid and attempt to maintain a normal blood pressure, while the kidney has a dominant role in the prevention of further losses and restoration of the volume deficit.

A. Cardiovascular Homeostatic Response to Sodium Depletion

Maintenance of an effective circulation is the immediate goal of the cardiovascular adjustments resulting from sodium depletion. It is well established that the hemodynamic consequences of sodium depletion are dependent on the *magnitude* and the *rapidity* of the appearance of the sodium deficit.

The classical studies of Elkinton, Danowski, and co-workers[2,3] have demonstrated that acute severe sodium depletion (greater than 300 to 500 mEq) results in marked contraction of the ECF volume (20 to 40%), significant reduction in mean blood pressure (30 to 80%) and cardiac output (60 to 80%), and in a 50 to 200% increase in total peripheral vascular resistance (PVR). There is tachycardia, prolongation of the circulation time, and decreased jugular venous pressure as well.[2,3]

These systemic hemodynamic changes result from the rapid decrease in ECF osmolality, the direct result of the sodium chloride loss. This is followed by movement of ECF water into the cells, which is soon followed by a decline in PV. As part of the systemic peripheral vasoconstriction, the renal circulation is also affected and decreased renal function, oliguria, and azotemia may appear.

On the other extreme, sodium depletion not exceeding 200 mEq of sodium and occurring within a few days to a week results in a much lesser circulatory impact.[4] There is an increase in pulse rate and a slight to modest decrease in blood pressure when changing to the upright posture (orthostatic test for ECF volume depletion).

When marked sodium depletion (such as a loss of 500 mEq or more) occurs chronically, as the classical experiments of McCance[5] have shown, the circulatory hemodynamic changes are much less impressive (tachycardia and modest decreases in blood pressure following postural change). The latter observation has two important practical clinical implications. The first is that the circulation can be effectively maintained in the presence of large sodium losses and ECF volume depletion, if the incurred losses are sustained in a chronic manner. The second one is that the presence of mild orthostatic increases in pulse rate and decline in arterial blood pressure indicates *either* moderate acute sodium loss or severe chronic sodium deficit.

Extensive studies in recent years have characterized the important pathogenetic roles of the renin-angiotensin and autonomic nervous systems in the mediation of the circulatory responses to sodium depletion.

The *renin-angiotensin* system plays a central role in the support of the blood pressure in sodium depletion. Plasma renin activity (PRA) has been found elevated in sodium depletion in man and experimental animals.[4,6-11] As expected, angiotensin II levels have also been found to be elevated in sodium depleted states.[12-14] It is not known yet if angiotensin III (a polypeptide fragment generated in vivo from angiotensin II[15]) may mediate the response of angiotensin II and thus also be of significance in the cardiovascular homeostatic response to sodium deficit.

The importance of the physiologic role of angiotensin II in blood pressure homeostasis has been underscored by recent studies in salt depleted humans and animals. The initial studies demonstrated a significant reduction in blood pressure and PVR when either an inhibitor of the converting enzyme (the enzyme that converts angiotensin I to angiotensin II) or an inhibitor of the vasoconstrictive effect of angiotensin II were given to anesthetized animals.[9,16,17] The hypotensive effect was not seen in the sodium-replete control animals. Subsequent studies in nonanesthetized man and animals have confirmed the above observations.[10,11,18,19] Indeed, even in other states characterized by enhanced PRA and angiotensin II levels (decompensated cirrhosis), the blockade of angiotensin II by saralasin administration produced a rapid, marked, and reversible fall in blood pressure in salt-restricted patients with cirrhosis.[20] This study provides evidence that some patients with cirrhosis, particularly when depleted of salt, might be extremely sensitive from the circulatory standpoint to interference with the supporting vascular effects of the renin-angiotensin system.

The cause of activation of the renin-angiotensin system in sodium depletion with its resultant increased peripheral vasoconstriction is not completely known. The possible mechanisms resulting in increased renin secretion include the stimulation of an intrarenal vascular receptor, the involvement of catecholamines, or alterations in the delivery of sodium to, or its transport at, the macula densa locus.

A catecholamine-mediated renal nerve stimulation of renin secretion seems unlikely since PRA increases in denervated kidneys of sodium-depleted man[21] or dog.[22] Adrenergic stimulation, as a mediator of renin secretion (see below), however, cannot be disregarded in sodium depletion.[23] The presence of renin release following hemorrhage or suprarenal aortic constriction in the dogs with abolished sodium delivery to the macula densa reported by Blaine et al.[24] appears to minimize the role of the macula densa mechanism. Therefore, it appears that the available data support the stimulation of an intrarenal vascular receptor (since the renin response to sodium depletion can be abolished by papaverine[25]) as the most important mechanism for enhanced renin secretion in states of sodium deficit.

There is considerable evidence supporting the *activation of the adrenergic system* in sodium depletion. Not only are there increases in plasma levels and the urinary excretion rate of norepinephrine,[26-28] but also dopamine beta hydroxylase[29] (the enzyme that converts dopamine to norepinephrine in the adrenal medulla and sympathetic nerve endings) is elevated in sodium depletion.

In patients with idiopathic postural hypotension,[30-32] mild ECF volume contraction results in greater falls in arterial blood pressure than in control subjects volume-depleted to the same extent. Likewise, a similar response was reported in normal man treated with guanethidine, a catecholamine depleting drug.[33]

Figure 1 outlines the interrelationships between the renin-angiotensin and adrenergic systems in their efforts to restore arterial blood pressure in sodium depletion. It can also be appreciated that stimulation of the adrenergic system can itself stimulate renin secretion,[23] and that angiotensin II can influence the adrenergic system by facilitating the release of norepinephrine at the sympathetic nerve endings[34] or by enhancing the peripheral vasoconstrictor response of the catecholamines.[35]

B. Renal Response to Sodium Depletion

Because the kidney is the most important organ in the regulation of sodium excretion, its participation is crucial in the normal homeostatic response to sodium depletion.

1. Renal Hemodynamics

Moderate to severe sodium depletion is associated with a decrease in renal blood flow (RBF), renal plasma flow (RPF), and glomerular filtration rate (GFR), while

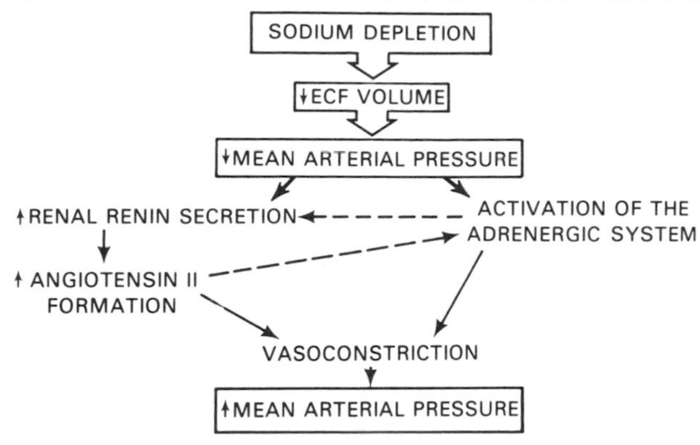

FIGURE 1. Circulatory homeostatic response in sodium depletion. Outline of the factors that are activated to maintain arterial blood pressure in sodium depletion. Note the interrelationships between the renin-angiotensin and adrenergic systems (---).

filtration fraction (FF) and renal vascular resistance (RVR) increase.[4,36-39] Sodium restriction in man results in a preferential decrease of renal blood flow through cortical nephrons.[39a] Mild sodium deficiency, on the other hand, results only in small but significant decreases in renal hemodynamics.[4,36-39] Lesser sodium losses (not exceeding 175 mEq in 5 days), however, did not significantly reduce GFR in middle-aged normal volunteers who exhibited a normal renal response to dietary sodium restriction studied in our laboratory.[40] In contrast, when a single dose of furosemide was given to salt restricted young normal volunteers, we observed a small but significant decrease in GFR (138 ± 9 to 125 ± 12 mℓ/min; $p < 0.025$). The cumulative sodium loss in these subjects, however, exceeded 200 mEq of sodium (213 ± 34 mEq in 4 days).[41]

2. *Renal Excretion of Sodium*

The normal renal response to sodium deficiency is a rapid and predictable decrease in sodium excretion. Several investigators have described in normal man an exponential decrease in urinary sodium excretion in response to sodium restriction.[40,42-48] When normal volunteers are placed on a low sodium intake renal sodium excretion decreases each successive day to about one half the value of the preceding day (Figure 2). It follows that, depending on the baseline level of sodium intake and excretion, sodium balance will be achieved within 3 to 5 days of salt restriction and the sodium loss will vary accordingly. For example, let us assume that two normal individuals have similar renal capacity for sodium conservation, and are in balance at baseline taking different sodium intakes (100 and 240 mEq/day, respectively). The response of these individuals to dietary sodium restriction (10 mEq/day) will differ. The first subject will achieve sodium balance by day three and his cumulative sodium loss will be less than 60 mEq/day); on the other hand, the second subject will have a cumulative sodium loss of 185 mEq and his urinary sodium will match intake by day five of dietary sodium restriction.

When the sodium loss occurs more rapidly (dietary sodium restriction plus diuretics[45] or induced sweating[5]) the renal conservation of sodium is also achieved more rapidly. Age is known to influence the renal conservation of sodium.[47] For example, the half-time for the reduction in renal sodium excretion from baseline in subjects under age 30 years was 18 ± 1 hr (N = 34), significantly faster than for subjects aged 30 to 59 years (23 ± 1 hr, N = 47), and for those over 60 years of age (31 ± 3 hr, N = 8).[47] Thus, older patients may respond to salt restriction at a rate of half of that observed in young subjects.

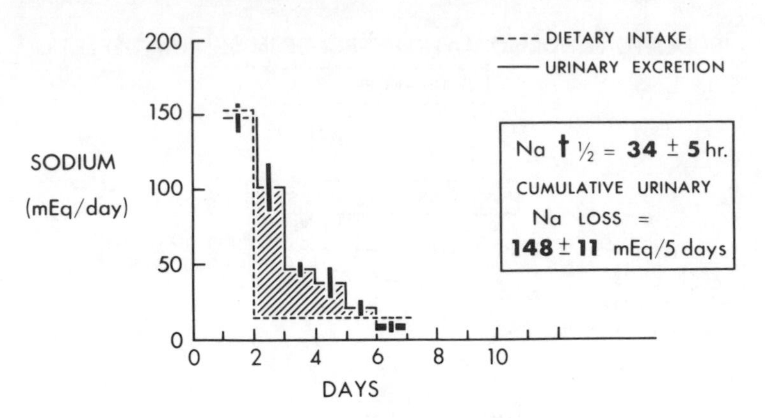

FIGURE 2. Renal response to dietary sodium restriction (10 mEq/day) in four normal male volunteers (range of GFR 99 to 126 mℓ/min) aged 48 to 53 years. Note that these patients were in sodium balance (intake of 150 mEq/day) prior to salt restriction, and that on the average they achieved sodium balance in 4 days (Sodium t ½ 34 ± 5 hr). The shaded area represents urinary sodium loss in excess of dietary intake. The vertical bars represent 1 ± SE. Their cumulative sodium loss ranged from 127 to 173 mEq/5 days. (Adapted from Vaamonde, C. A., Oster, J. R., Lohavichan, C., Carroll, K. E., Jr., Sebastianelli, M. J., and Papper, S., *Proc. Soc. Exp. Biol. Med.*, 146, 936, 1974. With permission.)

The mechanism(s) responsible for the decrease in sodium excretion that follows sodium loss from the body have been the subject of extensive investigation and, although a clear picture has emerged in recent years, some uncertainties remain. Consideration should be given to a decreased filtered load of sodium, stimulation of the renin-angiotensin-aldosterone system, peritubular capillary physical forces (increase in FF and RVR), and activation of the adrenergic system.

Since renal sodium conservation begins early in the course of sodium depletion, at a time when systemic (blood pressure and cardiac output) and renal hemodynamics and ECF volume may be still normal or only minimally decreased,[4,36-39,45] it appears that the initial decrease in renal sodium excretion may occur with an unchanged *filtered load of sodium*. This, or course, points towards mechanism(s) of sodium conservation resulting from *increased tubular sodium reabsorption*.

Several segments of the nephron have been shown to exhibit an increase in the transport capacity for sodium in experimental sodium deficiency.[49-53] In the rat and dog with severe or moderated sodium depletion enhanced sodium reabsorption was demonstrated by micropuncture or clearance studies in the proximal tubule,[49-52] ascending limb of Henle's loop[52] and collecting duct.[50,53] Since in mild sodium depletion some investigators[54] did not find increased fractional sodium reabsorption in the proximal tubule, it is possible that the involvement of different nephron segments in the sodium conservation response may be chronologically different depending on the rapidity and severity of the sodium loss.

What are the factors that influence the renal tubules to increase sodium reabsorption in sodium depletion?

A great deal of experimental and clinical evidence in this context supports a prominent role for *aldosterone* in renal sodium conservation. Sodium restriction is accompanied by an increase in plasma aldosterone concentration or urinary excretion rates.[12,13,55,56] Clinically, it is well known that patients with Addison's disease cannot reduce renal sodium excretion to the minimal levels as normal subjects do when sodium depleted.[57-59] Likewise, normal volunteers treated with spironolactone also show an impaired renal sodium conservation ability.[46] The increased aldosterone secretion most

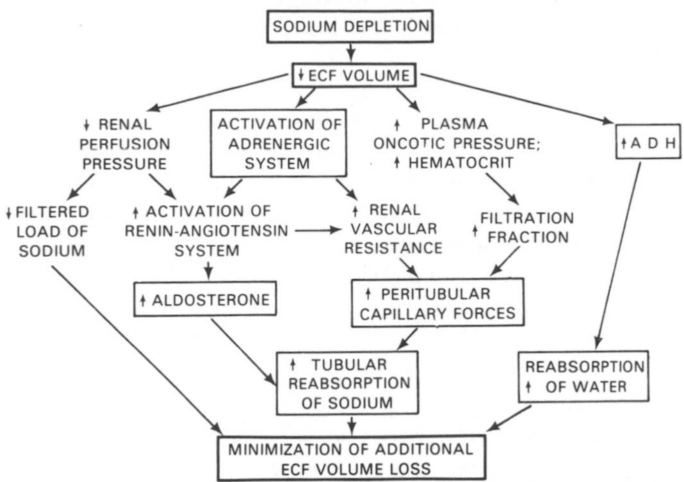

FIGURE 3. Renal homeostatic response to sodium depletion. Outline of the renal mechanisms that favor renal sodium and water conservation. Note that for reasons of simplicity the stimulation of thirst by the renin-angiotensin system is not shown. Note also, that because of its unproven status, the pathway of suppression of a natriuretic hormone during sodium depletion has not been included in the scheme leading to sodium conservation.

likely results from increased angiotensin II and III stimulation,[7,13,59,60] but a decrease in the metabolic clearance rate of aldosterone might also contribute.[13] On the other hand, it has been pointed out that patients with Addison's disease receiving a fixed dosage of both glucocorticoid and mineralocorticoid replacement are capable of regulating sodium excretion,[61] and that the decrease in sodium excretion observed in normal man in response to salt depletion starts before aldosterone excretion increases.[56] The data suggest, therefore, that other factor(s) besides aldosterone are of importance in mediating the renal response to sodium depletion. A role for suppression of natriuretic hormone is unproven.

Alterations in the *peritubular capillary physical forces* should be considered first. We have already alluded to the fact that because of a proportionally greater decrease in RPF than in GFR[4,38,39] the FF increases in sodium depletion. This circumstance should facilitate enhanced proximal tubular sodium reabsorption.[62,63] Likewise, the increased RVR associated with sodium depletion[4,39] will also favor increased reabsorption of sodium in the proximal tubule by altering the peritubular capillary physical forces.

The activation of the *adrenergic system* in sodium deficiency directly[23,33] or indirectly by its own stimulation of the renin-angiotensin system[15,39] will affect RVR (Figure 1). In Figure 3 an attempt has been made to integrate all the factors that determine decreased renal sodium excretion in response to sodium depletion. From the above description of the normally complete and relatively rapid renal response to sodium depletion, it follows that sodium deficiency of extrarenal origin will be accompanied by a very low urine sodium concentration (less than 10 mEq/ℓ), while the concentration will be higher (greater than 30 mEq/ℓ) when the kidney itself is responsible for the sodium deficit. This different response provides us with a simple and practical diagnostic tool.

3. Renal Excretion of Water

It is well known that water excretion is decreased during states of sodium depletion.[64,65] This undoubtedly is an effort on the part of the kidney to further limit volume loss (Figure 3). The mechanisms involved in this response include both the release of antidiuretic hormone (ADH)[66-71] and ADH-independent water reabsorption.[72,73]

At the beginning of sodium depletion, there is an increase in plasma ADH concentration initially proportional to the decrease in plasma volume. When the decrease in PV exceeds 10% there is an acceleration in the rate of release of ADH by the posterohypophysis as well as a lowering of the threshold for ADH release.[66,67] The exact mechanism for ADH release in sodium depletion is not well defined, but it may result from stimulation of volume baroreceptors with a consequent decrease in parasympathetic activity. Initially, the stimulation site may be located at the low pressure atrial receptors.[69,70] Subsequently, if sodium depletion progresses to the point of reducing arterial blood pressure and cardiac output, the activity of the carotid arterial baroreceptors will be involved in the release of ADH.[70,71]

In the absence of ADH, water excretion can also be abnormal. This has been demonstrated by Berliner and Davidson[72] in the dog undergoing a maximal water diuresis (physiological inhibition of ADH). When the GFR of one kidney was reduced by renal artery clamping, the urine produced by this kidney became hypertonic to plasma while the contralateral kidney continued producing a very dilute urine (no circulating ADH). This was attributed to a decrease in the delivery of filtrate to the distal diluting site of the clamped kidney, with resulting impaired free-water generation and thus formation of hypertonic urine. Similar results were obtained in sodium-depleted animals devoid of ADH.[73] The possibility that ADH release may also be under the influence of the renin-angiotensin system[74] has not been confirmed in more recent studies.[75,76]

Regardless of the exact mechanism responsible for increased ADH activity in sodium depletion, the clinical consequences of the impaired ability to excrete water is that unless careful restraint is exercised hyponatremia may be produced or accentuated by giving excessive hypotonic or salt-free solutions to patients with severe sodium depletion.

4. Other Renal Function Alterations Observed in Sodium Depletion

In the sodium-depletion states there is commonly a fall in urea clearance that is proportionally greater than the decrease in GFR, reflecting enhanced fractional reabsorption of water and urea.[36] These changes are reflected in enhanced urea retention and increasing BUN. This will occur more rapidly if renal function is already impaired. The disproportionate increase of BUN over creatinine offers another practical clinical clue for the diagnosis of sodium depletion. Proteinuria may appear or be accentuated in sodium depletion and renin may be in part responsible for this, at least in experimental animals.[77] Maximal urine osmolality is usually minimally affected by sodium depletion, while T^cH_2O formation is reduced, apparently due to a decrease in the amount of sodium available for transport at the distal concentrating site.[78] Sodium depletion may limit renal acidification by impairment of ammonium excretion.[79] In addition, we have shown in normal volunteers that following sodium depletion a decreased distal delivery of sodium limited maximal hydrogen ion secretion despite the probable enhancement of the avidity for the hydrogen and sodium ion exchange.[41]

C. Extrarenal Response to Sodium Depletion

The skin, which under normal conditions has only a negligible role in sodium homeostasis, will also contribute to sodium conservation, while the thirst mechanism will be intensively activated during sodium depletion.

Thirst — It is a common experience that sodium depletion results in increasing thirst[64,80,81] and salt appetite.[44,82] It appears that angiotensin II may stimulate thirst, since its blockade abolishes the increase in thirst.[81] This response of sodium depletion, however, has not been found by all investigators.[83] It has been noted by Baker et al.[80]

Table 1
SOME CLINICAL HINTS RESULTING FROM APPLICATION OF KNOWN PATHOPHYSIOLOGIC MECHANISMS IN SODIUM DEPLETION

1. Large sodium losses and significant ECF volume depletion may be associated with little compromise of the circulation when these losses take place over a period of days or weeks.
2. Mild orthostatic tachycardia and hypotension indicates either moderate acute sodium loss or severe chronic sodium deficit.
3. Sodium depletion of extrarenal origin results in the production of urine with a low sodium concentration (<10 mEq/ℓ), while the sodium concentration is higher (>30 mEq/ℓ) when the kidney itself is responsible for the sodium loss.
4. It is virtually impossible to produce sodium depletion by dietary sodium restriction alone in normal man under normal environmental conditions.
5. The rapidity with which normal sodium conservation is reached following sodium restriction depends on the prior level of sodium intake and the presence of additional sodium losses.
6. The absence of modest weight loss (about 1 kg) in response to dietary sodium restriction indicates non-compliance with the restricted diet or inappropriate high intake (or prescription) of fluids. In the latter situation modest hyponatremia may appear.
7. Older patients may exhibit an impaired response to dietary salt restriction.
8. Dry mucous membranes, decreased skin turgor and sweating (dry axilla and groin areas), and increased thirst associated with hyponatremia suggest the presence of sodium depletion.
9. Some patients with sodium depletion and severely decompensated cirrhotics when depleted of salt as well, might be very sensitive to interference with their circulatory homeostatic responses by administration of angiotensin II blocking drugs. This circumstance may result in severe hypotension and vascular collapse.
10. Likewise, patients who are receiving catecholamine-depleting drugs (antihypertensive agents, i.e., guanethidine, may be at an increased risk when sodium depleted.
11. Hyponatremia may be induced or accentuated by giving excessive amounts of hypotonic or salt-free solutions to patients with severe sodium depletion.

that thirst may persist even in the presence of hyponatremia, a circumstance that normally results in suppression of thirst, thus providing another clinical clue for the diagnosis of sodium depletion.

Sweat — A decrease in the volume of sweat and in the concentration of sodium in the sweat has been reported in sodium depletion in man, usually during experiments of acclimatization to heat.[84-87] This response of the skin — an organ not involved in body fluid volume regulation, but rather in caloric homeostasis — can be taken as another effort to decrease additional volume loss in the body's goal to restore volume homeostasis. The decrease in sweat sodium concentration is mediated by the increased circulating levels of aldosterone,[84,86] since it can be reproduced by DOCA administration[85] and prevented by spironolactone.[87] ADH is not involved in the reduction of sweat volume during sodium depletion, since the sweat gland is not responsive to changes in circulating ADH levels.[88] Table 1 summarizes some clinical hints resulting from the application of known pathophysiologic mechanisms related to sodium depletion.

III. CLINICAL ASPECTS OF SODIUM DEPLETION

Table 2 lists the symptoms, signs, and common laboratory findings of sodium depletion. The clinical features of sodium depletion are dominated by its effects on the cardiovascular system. Some of the symptomatology, particularly that related to the central nervous system, is commonly found in other clinical states associated with hyponatremia, particularly of the dilutional type,[89] making it occasionally difficult to separate the two major types of hyponatremias.

Factors that influence the appearance of these symptoms and signs are the rapidity and magnitude of the sodium loss as well as the background health status of the patient

Table 2
CLINICAL FEATURES IN SODIUM DEPLETION

Symptoms

Increased thirst
Decreased sense of flavor and taste

Anorexia, nausea and vomiting
Abdominal cramps and pain
Paralytic ileus (in severe depletion)

Fatigue, weakness
Muscle cramps

Mental dullness, nightmares
Mental apathy
Light-headness, headaches

Signs

Orthostatic decreases in blood pressure
Orthostatic rises in pulse rate
Decreased pulse volume
Decreased jugular venous pressure

Dry mucous membrane
Decreased skin turgor
Decreased sweating
Sunken eyes and cheeks

Peripheral vasoconstriction (cold extremities)

Laboratory Findings

Increased hematocrit
Increased serum total protein concentration
Increased serum uric acid concentration
Increased serum calcium concentration

Hyponatremia, normal serum sodium or hypernatremia
Serum osmolality, variable depending on sodium changes
K, Cl, HCO_3 concentrations, variable

BUN increases proportionally more than serum creatinine

Urine volume, variable depending on cause of sodium loss
Concentrated urine
Urine sodium concentration
  <10 mEq/ℓ, extrarenal sodium loss
  >30 mEq/ℓ, renal sodium loss
Decreased GFR
Proteinuria
Hyaline and granular casts

Normal or increased CSF pressure
Normal CSF protein

undergoing salt depletion. A loss of sodium not exceeding 150 mEq usually does not produce symptoms, while, depending on the circumstances mentioned above, losses in excess of 200 to 300 mEq do result in the development of symptoms. A loss of sodium in excess of 500 mEq should be considered as a moderately severe to severe sodium deficiency.[5,90]

Thirst is typical of sodium depletion and is probably the result of ECF volume contraction and its consequences (Figure 3), since it is not observed in most cases of inappropriate ADH syndrome. In the latter situation there is hyponatremia with increased body water and preservation of the ECF volume. Frequently the sense of flavor and taste for foods, liquids, and even cigarettes is decreased or lost as vividly described by McCance.[5]

Anorexia, nausea, and vomiting are frequently associated with sodium depletion, and abdominal cramps and pain may further limit food and fluid intake. Paralytic ileus is seen only in severe cases of sodium depletion, but gastric emptying time remains normal.[91] Diarrhea is uncommon in salt depletion; thus, when present, it should suggest a cause of sodium loss rather than a symptom.

Patients feel weak, with easy fatigability. Muscle cramps are common in sodium depletion states. These may be localized (fingers, legs) or generalized, very painful or most frequently tolerable. They may be also accompanied by muscular twitches and occasionally by tremors.

The rapidity of the development of the sodium loss is of importance in the appearance and magnitude of the central nervous system symptoms and signs associated with sodium deficiency. Mental symptoms reflect the effect of hyponatremia on brain water distribution.[89] This may lead to overhydration of brain cells if hyponatremia is of rapid onset. From the data available in experimental studies, it appears, however, that given time an adjustment of the fluid content of the brain takes place.[89] Neurological symptoms, thus, result from a balance between gain in brain water vs. loss of brain electrolytes.

Recovery from sodium depletion is usually dramatic with many of the symptoms disappearing quite rapidly with sodium repletion.

The clinical signs of sodium deficit are, as mentioned above, dominated by the effects of sodium loss on the cardiovascular system. Since blood pressure may not be initially decreased, evaluation of orthostatic changes in blood pressure and pulse rate are mandatory in patients evaluated for volume depletion or hyponatremia (Table 2). During this phase postural hypotension and elevation of pulse rate may be detected prior to the appearance of decrease in the supine blood pressure. A normal pressure in a previously hypertensive patient may represent the presence of a decreased blood pressure secondary to volume contraction.

The presence of dry mucous membranes (except in mouth-breathing patients), decreased skin turgor, sunken eyes and cheeks, and decreased sweating (the axilla and groin are the preferred areas to look for this sign) suggests significant sodium depletion.

Table 2 also lists the most common laboratory findings found in patients with sodium deficiency. There is evidence of hemoconcentration (increased hematocrit and total serum protein concentration) with the BUN increasing disproportionately to the elevation of serum creatinine. This rise in BUN seldom exceeds 100 mg/dℓ, unless acute renal insufficiency results from vascular collapse induced by severe sodium depletion. Because of hemoconcentration and decreased renal excretion,[92] serum uric acid and calcium levels may also be elevated in sodium depletion.

Although hyponatremia frequently accompanies sodium depletion, as alluded to before, depending on concomitant water losses and type of fluid replacement a normal or even an increased serum sodium concentration[93] may be found. Likewise, many sodium depletion states are associated with concomitant losses of potassium and bicarbonate.

Table 3
DIAGNOSES OF SODIUM DEPLETION

	Hyponatremia of sodium depletion	Dilutional hyponatremia
Hyponatremia	Present	Present
Recent weight loss	Present	Absent
Thirst	Present	Absent
Orthostatic hypotension and tachycardia	Present	Absent
Decreased CVP	Present	Absent
Poor skin turgor and dry mucous membranes	Present	Absent
Decreased sweat	Present	Absent
Hemoconcentration	Present	Absent
Increased BUN	Present	Normal[a]
Increased BUN/creatinine	Present	Absent

[a] May also be increased.

In general, urine volume is decreased and the osmolality of the urine is greater than plasma osmolality. When the cause of the sodium loss is of renal origin, the above findings are not present. As described before, urinary sodium concentration is variable but its assessment in patients with hyponatremia or sodium loss is very helpful in determining the origin of the sodium deficit: extrarenal losses are characterized by a low sodium concentration, while sodium depletion secondary to renal sodium loss is accompanied by a urine sodium concentration usually greater than 30 mEq/ℓ. Occasionally, there may be slight proteinuria with numerous hyaline and granular casts found in the urine sediment.

In Table 3, some of the guidelines that permit a differentiation of the patient with sodium depletion and hyponatremia from the commonly encountered patient with dilutional hyponatremia are shown. It should be emphasized that many patients present with confusing features or actually have sodium and water losses in various combinations, or have received excess water replacement in the presence of sodium depletion and impaired water excretion. All the above circumstances indicate that in practical terms the differential diagnosis is occasionally less simple than the one outlined in Table 3.

IV. CLASSIFICATION OF SODIUM DEPLETION STATES

There are a great variety of disorders that can cause sodium depletion. Because of this, sodium losses are best divided into those of extrarenal origin (Table 4) or renal origin (Table 5). As we have already emphasized, the urinary sodium concentration allows for a rapid and clinically accurate differentiation (Table 2).

V. EXTRARENAL CAUSES OF SODIUM DEPLETION

Extrarenal sodium losses represent the most common causes of sodium depletion (Table 4). These may be due to losses thoughout the gastrointestinal tract, the skin, or other mechanisms. A review of the daily volume and salt content of the body fluids involved (Table 6) underlines the predominance of the gastrointestinal tract as a major cause of sodium depletion. External losses or internal sequestration of sodium are the pathogenic mechanisms responsible for the extrarenal sodium depletion states.

Table 4
EXTRARENAL CAUSES OF SODIUM DEPLETION

(Urinary sodium <10 mEq/ℓ)

A. Gastrointestinal sodium loss
- External loss of sodium
  1. Vomiting
  2. Diarrhea
  3. Small bowel suction
  4. Intestinal fistulae
  5. Villous adenoma
- Sequestration of sodium
  1. Small bowel obstruction
  2. Pancreatitis
  3. Peritonitis

B. Skin sodium loss
- External loss of sodium
  1. Heat exposure
  2. Cystic fibrosis
  3. Adrenal insufficiency
- Sequestration of sodium
  1. Burns
  2. Inflammatory

C. Other extrarenal causes of sodium loss
- External loss of sodium
  1. Severe hemorrhage
  2. Removal of large volumes of serous cavity fluid
- Sequestration of sodium
  1. Extensive limb trauma, retroperitoneal space
  2. Peripheral vaso/venodilation

Table 5
RENAL CAUSES OF SODIUM DEPLETION

(Urinary sodium >30 mEq/ℓ)

A. Intrinsic renal disorders
- Chronic renal failure
  - Mild form of salt wasting
  - Massive, overt, salt wasting nephropathy
  - Medullary cystic disease
  - Nephrocalcinosis
  - Interstitial nephritis
- Acute renal failure
- Postobstructive diuresis

B. Extrinsic renal processes
- Solute diuresis
  - Endogenous (bicarbonate, urea, glucose)
  - Exogenous (mannitol, dextran urea, glycerol, radiocontrast material)
- Diuretic administration
- Mineralocorticoid deficiency
- Fasting

A. Gastrointestinal Sodium Loss

The gastrointestinal sodium losses are the major route of extrarenal exit of sodium from the body and constitute the single, most common cause of sodium depletion in practice (Tables 4 and 6).

Table 6
VOLUME AND ELECTROLYTE CONTENT OF BODY FLUIDS[a]

Electrolyte concentrations

	Daily volume (mℓ)	Sodium (mEq/ℓ)	Potassium (mEq/ℓ)	Bicarbonate (mEq/ℓ)
Saliva	500—2000	20—80	15—25	30—50
Gastric	300—3000	30—90	5—15	1
Bile	200—1000	120—170	5—12	50
Pancreatic	200—1000	115—160	3—10	70—110
Small bowel	500—4000	70—160	4—10	20—40
Ileostomy[b]	100—1000	90—140	6—30	25—30
Cecostomy	500—1000	50—120	15—30	15—30
Feces	100—200	30—40	70—90	—[c]
Sweat	500—4000	15—80	6—12	0

[a] Sources are References 94 to 99. All concentrations are in mEq/ℓ of water, gastrointestinal fluid, or sweat.

[b] Recent ileostomy (concentrations decrease in adapted ileostomy).

[c] Normal stool electrolyte concentrations are such that the sum of Na^+ and K^+ exceeds that of HCO_3^- and Cl^-. The anion gap of the normal stool consists of various organic acids (partly derived from bacterial metabolism of dietary carbohydrate) and represent potential HCO_3^- after hepatic oxidation.[99]

Diarrhea is the most common gastrointestinal disturbance causing sodium depletion. The amounts of sodium, potassium, chloride, and bicarbonate loss are very variable depending on the type of diarrhea and its duration. Generally, however, as the stool volume increases so does its sodium concentration, while potassium concentration diminishes. At fecal volumes of about 3 ℓ/day the electrolyte composition is similar to that of normal plasma.[100] Thus in severe diarrhea, the loss of sodium and chloride in the stool greatly exceed that of potassium.[100,101] While acute diarrhea tends to produce hyperchloremic metabolic acidosis, patients with chronic diarrhea usually develop hypochloremic metabolic alkalosis.

Any of the mechanisms resulting in diarrhea increase the external losses of sodium through the gastrointestinal tract. These include: (1) osmotic retardation of sodium and water reabsorption during administration of magnesium salts or in patients with lactase deficiency; (2) perturbations of sodium and water transport in infective diarrhea, such as classically described in cholera[101] or in villous adenomas,[102,103] and (3) increased gut transit time as described in the carcinoid syndrome. It should be remembered that in some cases of villous adenoma the increased stool formation is not perceived by the patient as diarrhea but rather as enhanced rectal mucous secretion.

Protracted vomiting can induce sodium loss through external loss with the vomitus or by way of the kidney. The concentration of sodium in the vomitus or nasogastric aspirate is variable, ranging from 20 to 170 mEq/ℓ depending on the circumstances,[104-105] but averaging about 50 to 60 mEq/ℓ in patients with pyloric obstruction secondary to peptic ulcer or gastric carcinoma.[105] When metabolic alkalosis results from severe vomiting, enhanced urinary excretion of sodium bicarbonate can be observed occasionally despite the usually associated ECF volume contraction,[106,107] compounding the sodium and potassium losses.

Ileostomy represents the commonest source of loss of sodium, particularly in recently surgically created ileostomies (Table 6). It should be noted that electrolyte losses, however, diminish in chronic or adapted ileostomies.[108]

Small bowel obstruction, pancreatitis, and peritonitis can lead to sequestration of large amounts of sodium-containing fluid causing ECF volume contraction and a clinical picture with some similarities to that of sodium depletion due to external sodium losses.

B. Skin Sodium Loss

Excessive losses of sodium through the skin are related to three mechanisms: (1) enhanced thermal-stimulated loss, such as in copious sweating in high ambient temperatures, particularly in nonacclimatized subjects; (2) increased sodium sweat loss due to inability of the sweat gland to achieve normal conservation of sodium, such as in Addison's disease[85] and in patients with cystic fibrosis,[109,110] and (3) sequestration of sodium in disorders where the normal structure and function of the skin is lost, such as in burns and inflammatory skin processes. Since in cystic fibrosis and Addison's disease the concentration of sodium in the sweat is high, ranging from 70 to 130 mEq/ℓ, patients with these disorders can lose significant amounts of sodium when exposed to high humidity and temperature.

C. Other Extrarenal Causes of Sodium Loss

Sodium can also be lost from the circulation by sequestration in traumatized extremities, retroperitoneal spaces or translocated to stagnant areas of the circulation (peripheral or splanchnic vasodilation), by removal of large volumes of fluid accumulated in the body's serous cavities or by massive bleeding.

VI. RENAL CAUSES OF SODIUM DEPLETION

The failure of the kidney to conserve sodium normally may result in sodium depletion. The normal response to sodium restriction has already been discussed. Interference with this crucial renal process may be the result of intrinsic renal diseases or factors that affect renal sodium reabsorption in the absence of renal disease (extrinsic renal sodium depletion) (Table 5).

A. Intrinsic Renal Sodium Loss

When patients with chronic renal failure are given a low intake of sodium they fail to conserve sodium normally.[111-117] The deleterious consequences of dietary restriction in these patients, usually prescribed because of hypertension or heart failure, have been recognized for many years.[111-113]

Two forms of this salt-wasting state are seen in chronic renal failure. The most commonly observed variety is usually clinically inapparent (potential salt wasting) and is only evident when the patient is salt restricted. The second form or overt variety, sometimes named salt-losing nephropathy, is usually clinically apparent and manifests itself as severe sodium depletion. Figure 4 illustrates the clinically inapparent variety. Almost all patients with chronic renal failure exhibit this type of salt wastage if appropriately tested (Figure 4). The amount of sodium wasted (sodium lost in the urine in excess of dietary intake) is modest, usually not exceeding 30 to 50 mEq/day in the majority of patients. It should be emphasized, however, that these patients are capable of achieving salt balance, although their renal response takes longer than in normal patients.[115,117]

In the study of Coleman et al.,[115] for example, 10 of 14 patients persisted in negative, albeit small, salt balance at the end of the 2- to 3-week period of salt restriction. Likewise, only in 2 of 14 patients was the urinary sodium concentration below 10 mEq/ℓ. The consequence of this sluggish renal sodium conservation ability is a moderately negative sodium balance. In the study of Coleman et al.[115] the cumulative sodium loss

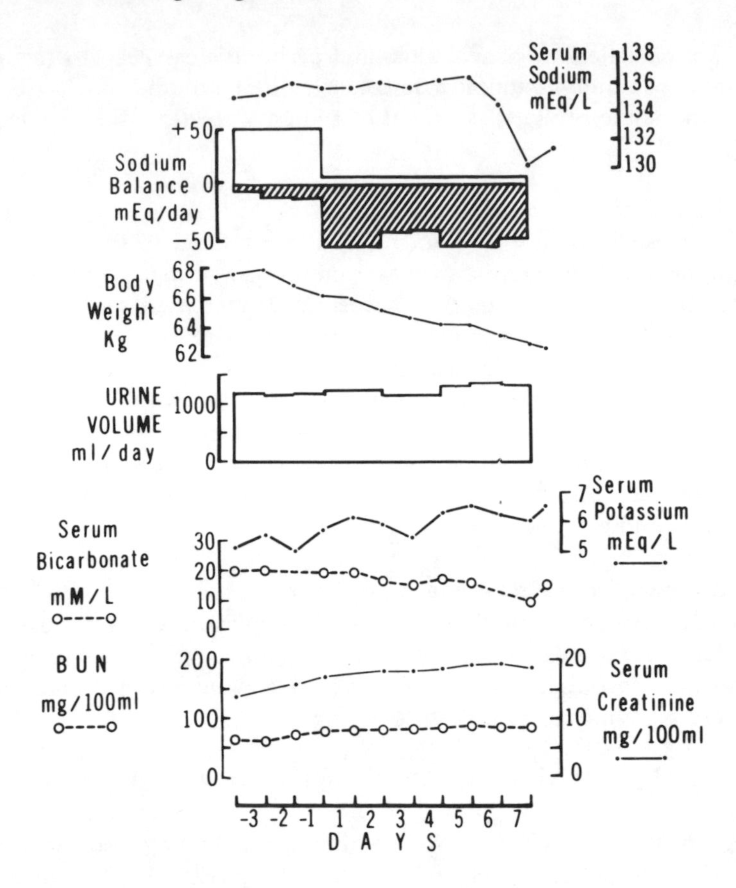

FIGURE 4. Renal sodium loss in a patient with chronic renal failure. The patient was a 44-year-old man with chronic glomerulonephritis and an endogenous creatinine clearance of 2 mℓ/min. The patient was maintained on a diet of 2000 kcal/day, with an intake of 30 g/day of protein of high biologic value and 50 and 25 mEq/day of sodium and potassium, respectively. He had no edema and was on sodium balance prior to salt restriction. Following institution of a low sodium diet (10 mEq/day), a negative sodium balance (the shaded area in the top panel) of approximately 40 to 50 mEq a day became apparent. This patient, in marked contrast to normal subjects of comparable age (Figure 2) continued to lose sodium at the same rate, without showing any evidence of sodium conservation throughout the period of observation. His cumulative sodium loss was 300 to 350 mEq in 7 days in comparison to a normal cumulative loss of about 120 to 150 mEq in 4 to 5 days (Figure 2). He lost weight, became volume depleted, tachycardic, moderately hypotensive, and more azotemic. Note that hyponatremia only appeared after almost a week of salt deprivation and that oliguria did not occur. There was a drop in serum bicarbonate and an increase in serum potassium. Note that the absence of hyponatremia and/or oliguria does not exclude the presence of significant sodium depletion. All these changes reversed rapidly after sodium repletion. (From Vaamonde, C. A., *J. Fla. Med. Assoc.*, 56, 405, 1969. With permission.)

averaged 298 mEq in 2 to 3 weeks. On the other hand, in the overt, sometimes massive salt wasting, the amount of sodium loss in the urine may exceed 200 mEq/day. This extraordinary salt-wasting state is much rarer than the mild form, and has been reported in patients with chronic renal failure secondary to certain tubulointerstitial disorders, such as medullary cystic disease, nephrocalcinosis, and chronic pyelonephritis.[119-125] Renal conservation, however, is even possible in these patients, provided the reduction in dietary sodium intake is done in a very slow stepwise (over 4 to 14 weeks period) fashion.[126] In the words of the authors,[126] performance of a sodium-reduction protocol in patients with severe salt wasting is fraught with danger and requires the cooperation of the patient in a difficult dietary regimen necessitating prolonged hospitalization in a metabolic ward.

Table 7
PATHOPHYSIOLOGY OF SODIUM WASTING IN CHRONIC RENAL FAILURE

1. Anatomical damage to nephrons
2. Increased solute diuresis per remaining nephrons
3. Hyperperfusion of remaining nephrons
4. Alteration of peritubular capillary forces
5. Hormonal perturbations
   Aldosterone
   Natriuretic hormone

Other more transient forms of sodium wasting have been reported during the recovery phase of oliguric acute renal failure, during the course of nonoliguric acute renal failure or during postobstructive diuresis.[127-130]

The study of the *pathophysiology* of salt wasting in chronic renal failure has been the subject of extensive effort and argument through the last three decades.[131-134] It is not known, for example, if the massive salt wasting (salt-losing nephropathies) is simply an extreme of the milder and almost universal salt-wasting state of chronic uremia or represents a different entity in itself.

Whereas the exact nature of salt wasting in chronic renal diseases remains undefined, withdrawal of dietary sodium in these patients may lead to negative sodium balance, contraction of ECF volume, vascular collapse, and further deterioration of renal function.[111-113,135] It is, therefore, good clinical practice to challenge patients with chronic renal failure who present with a questionable or apparent salt loss or an otherwise unexplained decline in renal function with intravenous sodium replacement (isotonic saline). Experience has repeatedly shown that many of these patients exhibit clinical improvement, concomitant with enhancement of renal function after gaining 1 or 2 kg of body weight. Mild, potential salt wasting transiently decreases when the chronic renal failure patients develop congestive heart failure or severe hypoalbuminemia and the nephrotic syndrome.

Several mechanisms have been proposed to explain enhanced salt wasting in chronic renal failure (Table 7).

Data favoring *anatomic damage* to the nephrons responsible for sodium reabsorption include the higher salt wasting at comparable GFR observed in tubulointerstitial disorders compared to glomerular diseases,[114] and supportive morphologic evidence.[133] Opposing this view are the studies of Bricker and co-workers[132] in a model of unilateral renal disease demonstrating the ability of the severely damaged kidney to conserve sodium effectively in the presence of a normal contralateral kidney, and the clinical data demonstrating normal salt conservation in some patients with renal disease.[112,136,137] Perhaps salt wasting accompanying nonoliguric acute renal failure[127] or postobstructive diuresis[129] may best be explained by transient nephron damage. Experimental studies give some support to the presence of a tubular sodium transport defect in bilateral[138,139] or unilateral[140] obstruction. It is, however, apparent that other mechanisms are involved in the genesis of the exaggerated sodium loss observed in some patients after the relief of urinary tract obstruction.

The salt-wasting state in chronic renal failure was first attributed to an *increased solute load per remaining nephrons.*[112,131] Clinical[141] and experimental data,[142] however, tend to exclude this as the only mechanism responsible for the decreased tubular reabsorption of sodium in chronic uremia.

Since single nephron GFR is usually increased in the residual nephrons of animals with experimental uremia,[143,144] it has been proposed that *hyperperfusion of these func-*

tionally remaining nephrons may be implicated in the sodium transport defect. At the same increased single nephron GFR, however, fractional excretion of sodium is variable[143,144] and when a uniform, generalized decrease in single nephron GFR is induced in experimental glomerulonephritis,[145] there is still enhanced fractional sodium excretion, indicating that other factors may be operative, These may include a *perturbation of the peritubular capillary physical forces,*[146,147] with the oncotic gradient being reduced because of a lowered FF and the peritubular capillary hydraulic pressure being increased,[147] both favoring a decrease in proximal tubular sodium reabsorption.

Hormonal influences have also been suggested to explain the enhanced salt excretion per nephron in uremia. The majority of patients with chronic renal failure exhibit normal *aldosterone* levels, and these levels do respond normally to physiologic stimuli.[117,148] Furthermore, administration of mineralocorticoids does not affect substantially the pattern of sodium excretion in most of these patients.[115,149] Thus, in the majority of patients with chronic renal failure, a defect in the synthesis or the renal tubular response to aldosterone appears not to be a common event. Note should be made here, however, of the existence of a group of patients with chronic renal failure, usually of a mild to moderate degree, that exhibit the syndrome of hyporeninemic hypoaldosteronism. This important group will be discussed below.

Finally, in the last few years the possible role of a *natriuretic hormone* influencing tubular sodium reabsorption in uremia has been extensively studied.[150-152] Confirmation of the existence of this natriuretic factor, as well as its purification and isolation, awaits further studies. Likewise, its relationship to the humoral factor that decreases renal tubular sodium reabsorption in unanesthetized volume expanded normal rats[153] also remains to be elucidated.

From the above discussion it seems clear that a single pathogenetic mechanism cannot explain the perturbation of sodium transport in uremia (Table 7). It is conceivable, however, that at different times in the evolution of chronic renal disease or in different models of disease these mechanisms may interact causing renal salt wasting.

B. Extrinsic Renal Processes

Sodium depletion can also develop due to the action of a variety of mechanisms which have in common the fact that they may influence sodium transport in otherwise normal kidneys. These extrinsic renal causes of sodium depletion are listed in Table 5.

1. Osmotic Diuresis

Solute diuresis is the most common cause of extrinsic renal sodium wasting. Endogenous solutes, such as glucose, urea, or bicarbonate and those of exogenous origin, such as mannitol, dextran, urea, glycerol, radiocontrast material, etc., can induce a large diuresis with significant salt loss.[154-157] Depending on other factors (state of hydration, sodium balance) the concentration of sodium changes from its baseline value towards a level approximating 50 to 80 mEq/ℓ at the height of the diuresis. It follows that if this high solute diuretic state persists for any extended period of time large quantities of sodium are lost in the urine. Table 8 illustrates the dramatic effect of mannitol diuresis on sodium excretion in a dehydrated salt restricted normal man.[158] Despite adequate sodium conservation before mannitol administration, large quantities of sodium were lost in the urine following the induction of the osmotic diuresis.

Clinical examples of sodium wasting due to solute diuresis can be found in diabetic ketoacidosis or hyperosmolar nonketotic coma,[154,156] in comatose patients with barbiturate overdose treated with mannitol, in hyperosmolar high-protein tube feeding,[159] etc. Renal sodium wasting also occurs in patients with proximal renal tubular acidosis or severe metabolic alkalosis.[107,160] The poorly reabsorbable characteristic of the anion bicarbonate is responsible for the enhanced sodium excretion.

Table 8
EFFECT OF MANNITOL DIURESIS ON SODIUM EXCRETION IN A SALT RESTRICTED HYDROPENIC NORMAL MAN

Time (min)	V (mℓ/min)	UOsm (mOsm/ kg H_2O)	UNa (mEq/ℓ)	UNaV (μEq/min)	Cumulative Na loss (mEq)	CIN (mℓ/min)	Serum Na (mEq/ℓ)
Baseline (−109 to 1)	0.4	762	15	7	0.7	99	138
0							
Start mannitol 10% at 16 mℓ/min							
0—30[a]	2.3	616	65	152	5	125	134
66—74	7.6	459	73	553	21	82	
94—105	10.6	430	73	768	43	95	131
125—135	13.6	413	72	897	71	89	128

[a] For simplicity only 4 of the 10 urine collection periods are shown. Note that if a steady state of solute diuresis of this magnitude is assumed, in 24 hr the loss of sodium would reach 760 mEq. Adapted from Vaamonde et al.[158]

2. Diuretics

All diuretic agents because of their effects on renal sodium reabsorption are capable of inducing salt wasting. The loop diuretics (furosemide, ethacrynic acid), however, alone or in combination with other diuretics, are responsible for most of the clinically reported cases of diuretic-induced salt depletion.

3. Mineralocorticoid Deficiency

A variety of disorders characterized by deficiency of mineralocorticoid activity may be associated with salt wasting (Table 9).[161] It is well established that renal sodium wasting is a major feature of untreated *Addison's disease.*[57] The effect of renal salt loss on ECV volume and serum sodium concentration depends on the patient's concomitant salt intake and extrarenal salt losses. Prominent hyperplasia of the juxtaglomerular apparatus has been observed in experimental adrenocortical insufficiency and plasma renin activity (PRA) is increased. A recent study, however, demonstrated that although mineralocorticoid deficiency stimulates renin release, adequate glucocorticoid production is crucial in preventing renin substrate depletion,[162] suggesting that the homeostatic role of the renin-angiotensin system may be impaired in severe adrenal insufficiency. Nevertheless, the increase of PRA in Addisonian patients stands in marked contrast to the low levels found in hyporeninemic hypoaldosteronism (Table 9). In addition, patients with Addison's disease have impaired ability to excrete free water, and this defect also contributes to the development of hyponatremia. This abnormality in water excretion appears to be due to the deficiency of both mineralocorticoid and glucocorticoid hormones. The impaired diluting ability secondary to mineralocorticoid deficiency appears to be corrected by ECV repletion.[163] In experimental animals with isolated glucocorticoid deficiency, however, the defect in water excretion may be related to impaired ADH suppression caused by hemodynamic changes.[164]

Deficiency of enzymes involved in the biosynthetic pathways for mineralocorticoids may result in salt wasting, hyperkalemia, and acidosis. Defects in the early steps cause decreased production of all adrenal steroids and lethal disease.[165] Deficiencies in the intermediate steps are responsible for the various *adrenogenital syndromes.* The only one of these which is characterized by mineralocorticoid deficiency is the complete form of 21-hydroxylase deficiency ("salt losing" form) which is associated with de-

Table 9
SYNDROMES OF MINERALOCORTICOID (MC) HORMONE DEFICIENCY

1. Primary MC deficiency (low aldosterone, high PRA)
   a. Addison's disease[a]
   b. Defectual biosynthetic pathways
      Early steps (desmolase, 3β-ol-dehydrogenase deficiencies)[a]
      Intermediate steps (complete 21-hydroxylase deficiency: adrenogenital syndrome with "salt-wasting")[a]
      Final steps (corticosterone methyl oxidase deficiencies)[a]
   c. Heparin[a]
2. Secondary MC deficiency (low PRA, low aldosterone)
   a. Syndrome of hyporeninemic hypoaldosteronism (SHH)[a]
3. Pseudohypoaldosteronism (normal or high aldosterone[b])
   a. Pseudohypoaldosteronism of infancy[a]
   b. Renal tubular unresponsiveness to MC hormones associated with renal disease ("salt-wasting" syndromes)[a]
   c. Others[c]

[a] Characterized by mild to severe salt wasting.
[b] The term pseudohypoaldosteronism denotes those hyperkalemic conditions usually characterized by normal or elevated levels of plasma aldosterone and/or renal tubular resistance to MC hormones.
[c] There are some patients with SHH who have relative renal tubular resistance to MC hormones.

Adapted from Vaamonde, C. A., Perez, G. O., and Oster, J. R., *Mineral Electrolyte Metabolism*, 5, 121, 1981. With permission.

creased cortisol and aldosterone secretion as well as an increase in adrenal androgens. A congenital defect in the final steps required for the biosynthesis of aldosterone results in hypoaldosteronism but normal production of cortisol and adrenal androgen.[166] Administration of *heparin* rarely is associated with damage to the adrenal cortex and subsequent hypoaldosteronism.[167]

The most common form of isolated mineralocorticoid deficiency in the adult is the *syndrome of hyporeninemic hypoaldosteronism (SHH).*[161,168-170] As already mentioned, considerable interest in this syndrome has evolved in the last few years and what was originally thought to be a rarity is now recognized frequently. The presence of unexplained persistent hyperkalemia, in the range of 5.5 to 6.5 mEq/ℓ, in disproportion to the decrement of renal function is the usual diagnostic clue. Hypoaldosteronism appears to be an acquired defect secondary to insufficient stimulation of the adrenal gland by the renin-angiotensin system. A defect in one of the steps of aldosterone biosynthesis, however, has also been described in patients with SHH.[171] Although this cannot explain low PRA, it is possible that they coexist as associated abnormalities. In most patients, hyporeninemia is associated with chronic renal disease usually of the tubulointerstitial type. Most patients exhibit mild to moderate chronic renal insufficiency (typical serum creatinine concentrations 2 to 4 mg/dℓ) and normal glucocorticoid production and response to ACTH stimulation. At least a third or more of the patients have diabetes mellitus.[169,170,172] The majority of patients are in the fifth to seventh decade of life, and many are hypertensive. As expected from the aldosterone deficit, these patients manifest hyperkalemia and hyperchloremic acidosis.[173,174] Most patients remain asymptomatic, and hyperkalemia-related electrocardiographic manifestations are uncommon. Most of these patients are capable of achieving salt balance in a restricted sodium diet, but they may do so with considerable loss of body weight and decrease in renal function, particularly when they also receive diuretics.[175] The natural course, particularly its relationship to the progression of renal failure, is unknown.

Patients with *pseudohypoaldosteronism* of infancy (congenital "salt-wasting" syndrome) are usually male and present before age one with failure to thrive, electrolyte imbalance, and volume contraction.[176] Plasma aldosterone and PRA are elevated and unresponsiveness to MC has been demonstrated in the kidney, colon, sweat, and salivary glands. As already described, a pseudohypoaldosteronism state is present in patients with renal disease and overt renal salt wasting[119-125] and in some patients with SHH.[168,169]

4. Fasting

The mild augmented renal excretion of sodium observed in patients undergoing caloric deprivation results from sodium accompanying the urinary losses of phosphate and other organic anions that occur during fasting.[177,178] A relative state of renal tubular resistance to aldosterone (mediated by the elevated levels of glucagon present in fasting) has also been proposed.[178]

Although patients with severe *myxedema* may exhibit a sluggish renal response to sodium restriction[40] and an enhanced fractional excretion of sodium after salt loading,[179] severe clinical perturbations of sodium balance have not been described in hypothyroid patients.[180] This contrasts with the findings in rats with experimentally induced thyroid deficiency. Species or experimental protocol differences may account for the discrepancy.[180]

VII. TREATMENT

Prevention of sodium depletion requires both physician's knowledge of the conditions in which it occurs (Tables 4 and 5) and, particularly, recognition of patients who are prone to develop sodium depletion because their homeostatic mechanisms for sodium conservation may be impaired (Table 1). Those patients at risk have either chronic renal disease, salt wasting nephropathy, adrenocortical insufficiency, autonomic deficiency, cystic fibrosis, advanced age, or have received drugs such as: potent diuretics, catecholamine-depleting drugs, or perhaps angiotensin blockers.

Early recognition remains the key to a successful therapy. Since dilutional hyponatremia is the more common type of hyponatremia seen in practice and therapy of this condition is in general quite different from correction of sodium depletion, rapid differentiation between the two is necessary. Some guidelines were summarized in Table 3.

Treatment of sodium depletion requires an *estimation of the sodium loss* incurred. This can be better done by direct measurement of losses, but in practical terms, this is estimated from tables, such as Table 6, or by the clinical consequences of the sodium deficient state.

The initial goal of therapy is the *rapid restoration of ECF volume.* This should be done by intravenous infusion of isotonic sodium chloride solution. Other sodium salts should not be used because of the relative renal tubular impermeability of accompanying ions, i.e., bicarbonate. Once an adequate response of the circulation has occurred, attention can then be focused to the correction of other abnormalities which may be present (osmolality, sodium, potassium, acid-base, etc.). Addition of bicarbonate, hypotonic solutions, etc. can then be prescribed. Because many volume-depleted patients may also have serious limitations of their cardiovascular system, volume replacement should be carried out under close monitoring. The absence, however, of monitoring facilities (i.e., intensive care unit) or of complete laboratory work-up, however, should never delay the initiation of therapy leading to restoration of ECF volume.

Finally, *correction of the specific cause of sodium depletion* should be undertaken. Mineralocorticoids should be given to patients with hypoaldosteronism (Table 9), sodium and potassium salts should be added to mannitol infusions in patients with postobstructive or forced diuresis (overdose), potent diuretics should be carefully used and their dosage adjusted frequently.

As mentioned earlier, recovery from sodium depletion is usually dramatic with many of the symptoms dissappearing quite rapidly with sodium repletion.

ACKNOWLEDGMENT

Part of the work included in this chapter was supported by NIH grants HE-07665 and HE-12544, and by Veterans Administration designated research funds VA-8943.

REFERENCES

1. **Welt, L. G.,** *Clinical Disorders of Hydration and Acid-Base Equilibrium,* 2nd ed., Little, Brown, Boston, 1959, 133.
2. **Elkington, J. R., Danowski, T. S., and Winkler, A. W.,** Hemodynamic changes in salt depletion and in dehydration, *J. Clin. Invest.* 25, 120, 1946.
3. **Elkington, J. R. and Danowski, T. S.,** *The Body Fluids. Basic Physiology and Practical Therapeutics,* Williams & Wilkins, Baltimore, 1955, 174.
4. **Romero, J. C., Staneloni, R. J., Dufau, M. L., Dohmen, R., Binia, A., Kliman, B., and Fasciolo, J. C.,** Changes in fluid compartments, renal hemodynamics, plasma renin and aldosterone secretion induced by low sodium intake, *Metabolism,* 17, 10, 1968.
5. **McCance, R. A.,** Experimental sodium chloride deficiency in man, *Proc. R. Soc. London,* 119, 245, 1936.
6. **Brown, J. J., Davies, O. L., Lever, A. F., and Robertson, J. I.,** Influence of sodium deprivation and loading on the plasma renin in man, *J. Physiol. London,* 173, 408, 1964.
7. **Binnion, P. F., Davis, J. O., Brown, T. C., and Olichney, M. J.,** Mechanisms regulating aldosterone secretion during sodium depletion, *Am. J. Physiol.,* 208, 655, 1965.
8. **Laragh, J. H., Sealey, J. E., and Sommers, S. C.,** Patterns of adrenal secretion and urinary excretion of aldosterone and plasma renin activity in normal man and hypertensive subjects, *Circ. Res.,* 18, 19 (Suppl. 1), 158, 1966.
9. **Coleman, T. G., Cowley, A. W., and Guyton, A. C.,** Angiotensin and the hemodynamics of chronic salt deprivation, *Am. J. Physiol.,* 229, 167, 1975.
10. **Sancho, J., Re, R., Burton, J., Barger, A. C., and Haber, E.,** The role of the renin-angiotensin-aldosterone system in cardiovascular homeostasis in normal human subjects, *Circulation,* 53, 400, 1976.
11. **Samuels, A. I., Miller, E. D., Fray, J. C., Haber, E., and Barger, A. C.,** Renin-angiotensin antagonists and the regulation of blood pressure, *Fed. Proc.,* 35, 2512, 1976.
12. **Brown, J. J., Fraser, R., Lever, A. F., Love, D. R., Morton, J. J., and Robertson, J. I.,** Raised plasma angiotensin II and aldosterone during sodium restriction in man, *Lancet,* 2, 1106, 1972.
13. **Tuck, M. L., Dluhy, R. G., and Williams, G. H.,** Sequential response to the renin angiotensin-aldosterone axis to acute postural change: effect of dietary sodium, *J. Lab. Clin. Med.,* 86, 754, 1975.
14. **Walker, W. G., Moore, M. A., Horvath, J. S., and Whelton, P. K.,** Arterial and venous angiotensin in normal subjects, *Circ. Res.,* 38, 477, 1976.
15. **Freeman, R. H., Davis, J. O., Lahmeier, T. E., and Spielman, W. S.,** Evidence that des-asp′-angiotensin II mediates the renin-angiotensin response, *Circ. Res.,* 38 (Suppl. 2), 99, 1976.
16. **Johnson, J. A. and Davis, J. O.,** Effect of a specific antagonist of angiotensin II on arterial pressure and adrenal steroid secretion in the dog, *Circ. Res.,* 32 (Suppl. 1), 159, 1973.
17. **Mimran, A., Guiod, L., and Hollenberg, N. K.,** The role of angiotensin in the cardiovascular and renal response to salt restriction, *Kidney Int.,* 5, 348, 1974.
18. **Posternak, L., Brunner, H. R., Gavras, H., and Brunner, D. B.,** Angiotensin blockade in normal man, *Kidney Int.,* 11, 197, 1977.
19. **Anderson, R. J. and Limas, S. L.,** Sodium depleted states, in *Sodium and Water Homeostasis,* Brenner, B. M. and Stein, J. H., Eds., Churchill Livingstone, New York, 1978, 154.
20. **Schroeder, E. T., Anderson, G. H., Goldman, S. H., and Streeten, D. H. P.,** Effect of blockade of angiotensin II on blood pressure, renin and aldosterone in cirrhosis, *Kidney Int.,* 9, 511, 1976.

21. **Blaufox, M.D., Lewis, E. J., Jagger, P., Lauler, D., Heckler, R., and Merill, J. P.,** Physiologic responses of the transplanted human kidney, *N. Engl. J. Med.*, 280, 62, 1966.
22. **Brennan, L. A., Mulcahy, J. J., Carretero, O. A., Malvin, R. L., Geis, P., and Kaye, M.,** Effect of chronic sodium depletion in dogs with denervated kidneys and hearts, *Am. J. Physiol.*, 227, 1289, 1974.
23. **Schrier, R. W.,** Effects of adrenergic nervous system and catecholamines on systemic and renal hemodynamics, sodium and water excretion and renin secretion. *Kidney Int.*, 6, 291, 1974.
24. **Blaine, E. H., Davis, J. O., and Witty, R. T.,** Renin release after hemorrhage and after suprarenal aortic constriction in dogs without sodium delivery to the macula densa, *Circ. Res.*, 27, 1081, 1970.
25. **Gotshall, R. W., Davis, J. O., Shade, R. E., Spielman, W., Johnson, J. A., and Braverman, B.,** Effects of renal denervation on renin-release in sodium-depleted dogs, *Am. J. Physiol.*, 225, 344, 1973.
26. **Kelsch, R. C., Light, G. S., Luciano, J. R., and Oliver, W. J.,** The effect of prednisone on plasma norepinephrine concentration and renin activity in salt-depleted man, *J. Lab. Clin. Med.*, 77, 267, 1971.
27. **Cuche, J. L., Kuchel, O., Barbeau, A., Beucher, R., and Genest, J.,** Relationship between the adrenergic nervous system and renin during upright posture: a possible role for 3,4-dehydroxyphenethylamine, *Clin. Sci.*, 43, 481, 1972.
28. **Alexander, R. W., Gill, J. R., Yamobe, H., Lovenberg, W., and Keiser, H. R.,** Effects of dietary sodium and of acute saline infusion on the interrelationship between dopamine excretion and adrenergic activity in man, *J. Clin. Invest.*, 54, 194, 1974.
29. **North, R. H. and Mulrow, R. J.,** Serum dopamine β-hydroxylase as an index of sympathetic nervous system activity in man, *Circ. Res.*, 38, 1, 1976.
30. **Stead, E. A. and Ebert, R. V.,** Postural hypotension. A disease of the sympathetic nervous system, *Arch. Intern. Med.*, 64, 546, 1941.
31. **Shear, L.,** Renal function and sodium metabolism in idiopathic orthostatic hypotension, *N. Engl. J. Med.*, 268, 347, 1963.
32. **Ibrahim, M. M., Tarazi, R. C., Dustan, H. P., and Bravo, E. L.,** Idiopathic orthostatic hypotension: circulatory dynamics in chronic autonomic insufficiency, *Am. J. Cardiol.*, 34, 288, 1974.
33. **Gill, J. R. and Bartter, F. C.,** Adrenergic nervous system in sodium metabolism. II. Effects of guanethidine on the renal response to sodium deprivation in normal man, *N. Engl. J. Med.*, 275, 1466, 1966.
34. **Panisset, J. C. and Bourdois, P.,** Effect of angiotensin on the response to noradrenalin and sympathetic nerve stimulation, and on ^3H noradrenalin uptake in cut mesenteric vessels, *Can. J. Physiol. Pharmacol.*, 46, 125, 1968.
35. **Roth, R. H.,** Action of angiotensin on vascular adrenergic nerve endings: enhancement of norepinephrine biosynthesis, *Fed. Proc.*, 31, 1344, 1972.
36. **McCance, R. A. and Widdowson, E. M.,** The secretion of urine in man during experimental salt deficiency, *J. Physiol. London*, 91, 222, 1937.
37. **Leaf, A. and Couter, W. T.,** Evidence that renal sodium excretion by normal human subjects is regulated by adrenal cortical activity, *J. Clin. Invest.*, 28, 1067, 1949.
38. **Wiggins, W. S., Manry, C. H., Lyons, R. H., and Pitts, R. F.,** The effect of salt loading and salt depletion on renal function and electrolyte secretion in man, *Circulation*, 3, 275, 1951.
39. **Hollenberg, N. K., Adams, D. F., Rashid, A., Epstein, M., Abrams, H. L., and Merrill, J. P.,** Renal vascular response to salt restriction in normal man, *Circulation*, 43, 845, 1971.

39a. **Hollenberg, N. K., Epstein, M., Guttman, R. D., Conroy, M., Basch, R. I., and Merrill, J. P.,** Effect of sodium balance on intrarenal distribution of blood flow in normal man, *J. Appl. Physiol.*, 28, 312, 1970.

40. **Vaamonde, C. A., Oster, J. R., Lohavichan, C., Carroll, K. E., Jr., Sebastianelli, M. J., and Papper, S.,** Renal response to sodium restriction in myxedema, *Proc. Soc. Exp. Biol. Med.*, 146, 936, 1974.
41. **Perez, G. O., Oster, J. R., and Vaamonde, C. A.,** The effect of sodium depletion on the renal response to short-duration NH_4Cl acid loading, *Proc. Soc. Exp. Biol. Med.*, 154, 562, 1977.
42. **Black, D. A., Platt, R., and Stanbury, S. W.,** Regulation of sodium excretion in normal and salt-depleted subjects, *Clin. Sci.*, 9, 205, 1950.
43. **Renwick, R., Robson, J. S., and Stewart, C. P.,** Observations upon the withdrawal of sodium chloride from the diet in hypertensive and normotensive individuals, *J. Clin. Invest.*, 34, 1039, 1955.
44. **Strauss, M. B.,** *Body Water in Man*, Churchill, London, 1957.
45. **Strauss, M. B., Lamdin, E., Smith, W. P., and Bleifer, P. I.,** Surfeit and deficit of sodium, *Arch. Intern. Med.*, 102, 527, 1958.
46. **Ross, E. J. and Winternitz, W. W.,** The effect of an aldosterone antagonist on the renal response to sodium restriction, *Clin. Sci.*, 20, 143, 1960.
47. **Epstein, M. and Hollenberg, N. K.,** Age as a determinant of renal sodium conservation in normal man, *J. Lab. Clin. Med.*, 87, 411, 1976.

48. **Kirkendall, W. M., Connor, W. E., Abboud, F., Rastagi, S. P., Anderson, T. A., and Fry, M.,** The effect of dietary sodium chloride on blood pressure, body fluids, electrolytes, renal function and serum lipids of normotensive man, *J. Lab. Clin. Med.,* 87, 418, 1976.
49. **Weiner, M. W., Weinman, E. J., Kashgarian, M., and Hayslett, J. P.,** Accelerated reabsorption in the proximal tubule produced by volume depletion, *J. Clin. Invest.,* 40, 1379, 1971.
50. **Stein, J. H., Osgood, R. W., Boonjaren, S., Cox, J. H., and Ferris, T. F.,** Segmental sodium reabsorption in rats with mild and severe volume depletion, *Am. J. Physiol.,* 227, 351, 1974.
51. **Mohammad, G., DiScala, V., and Stein, R. M.,** Effects of chronic sodium depletion on tubular sodium and water reabsorption in the dog, *Am. J. Physiol.,* 227, 469, 1974.
52. **Chou, S., Ferder, L. F., Levin, D. L., and Porush, J. G.,** Evidence for enhanced distal tubule sodium reabsorption in chronic salt-depleted dogs, *J. Clin. Invest.,* 57, 1142, 1976.
53. **Diezi, J., Michoud, P., Aceves, J., and Giebisch, G.,** Micropuncture study of electrolyte transport across papillary collecting duct of the rat, *Am. J. Physiol.,* 224, 623, 1973.
54. **Willis, L. R., Schneider, E. G., Lynch, R. E., and Knox, F. G.,** Effect of chronic alteration of sodium balance on reabsorption by proximal tubule of the dog, *Am. J. Physiol.,* 223, 34, 1972.
55. **Luetscher, J. A., Jr. and Axelrad, B. J.,** Increased aldosterone output during sodium deprivation in normal men, *Proc. Soc. Exp. Biol. Med.,* 87, 650, 1954.
56. **Crabbe, J., Ross, E. J., and Thorn, G. W.,** The significance of excretion of aldosterone during dietary sodium deprivation in normal man, *J. Clin. Endocrinol. Metab.,* 18, 1159, 1958.
57. **Lipsett, M. B. and Pearson, O. H.,** Sodium depletion in adrenalectomized humans, *J. Clin. Invest.,* 37, 1395, 1958.
58. **Brown, J. J., Fraser, R., Lever, A. F., Robertson, J. I. S., James, V. H. T., McCusker, J., and Wynn, V.,** Renin, angiotensin, corticosteroids, and electrolyte balance in Addison's disease, *Q. J. Med.,* 37, 97, 1968.
59. **Higgins, J. T., Jr. and Mulrow, P. J.,** Fluid and electrolyte disorders of endocrine diseases, in *Clinical Disorders of Fluid and Electrolyte Metabolism,* 3rd ed., Maxwell, M. H. and Kleeman, C. R., Eds., McGraw-Hill, New York, 1980, 1291.
60. **Campbell, W. B., Schmitz, J. M., and Itskovitz, H. D.,** (Des-asp') angiotensin II: a study of its pressor and steroidogenic activities in conscious rats, *Endocrinology,* 100, 46, 1977.
61. **Rosenbaum, J. D., Papper, S., and Ashley, M. M.,** Variations in renal excretion of sodium independent of change in adrenal cortical hormone dosage in patients with Addison's disease, *J. Clin. Endocrinol. Metab.,* 15, 1459, 1959.
62. **Martino, J. A. and Earley, L. E.,** Demonstration of the role of physical factors as determinants of natriuretic response to volume expansion, *J. Clin. Invest.,* 46, 1963, 1967.
63. **Schrier, R. W. and de Wardener, H. E.,** Tubular reabsorption of sodium ion; influence of factors other than aldosterone and glomerular filtration rate, *N. Engl. J. Med.,* 285, 1231 and 1292, 1971.
64. **Cizek, L. J. and Huang, K. C.,** Water diuresis in the salt-depleted dog, *Am. J. Physiol.,* 167, 473, 1951.
65. **Vaamonde, C. A. and Papper, S.,** Maintenance of body tonicity, in *Pathophysiology: Altered Regulatory Mechanisms of Disease,* 2nd ed., Frohlich, E. D., Ed., Lippincott, Philadelphia, 1976, 265.
66. **Dunn, F. L., Brennan, T. J., Nelson, A. E., and Robertson, G. L.,** The role of blood osmolality and volume in regulating vasopressin secretion in the rat, *J. Clin. Invest.,* 52, 3212, 1973.
67. **Robertson, G. L. and Athar, S.,** The interaction of blood osmolality and blood volume in regulating plasma vasopressin in man, *J. Clin. Endocrinol. Metab.,* 42, 613, 1976.
68. **Gupta, P. D., Henry, J. P., Sinclair, R., and von Baumgarten, R.,** Responses of atrial and aortic baroreceptors to nonhypotensive hemorrhage and to transfusion, *Am. J. Physiol.,* 211, 1429, 1966.
69. **Henry, J. P., Gupta, P. D., Mechan, J. P., Sinclair, R., and Share, L.,** The role of afferents from the low-pressure system in the release of antidiuretic hormone during nonhypotensive hemorrhage, *Can. J. Physiol. Pharmacol.,* 46, 287, 1968.
70. **Schrier, R. W. and Berl, T.,** Mechanism of the antidiuretic effect associated with interruption of parasympathetic pathways, *J. Clin. Invest.,* 51, 2613, 1972.
71. **Anderson, R. J., Cadnapaphornchai, P., Harbottle, J. A., McDonald, K. M., and Schrier, R. W.,** Mechanism of effect of thoracic inferior vena cava constriction on renal water excretion, *J. Clin. Invest.,* 54, 1473, 1974.
72. **Berliner, R. W. and Davidson, D. G.,** Production of a hypertonic urine in the absence of pituitary antidiuretic hormone, *J. Clin. Invest.,* 36, 1416, 1957.
73. **Harrington, A. R.,** Hyponatremia due to sodium depletion in the absence of vasopressin, *Am. J. Physiol.,* 222, 768, 1972.
74. **Bonjour, J. P. and Malvin, R. L.,** Stimulation of ADH release by the renin-angiotensin system, *Am. J. Physiol.,* 218, 1555, 1970.
75. **Shade, R. E. and Share, L.,** Vasopressin release during nonhypotensive hemorrhage and angiotensin II infusion, *Am. J. Physiol.,* 228, 149, 1975.

76. **Cadnapaphornchai, P., Boykin, J., Harbottle, J. A., McDonald, K. M., and Schrier, R. W.,** Effect of angiotensin II on renal water excretion, *Am. J. Physiol.*, 228, 155, 1975.
77. **Deodhar, S. D., Cuppage, F. E., and Gableman, E.,** Studies on the mechanism of experimental proteinuria induced by renin, *J. Exp. Med.*, 120, 677, 1965.
78. **Stein, R. M., Levitt, B. H., Goldstein, M. H., Porush, J. G., and Levitt, M. F.,** The effects of salt restriction on renal concentrating operation in normal hydropenic man, *J. Clin. Invest.*, 41, 2102, 1962.
79. **Clarke, E., Evans, B. M., MacIntyre, I., and Milne, M. D.,** Acidosis in experimental electrolyte depletion, *Clin. Sci.*, 14, 421, 1955.
80. **Baker, G. P., Levitan, H., and Epstein, F. H.,** Sodium depletion and renal conservation of water, *J. Clin. Invest.*, 40, 867, 1961.
81. **Fitzsimons, J. T.,** The physiological basis of thirst, *Kidney Int.*, 10, 3, 1976.
82. **Richter, C. P.,** Total self-regulatory functions in animals and human beings, *Harvey Lect.*, 38, 63, 1942.
83. **Striker, E. M., Bradshaw, W. G., and McDonald, R. H.,** The renin-angiotensin system and thirst: a reevaluation, *Science*, 194, 1169, 1976.
84. **Dill, D. B., Hall, F. G., and Edwards, H. T.,** Changes in composition of sweat during acclimatization to heat, *Am. J. Physiol.*, 12, 412, 1938.
85. **Conn, J. W.,** The mechanism of acclimatization to heat, *Adv. Intern. Med.*, 3, 373, 1949.
86. **Streeten, D. H., Conn, J. W., Louis, L. H., Fajans, S. H., Seltzer, H. S., Johnson, R. D., Gittler, R. D., and Dube, A. H.,** Secondary aldosteronism; metabolic and adrenocortical responses of normal men to high environmental temperatures, *Metabolism*, 9, 1071, 1960.
87. **Ladell, W. S. and Shepard, R. J.,** Aldosterone inhibition and acclimatization to heat, *J. Physiol. London*, 160, 19P, 1962.
88. **Ratner, A. C. and Dobson, R. L.,** The effect of antidiuretic hormone on sweating, *J. Invest. Dermatol.* 43, 379, 1964.
89. **Arieff, A. I., Llach, F., and Massry, S. G.,** Neurological manifestations and morbidity of hyponatremia: correlations with brain water and electrolytes, *Medicine*, 55, 121, 1976.
90. **Nabarro, J. D., Spencer, A. G., and Stowers, J. M.,** Metabolic studies in severe diabetic ketosis, *Q. J. Med.*, 21, 225, 1952.
91. **Krnjevic, K., Kilpatrick, R., and Aungle, P. G.,** A study of some aspects of nervous and muscular activity during experimental human salt deficiency, *Q. J. Exp. Physiol.*, 40, 205, 1955.
92. **Suki, W. N., Hull, A. R., Rector, F. C., and Seldin, D. W.,** Mechanism of the effect of thiazide diuretics on calcium and uric acid, *Clin. Res.*, 15, 78, 1967.
93. **Ross, E. J. and Christie, S. B. M.,** Hypernatremia, *Medicine*, 48, 441, 1969.
94. **Edelman, I. S. and Leibman, J.,** Anatomy of body water and electrolytes, *Am. J. Med.*, 27, 256, 1956.
95. **Randall, H. T.,** Water and electrolyte balance in surgery, *Surg. Clin. N. Am.*, 32, 445, 1952.
96. **Gamble, J. L.,** *Clinical Anatomy, Physiology and Pathology of Extracellular Fluid*, Harvard University Press, Cambridge, 1954, 106.
97. **Wrong, O., Metcalfe-Gibson, A., Morrison, R. B. I., Ng, S. T., and Howard, A. V.,** In vivo dialysis of faeces as a method of stool analyses, *Clin. Sci.*, 28, 357, 1965.
98. **Spector, W. S., Ed.,** *Handbook of Biological Data*, W. B. Saunders, Philadelphia, 1956.
99. **Fordtran, J. S.,** Speculations on the pathogenesis of diarrhea, *Fed. Proc.*, 26, 1405, 1967.
100. **Sladen, G. E.,** Effects of chronic purgative abuse, *Proc. R. Soc. Med.*, 65, 288, 1972.
101. **Pierce, N. F., Banwell, J. G., Mitra, R. C., Caranasos, G. J., Keimowitz, R. I., Mondal, A., and Manji, P. M.,** Effect of intragastric glucose-electrolyte infusion upon water and electrolyte balance on Asiatic cholera, *Gastroenterology*, 55, 333, 1968.
102. **Carey, R. J. and Burbank, C. B.,** Mucus-secreting villous adenoma of the colon. Presenting as circulatory collapse, *N. Engl. J. Med.*, 267, 609, 1962.
103. **Pilch, Y. H., Kiser, W. S., and Bartter, F. C.,** A case of villous adenoma of the rectum with hyperaldosteronism and unusual renal manifestations, *Am. J. Med.*, 39, 483, 1965.
104. **Lesser, J. H. and Pereira, M. D.,** Electrolyte patterns in gastrointestinal secretions, *Ann. Surg.*, 138, 846, 1953.
105. **Lans, H. S., Stein, I. F., Jr., and Meyer, K. A.,** Electrolyte abnormalities in pyloric obstruction resulting from peptic ulcer or gastric carcinoma, *Ann. Surg.*, 135, 441, 1952.
106. **Needle, M. A., Kaloyanides, G. J., and Schwartz, W. B.,** The effects of selective depletion of hydrochloric acid on acid-base and electrolyte equilibrium, *J. Clin. Invest.*, 43, 1836, 1964.
107. **Kassirer, J. P. and Schwartz, W. B.,** The response of normal man to selective depletion of hydrochloric acid, *Am. J. Med.*, 40, 10, 1966.
108. **Lockwood, J. S. and Randall, H. T.,** Place of electrolyte studies in surgical patients, *Bull. N.Y. Acad. Med.*, 25, 228, 1949.

109. **Kessler, W. R. and Anderson, D. H.**, Heat prostration in fibrocystic disease of pancreas and other conditions, *Pediatrics*, 8, 648, 1951.
110. **Mangos, J. A.**, Microperfusion study of the sweat gland abnormality in cystic fibrosis, *Tex. Rep. Biol. Med.*, 31, 651, 1973.
111. **Peters, J. P., Wakeman, A. M., and Lee, C.**, Total acid-base equilibrium of plasma in health and disease. XI. Hypochloremia and total salt deficiency in nephritis, *J. Clin. Invest.*, 6, 551, 1929.
112. **Platt, R.**, Sodium and potassium excretion in chronic renal failure, *Clin. Sci.*, 9, 367, 1950.
113. **Nickel, J. F., Lowrence, P. B., Leifer, E., and Bradley, S. E.**, Renal function, electrolyte excretion and body fluids in patients with chronic renal insufficiency before and after sodium deprivation, *J. Clin. Invest.*, 32, 68, 1953.
114. **Gonick, H. C., Maxwell, M. H., Rubini, M. E., and Kleeman, C. F.**, Functional impairment in chronic renal disease. I. Studies of sodium-conserving ability, *Nephron*, 3, 137, 1966.
115. **Coleman, A. J., Arias, J., Carter, N. W., Rector, F. C., and Seldin, D. W.**, The mechanism of salt wastage in chronic renal disease, *J. Clin. Invest.*, 45, 1116, 1966.
116. **Polak, A.**, Sodium depletion in chronic renal failure, *J. R. Coll. Physicians London*, 5, 333, 1971.
117. **Schrier, R. W. and Regal, R. M.**, Influence of aldosterone on sodium, water and potassium metabolism in chronic renal disease, *Kidney Int.*, 1, 156, 1972.
118. **Vaamonde, C. A.**, Treatment of chronic renal failure, *J. Fla. Med. Assoc.*, 56, 405, 1969.
119. **Thorn, G. W., Koepf, G. F., and Clinton, M.**, Renal failure simulating adrenocortical insufficiency, *N. Engl. J. Med.*, 231, 76, 1944.
120. **Sawyer, W. H. and Solez, C.**, Salt-losing nephritis simulating adrenocortical insufficiency, *N. Engl. J. Med.*, 240, 210, 1949.
121. **Murphy, R. V., Coffman, E. W., Pringle, B. H., and Iseri, L. T.**, Studies of sodium and potassium metabolism in salt-losing nephritis, *Arch. Intern. Med.*, 90, 750, 1952.
122. **Stanbury, S. W. and Mahler, R. F.**, Salt-wasting renal disease, *Q. J. Med.*, 28, 425, 1959.
123. **Strauss, M. B.**, Clinical and pathological aspects of cystic disease of the renal medulla. An analysis of eighteen cases, *Ann. Intern. Med.*, 57, 373, 1962.
124. **Knowles, R. C., Levitin, H., and Bridges, A.**, Salt-losing nephritis with fixed urinary composition, *Am. J. Med.*, 22, 158, 1957.
125. **Walker, W. G., Jost, L. J., Johnson, J. R., and Kowarski, A.**, Metabolic observations on salt wasting on a patient with renal disease, *Am. J. Med.*, 39, 505, 1965.
126. **Danovitch, G. M., Bourgoignie, J. J., and Bricker, N. S.**, Reversibility of the salt-losing tendency of chronic renal failure, *N. Engl. J. Med.*, 296, 14, 1977.
127. **Anderson, R. J., Linas, S. L., Berns, A. S., Henrich, W. L., Miller, T., Gabow, P. A., and Schrier, R. W.**, Nonoliguric acute renal failure. *N. Engl. J. Med.*, 296, 1134, 1977.
128. **Eiseman, B., Vivion, C., and Vivian, V.**, Fluid and electrolyte changes following the relief of urinary obstruction, *J. Urol.*, 74, 222, 1955.
129. **Bricker, N. S., Shwayri, E. I., Reardan, J. B., Kellog, D., Merrill, J. P., and Holmes, J. H.**, An abnormality in renal function resulting from urinary tract obstruction, *Am. J. Med.*, 23, 554, 1957.
130. **Massry, S. G., Schainuck, L. I., Goldsmith, C., and Schreiner, G. E.**, Studies on the mechanisms of diuresis after relief of urinary tract obstruction, *Ann. Intern. Med.*, 66, 149, 1967.
131. **Platt, R.**, Structural and functional adaptation in renal failure, *Br. Med. J.*, 1, 1313, 1952.
132. **Bricker, N. S., Klahr, S., Lubowitz, H., and Rieselbach, R. E.**, Renal function in chronic renal disease, *Medicine*, 44, 263, 1965.
133. **Biber, T. U., Mylle, M., Baines, A. D., Gottschalk, C. W., Oliver, J. R., and MacDowell, M. C.**, A study by micropuncture and microdissection of acute renal damage in rats, *Am. J. Med.*, 44, 664, 1968.
134. **Bricker, N. S.**, On the meaning of the intact nephron hypothesis, *Am. J. Med.*, 46, 1, 1969.
135. **Levin, D. M. and Cade, R.**, Influence of dietary sodium on renal function in patients with chronic renal disease, *Ann. Intern. Med.*, 62, 231, 1965.
136. **Swales, J. D., Thurston, H., and Pohl, F. J.**, Sodium conservation in renal failure: studies utilizing diazoxide, *Clin. Sci.*, 43, 771, 1972.
137. **Grausz, H., Lieberman, R., and Earley, L. E.**, Effect of plasma albumin on sodium reabsorption in patients with nephrotic syndrome, *Kidney Int.*, 1, 47, 1972.
138. **Jaenike, J. R.**, The renal function defect of postobstructive nephropathy, The effects of bilateral ureteral obstruction in the rat, *J. Clin. Invest.*, 51, 2999, 1972.
139. **Yarger, W. E., Aynedjian, H. S., and Bank, N.**, A micropuncture study of postobstructive diuresis in the rat, *J. Clin. Invest.*, 51, 625, 1972.
140. **Jaenike, J. R.**, The renal response to ureteral obstruction. A model for the study of factors which influence glomerular filtration pressure, *J. Lab. Clin. Med.*, 76, 373, 1970.
141. **Yeh, B. P. Y., Tomko, D. J., Stacy, W. K., Bear, E. S., Haden, H. T., and Falls, W. F., Jr.**, Factors influencing sodium and water excretion in uremic man, *Kidney Int.*, 7, 103, 1975.

142. **Schultze, R. G., Shapiro, H. S., and Bricker, N. S.**, Studies on the control of sodium excretion in experimental uremia, *J. Clin. Invest.*, 48, 869, 1969.
143. **Hayslett, J. P., Kashgarian, M., and Epstein, F. H.**, Mechanism of change in the excretion of sodium per nephron when renal mass is reduced, *J. Clin. Invest.*, 48, 1002, 1969.
144. **Weber, H., Lin, K. Y., and Bricker, N. S.**, Effect of sodium intake on single nephron glomerular filtration rate and sodium reabsorption in uremia, *Kidney Int.*, 8, 14, 1975.
145. **Allison, M. E., Wilson, C. B., and Gottschalk, C. W.**, Pathophysiology of experimental glomerulonephritis in the rat, *J. Clin. Invest.*, 53, 1402, 1974.
146. **Wagnild, J. P., Gutmann, F. D., and Rieselbach, R. E.**, Influence of hydrostatic and oncotic pressure on sodium reabsorption in the unilateral pyelonephritic dog kidney, *Clin. Sci.*, 47, 367, 1974.
147. **Maddox, D. A., Bennett, C. M., Deen, W. M., Glassock, R. J., Knutson, D., and Brenner, B. M.**, Control of proximal tubule fluid reabsorption in experimental glomerulonephritis, *J. Clin. Invest.*, 55, 1315, 1975.
148. **Wilkinson, R., Leutscher, J. A., and Dowdy, A. J.**, Studies on the mechanism of sodium excretion in uremia, *Clin. Sci.*, 42, 711, 1972.
149. **Slatopolsky, E., Elkan, J. O., Weerts, C., and Bricker, N. S.**, Studies on the characteristics of the control system governing sodium excretion in uremic man, *J. Clin Invest.*, 47, 521, 1968.
150. **Bourgoignie, J. J., Hwang, K. H., Espinel, C., Klahr, S., and Bricker, N. S.**, A natriuretic factor in the serum of patients with chronic uremia, *J. Clin. Invest.*, 51, 1514, 1972.
151. **Bourgoignie, J. J., Hwang, K. H., Ipakchi, E., and Bricker, N. S.**, The presence of a natriuretic factor in urine of patients with chronic uremia: the absence of the factor in nephrotic uremic patients, *J. Clin. Invest.*, 53, 1559, 1974.
152. **Fine, L. G., Bourgoignie, J. J., Hwang, K. H., and Bricker, N. S.**, On the influence of the natriuretic factor from patients with chronic uremia on the bioelectric properties and sodium transport of the isolated mammalian collecting duct, *J. Clin. Invest.*, 58, 590, 1976.
153. **Louis, F. and Favre, H.**, Natriuretic factor in rats acutely expanded by Ringer's versus albumin solution, *Kidney Int.*, 18, 20, 1980.
154. **Atchley, D. W., Loeb, R. F., Richards, D. W., Benedict, E. M., and Dresiall, M. E.**, On diabetic acidosis, *J. Clin. Invest.*, 12, 297, 1933.
155. **Mudge, G. H., Faulks, J., and Gilman, A.**, Effect of urea diuresis on renal excretion of electrolytes, *Am. J. Physiol.*, 158, 218, 1949.
156. **Seldin, D. W. and Tarail, J.**, The metabolism of glucose and electrolytes in diabetic acidosis, *J. Clin. Invest.*, 29, 552, 1950.
157. **Gennari, F. J. and Kassirer, J. P.**, Osmotic diuresis, *N. Engl. J. Med.*, 291, 714, 1974.
158. **Vaamonde, C. A., Presser, J. I., Vaamonde, L. S., and Papper, S.**, Renal concentrating ability in cirrhosis. III. Failure of hypertonic saline to increase reduced T^cH_2O formation, *Kidney Int.*, 1, 55, 1972.
159. **Gault, M. H., Dixon, M. E., Doyle, M., and Cohen, W. M.**, Hypernatremia, azotemia and dehydration due to high-protein tube feeding, *Ann. Intern. Med.*, 68, 778, 1968.
160. **Sebastian, A., McSherry, E., and Morris, R. C.**, On the mechanism of renal potassium wasting in renal tubular acidosis associated with the Fanconi syndrome, *J. Clin. Invest.*, 50, 231, 1971.
161. **Vaamonde, C. A., Perez, G. O., and Oster, J. R.**, Syndromes of aldosterone deficiency, *Mineral Electrolyte Metabolism*, 5, 121, 1981.
162. **Stockigt, J. R., Hewett, M. J., Topliss, D. J., Higgs, E. J., and Taft, P.**, Renin and renin substrate in primary adrenal insufficiency. Contrasting effects of glucocorticoid and mineralocorticoid deficiency, *Am. J. Med.*, 66, 915, 1979.
163. **Ufferman, R. C. and Schrier, R. W.**, Importance of sodium intake and mineralocorticoid hormone in the impaired water excretion in adrenal insufficiency, *J. Clin. Invest.*, 51, 1639, 1972.
164. **Berns, A. S., Pluss, R. G., Erickson, A. L., Anderson, R. J., McDonald, K. M., and Schrier, R. W.**, Renin-angiotensin system and cardiovascular homeostasis in adrenal insufficiency, *Am. J. Physiol.*, 233, F509, 1977.
165. **Migeon, C. J.**, Diagnosis and treatment of adrenogenital disorders, in *Endocrinology*, DeGroot, L. J., Cahill, G. F., Martini, L., Nelson, D. H., Odell, W. D., Potts, J. T., Jr., Sternberger, E., and Winegrad, A. I., Eds., Grune & Stratton, New York, 1979, 1203.
166. **Ulick, S.**, Diagnosis and nomenclature of the disorders of the terminal portion of the aldosterone biosynthetic pathway, *J. Clin. Endocrinol. Metab.*, 43, 92, 1976.
167. **Wilson, I. D. and Goetz, F. C.**, Selective hypoaldosteronism after prolonged heparin administration, *Am. J. Med.*, 36, 635, 1964.
168. **Perez, G., Siegel, L., and Schreiner, G. E.**, Selective hypoaldosteronism with hyperkalemia, *Ann. Intern. Med.*, 76, 757, 1972.
169. **Schambelan, M. and Sebastian, A.**, Hyporeninemic hypoaldosteronism, *Adv. Intern. Med.*, 24, 385, 1978.

170. **Knochel, J. P.**, The syndrome of hyporeninemic hypoaldosteronism, *Ann. Rev. Med.*, 30, 145, 1979.
171. **deLeiva, A., Christlieb, A. R., Melby, J. C., Graham, C. A., Day, R. P., Leutscher, J. A., and Zager, P. G.**, Big renin and biosynthetic defect of aldosterone in diabetes mellitus, *N. Engl. J. Med.*, 295, 639, 1976.
172. **Perez, G. O., Lespier, L., Jacobi, J., Oster, J. R., Katz, F. H., Vaamonde, C. A., and Fishman, L. M.**, Hyporeninemia and hypoaldosteronism in diabetes mellitus, *Arch. Intern. Med.*, 137, 852, 1977.
173. **Perez, G. O., Oster, J. R., and Vaamonde, C. A.**, Renal acidosis and renal potassium handling in selective hypoaldosteronism, *Am. J. Med.*, 57, 809, 1974.
174. **Perez, G. O., Oster, J. R., and Vaamonde, C. A.**, Renal acidification in patients with mineralocorticoid deficiency, *Nephron*, 17, 461, 1976.
175. **Perez, G. O., Lespier, L. E., Oster, J. R., and Vaamonde, C. A.**, Effect of alterations of sodium intake in patients with hyporeninemic hypoaldosteronism, *Nephron*, 18, 259, 1977.
176. **Oberfield, S. E., Levine, L. S., Carey, R. M., Bejar, R., and New, M. I.**, Pseudohypoaldosteronism: multiple target organ unresponsiveness to mineralocorticoid hormones, *J. Clin. Endocrinol. Metab.*, 48, 228, 1979.
177. **Sigler, M. H.**, The mechanism of the natriuresis of fasting, *J. Clin. Invest.*, 55, 377, 1975.
178. **Spark, R. F., Arky, R. A., Boulter, P. R., Saudek, C. D., and O'Brien, J. T.**, Renin, aldosterone and glucagon in the natriuresis of fasting, *N. Engl. J. Med.*, 292, 1335, 1975.
179. **Vaamonde, C. A., Sebastianelli, M. J., Vaamonde, L. S., Pellegrini, E. L., Watts, R. S., Klingler, E. L., Jr., and Papper, S.**, Impaired renal tubular reabsorption of sodium in hypothyroid man, *J. Lab. Clin. Med.*, 84, 451, 1975.
180. **Vaamonde, C. A. and Michael, U. F.**, The kidney in thyroid dysfunction, in *The Kidney in Systemic disorders*, 2nd ed., Suki, W. N. and Eknoyan, G., Eds., John Wiley & Sons, New York, 1981, 361.

Alterations of Serum Sodium Concentration

Chapter 11

HYPONATREMIA

Francisco Llach and Anthony W. Czerwinski

TABLE OF CONTENTS

I. INTRODUCTION

Hyponatremia is a frequent medical problem and can produce serious neurological manifestations and death. Plasma sodium (Na^+) and its attendant anions are the main determinants of plasma osmolality and changes in plasma Na^+ usually reflect abnormality in osmolality. The plasma osmolality is normally maintained within narrow limits by homeostatic changes in water intake and water excretion. In a normal physiological situation, water intake usually equals output. Most of the water output involves obligatory losses from the skin and respiratory tract as well as in the urine and stool. Evaporation from the surface of the skin and respiratory tract accounts for approximately 900 mℓ/day. These are the so-called insensible losses. Water is also excreted through the renal route and is usually related to solute excretion. A normal subject excretes approximately 600 mOsm of solute per day and the urine osmolality may reach a maximum of 1200 mOsm/kg; thus, the excretion of 600 mOsm usually requires a minimum urine volume of 500 mℓ/day. The amount of water lost through the gastrointestinal tract is usually small averaging 100 to 200 mℓ/day. This, of course, may be increased in the presence of vomiting or diarrhea.

In order to stay in water balance, water has to be taken to replace the losses; water intake usually comes from three sources: (1) ingested water, (2) water contained in foods, and (3) water produced by the oxidation of carbohydrates, protein, and fats. The ingested water is about 1400 mℓ and the water content of food and oxidation accounts for 1200, totaling 2600 mℓ/day.

Plasma osmolality usually ranges from 275 to 290 mOsm/kg which represents a variation of only 1 to 2%. A decrease in water intake results in an increase in plasma osmolality and is sensed by the hypothalamus, which stimulates the thirst center as well as increases antidiuretic hormone (ADH) secretion, resulting in increased water intake and appropriate decrease in renal water excretion. In hypernatremia and hyperosmolality, the major factor bringing plasma osmolality back to normal is the stimulation of thirst which results in an increase in water intake. An example of this is illustrated by patients with diabetes insipidus, who, despite the absence of ADH, manage to keep plasma osmolality near normal because the large intake of water compensates for the increased renal excretion.

The physiological response to an increase in water intake is a decrease in plasma Na^+ and osmolality which suppresses ADH secretion and thirst, leading to the formation of a dilute urine and the excretion of the excess water. In hypo-osmolality, a decrease in ADH levels and an increase in renal water excretion are the primary events bringing plasma osmolality back to normal.

II. WATER MOVEMENT BETWEEN THE INTRA- AND EXTRACELLULAR SPACE

The distribution of body fluids is determined by the number of osmotically active particles. The presence of an osmotic gradient is the major determinant of water distribution between the intra- and extracellular space; however, both compartments have an osmolality of the same magnitude, and since water can cross almost all membranes, all body fluids are in osmotic equilibrium.[1] The principal extracellular osmol is Na^+ and the principal intracellular osmol is potassium (K^+). It is the steady concentration of these solutes that holds water within these spaces. Sodium tends to move passively into the cell whereas K^+ has a tendency to escape out of the cell. This is prevented by the active transport of Na^+ out of the cell and K^+ into the cell, against a concentration gradient. This is the sodium-potassium pump which is located in the cell membrane and its activity appears to be regulated by an enzyme, sodium-potassium activated adenosine triphosphatase (ATPase). This enzyme catalyzes ATP into adenosine diphosphate and the energy generated is utilized for the active transport of Na^+ and K^+.[2] For further details, see Chapter 1.

III. PLASMA Na^+ AND OSMOLALITY

The osmolality of the plasma is the result of the sum of the osmolality of the individual solutes. Sodium and its attendant anions account for most plasma osmols, though other substances such as glucose and urea are also important. Sodium chloride dissociates roughly into 1.75 particles so that plasma Na^+ concentration has to be multiplied by 1.75. In addition, Na^+ is present only in the aqueous phase of the plasma so that to calculate plasma water Na^+ concentration, the actual plasma Na^+ concentration has to be divided by the factor of 0.93; thus, the following formula can be applied: osmolality of Na^+ salts = $1.75/0.93 \times$ plasma Na^+. However, in clinical medicine and for practical purposes, the osmotic effect can be estimated from the plasma Na^+ concentration multiplied by two.

The contribution of urea and glucose to total plasma osmolality can also be calculated from the formula:

$$\text{mOsm/kg} = \text{mmol}/\ell = \frac{\text{mg/d}\ell \times 10}{\text{mol wt}}$$

The molecular weight of glucose (180) and urea (28) are used as the basis for this calculation. Since urea is measured as blood urea nitrogen, two nitrogen atoms are taken. Thus, mOsm/kg of glucose = glucose × 10/180 = glucose/18. And mOsmol of urea nitrogen = BUN × 10/28 = BUN/2.8.

Thus the plasma osmolality can be estimated from the following formula:

$$\text{plasma osmolality} = 2 \times \text{plasma } Na^+ + \text{glucose}/18 + \text{BUN}/2.8$$

The *effective* plasma osmolality is the result of the number of osmotically active particles acting to bring water within the extracellular space. Since urea can cross the cellular membrane, it is an ineffective osmol. The above-mentioned formula can be reduced to: plasma osmolality = 2 × plasma Na^+ + glucose/18. In general, if the serum glucose is normal, glucose will add only 3 to 10 mOsm/kg; thus, the formula may further be simplified to plasma osmolality = 2 × plasma Na^+. In most clinical conditions, the plasma Na^+ concentration correlates well with osmolality so that when hyponatremia is discussed, hypo-osmolality is usually present.

IV. PLASMA AND BODY OSMOLALITY

Though total body water accounts for 50% of body weight, there is an individual and sexual variation in females vs. males. This is due to the higher percentage of adipose tissue in females which is free of water, resulting in a lower water content of females as compared with males.

The total body water (TBW) is contained in two major compartments: the intracellular and the extracellular. These two compartments are divided by the cell membrane, and the fluid content within these compartments is in an osmotic equilibrium. Thus plasma osmolality is the same as the osmolality of total body water (P osmol = TBW osmol). In this equation, plasma osmolality may be substituted for practical purposes by plasma Na^+. At the same time, TBW osmol is defined by the equation: extracellular

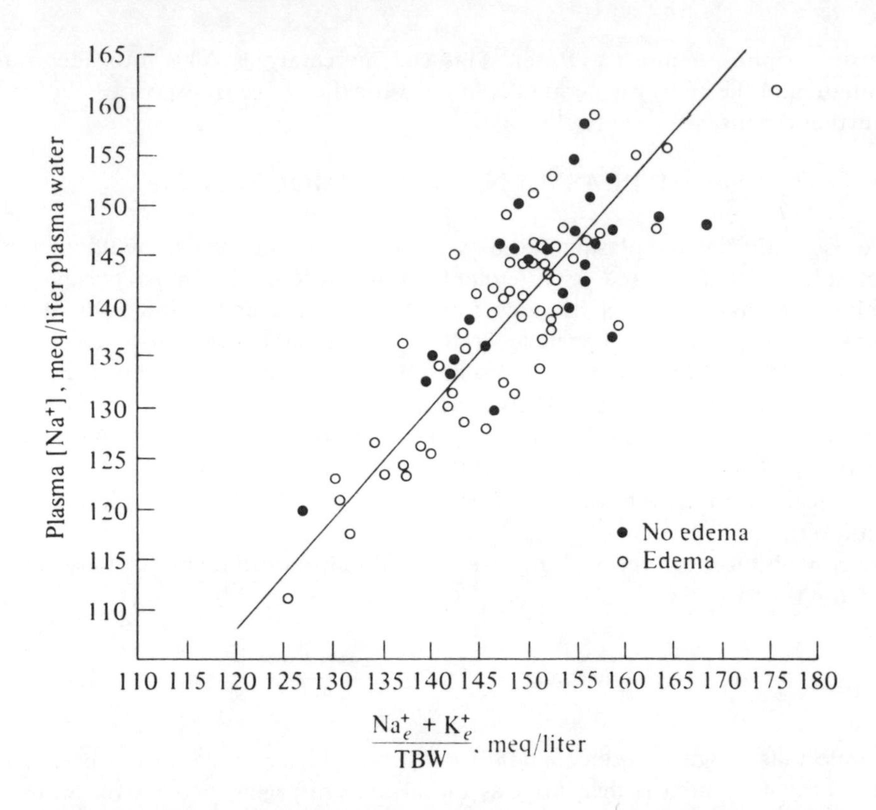

FIGURE 1. Relation between plasma water Na^+ concentration and the ratio of (Na_e^+ + K_e^+/TBW). (From Burton, D. R., *Clinical Physiology of Acid-Base and Electrolyte Disorders*, McGraw-Hill, New York, 1977, 28. With permission.)

solute + intracellular solute/TBW. Since Na^+ salts are the major extracellular solutes and K^+ salts are the primary intracellular solutes, the following equation results:

$$\text{plasma } Na^+ \propto \frac{Na_e + K_e}{TBW}^{3}$$

Na_e and K_e refer to total exchangeable Na and K in the body. The term exchangeable is used because a small fraction of the body's Na and K is bound and as such is osmotically inactive. In Figure 1, the significant correlation between plasma Na^+ and

$$\frac{Na_e + K_e}{TBW}$$

can be noted. This relationship observed in patients is maintained over a wide range of plasma concentrations regardless of the presence or absence of edema. It has important clinical implications which will be discussed. It can be appreciated that hyponatremia may be induced either by Na^+ or K^+ losses and/or water retention.

V. PHYSIOLOGICAL RESPONSE TO CHANGES OF PLASMA OSMOLALITY

An osmotic gradient is a major determinant of water distribution between the extra- and intracellular space. It follows that any variation in osmotic gradient between these compartments will result in the flow of water from the compartment of low to that of

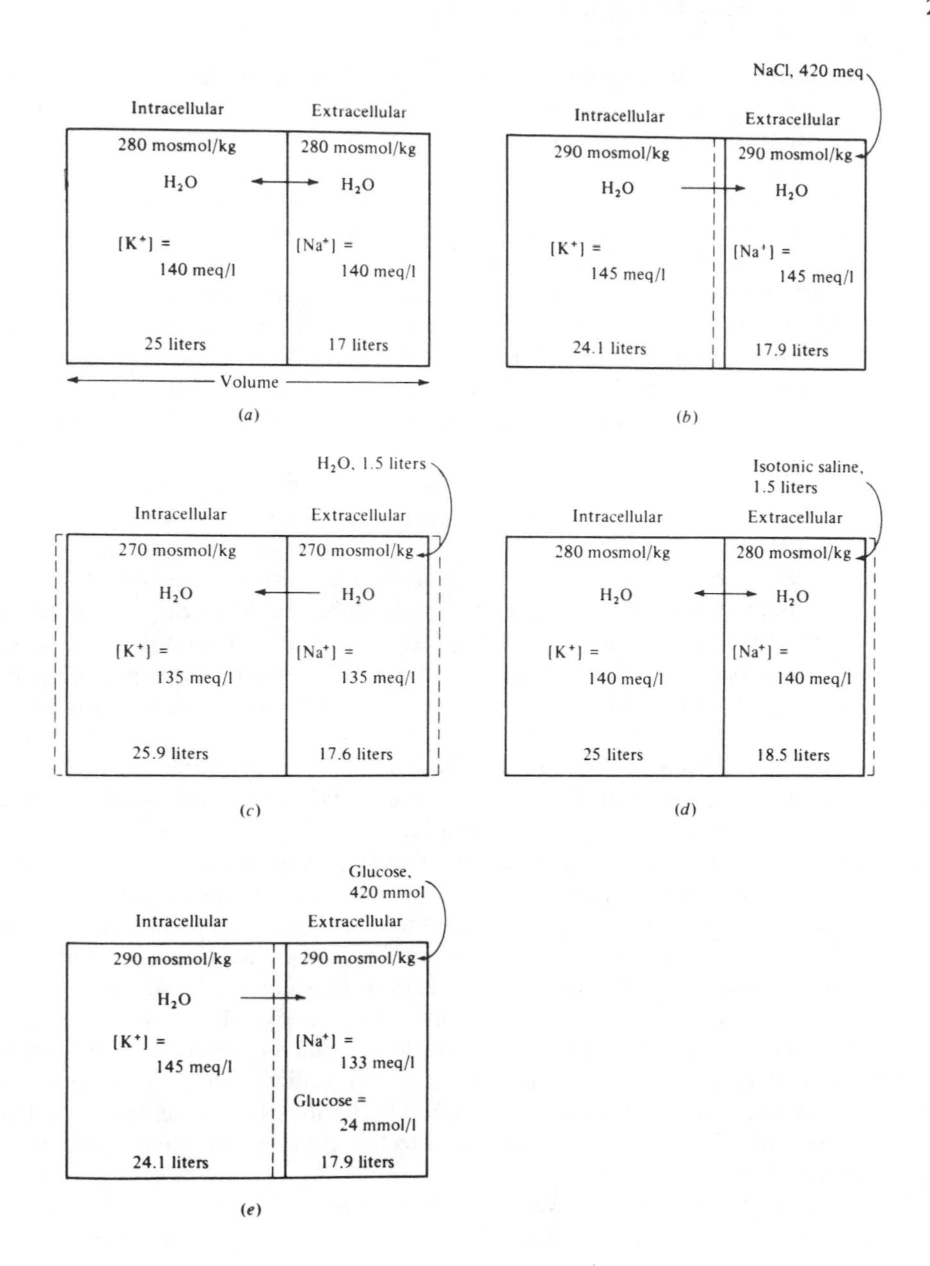

FIGURE 2. Body fluid osmolality and water distribution between body compartments: (*a*) normal state, (*b*) after hypertonic NaCl, (*c*) water (*d*) isotonic saline, and (*e*) glucose. Dash line represents new state after solution administration. (From Burton, D. R., *Clinical Physiology of Acid-Base and Electrolyte Disorders,* McGraw-Hill, New York, 1977, 381. With permission.)

high osmolality in such a way that osmotic equilibrium is reestablished. The changes in osmolality and water movement between compartments induced by the addition of hypertonic NaCl, H_2O, isotonic NaCl and hypertonic glucose are illustrated in Figure 2. The osmolality of the body fluids as well as the distribution of water between both compartments in a steady state are shown in Figure 2*a*. After administration of 420 mEq/ℓ NaCl (2*b*), the osmolality of the extracellular space increases and water moves out of the cells into the extracellular compartment until a new osmotic equilibrium is reached. Plasma Na^+ concentration increases from 140 to 145 mEq/ℓ and the end result is an increase in extracellular space and a decrease in the intracellular space. An important physiological and clinical point is apparent: the osmolar effect of NaCl is distributed throughout all body compartments despite the fact that Na^+, in itself, remains in the extracellular space.

The addition of a hypotonic solution such as 1.5ℓ of water (2*c*) induces an initial decrease in extracellular space osmolality which results in water moving from that space to the intracellular space. The new state is characterized by a low osmolality, a decrease in plasma Na^+ as well as volume expansion of all compartments.

The administration of isotonic NaCl (Figure 2*d*) does not change osmolality and therefore, there are no water shifts; there is only expansion of the extracellular space which eventually will be sensed by the kidney and the excess of sodium and water will be excreted. Thus, the volume and composition of the intracellular space remains unchanged.

The administration of 420 mmol of glucose (Figure 2*e*), an osmotic load identical to that of NaCl (2*b*), induces changes different from those observed after the saline load. There is an increase in extracellular space osmolality, water flows from the intra- to the extracellular compartment and, by dilution, the plasma Na^+ concentration decreases. The magnitude of the increase in osmolality is similar to that observed with saline (290 mOsm/kg), however, the plasma Na^+ concentration is lower. An important clinical point is apparent: in clinical conditions associated with hyperglycemia, the normal relationships between plasma osmolality and plasma Na^+ concentration are not present.

The above-induced changes in plasma osmolality illustrate several important clinical points. First, it can be appreciated that an increase in plasma Na^+ and osmolality results in intracellular dehydration, whereas hyponatremia and hypo-osmolality result in cellular overhydration. It is this overhydration, especially in brain cells with the development of brain edema that accounts for most clinical manifestations of hyponatremia. Second, from the above examples, it can be appreciated that plama Na^+ concentration may be high, low, and normal in the presence of an increase in extracellular fluid volume. Thus, plasma Na^+ is a measure of concentration and not of volume.

Third, hyperglycemia induces similar effects on plasma osmolality and water distribution from those observed by the administration of sodium chloride. However, the plasma Na^+ concentration is lower in the former condition. Thus, in patients with marked hyperglycemia, the presence of a normal plasma Na^+ concentration reflects lack of water shifts from the intra- to the extracellular space and may be evidence of severe intracellular dehydration.

It is important to mention that water movement does not occur when the hyperosmolality is due to an increase in solutes which can cross cell membranes; thus, in patients with renal failure and high levels of serum urea nitrogen (a low molecular weight solute able to cross cell membranes) an osmolar gradient is not established and thus there are no water shifts.

VI. GENERATION OF HYPONATREMIA

As we have mentioned, plasma Na^+ concentration is the major determinant of plasma osmolality and hyponatremia usually represents hypo-osmolality. Hyponatremia is associated with many different disease states. The mechanism by which hyponatremia is generated are multiple according to the various causes. Hyponatremia may be divided into two stages: (1) the mechanisms by which hyponatremia may develop; that is, the generation of hyponatremia, and (2) the mechanisms by which hyponatremia may persist despite therapeutic maneuvers; that is, the maintenance of hyponatremia.

For practical purposes, plasma Na^+ concentration is a reflection of plasma osmolality, which, in turn, is in equilibrium with the total body osmolality; thus the equation:

$$\text{plasma } Na^+ \propto \frac{Na_e + K_e}{TBW}$$

as was illustrated in Figure 1 holds over a very wide range of plasma Na^+ concentrations in hyponatremic patients. From this equation, it can be appreciated that hyponatremia can be produced either by Na^+ or K^+ loss or by increases in TBW, that is, water retention. Any K^+ losses from the extracellular fluid theoretically induce hyponatremia. As K^+ is lost, the extracellular K^+ concentration falls, K^+ moves out of the cells into the extracellular fluid, down a concentration gradient and in order to maintain electroneutrality, Na^+ enters the cells leading to a reduction in the plasma Na^+ concentration.[4-5] As is discussed later, there are certain clinical conditions associated with hypokalemia which will result in hyponatremia. A more common factor in the generation of hyponatremia is the loss of Na^+ via the kidney or the gastrointestinal tract. However, physiologically, it is rare to lose solute in excess of water; whenever there is electrolyte loss via vomiting or diarrhea, such losses are hypotonic or isotonic; with the loss of Na^+, water goes with it, resulting in fluid losses which will not change plasma Na^+ and osmolality. Fluid loss will lead to volume depletion and the decrease in blood volume stimulates the thirst center and results in an increase in water intake; this, in turn, induces a mild but significant decrease in plasma osmolality which suppresses both the thirst and ADH secretion, leading to the excretion of the water excess and the restoration of normal plasma osmolality. Thus, in the final analysis, an increase in water intake is necessary to cause hyponatremia; without water ingestion, hyponatremia cannot be generated.

VII. MAINTENANCE OF HYPONATREMIA

As mentioned, the physiological response to hypo-osmolality is inhibition of thirst and ADH secretion; the latter results in renal excretion of water excess. With no ADH present in the circulation, the urine osmolality decreases to 50 to 100 mOsm/kg and the kidney is able to excrete up to 15 to 20 ℓ of solute-free water per day; thus, a normal man has the ability to ingest large amounts of water without diluting his body fluid. This capability is dependent upon the capacity of the kidney to excrete large volumes of urine at an osmolality lower than plasma. With such a great capacity to excrete free water, it is unlikely that hyponatremia resulting from water retention will develop in a normal individual unless there is a defect in renal water excretion. The exception to this rule may be severe psychogenic polydipsia where the ingestion of large volumes of fluid may overcome the renal excretory capacity.

Since water retention and hyponatremia occur only if there is a defect in renal water excretion, it is appropriate now to review the various factors involved in the renal diluting system. The renal excretion of free water is dependent mainly on three factors. They are schematically depicted in Figure 3.

(1) The amount of fluid from the glomerular filtrate delivered to the distal segment of the nephron is important in the dilutional process. Proximal tubular fluid reabsorption is iso-osmotic and by the time this fluid reaches the loop of Henle, there has been an 80% reduction in its volume; therefore, although it does not directly contribute to urinary dilution, any decrease in the volume of this fluid delivered to the distal tubule will limit the rate of renal water excretion. A decrease in glomerular filtration and an increase in proximal tubular reabsorption are the most common factors resulting in a

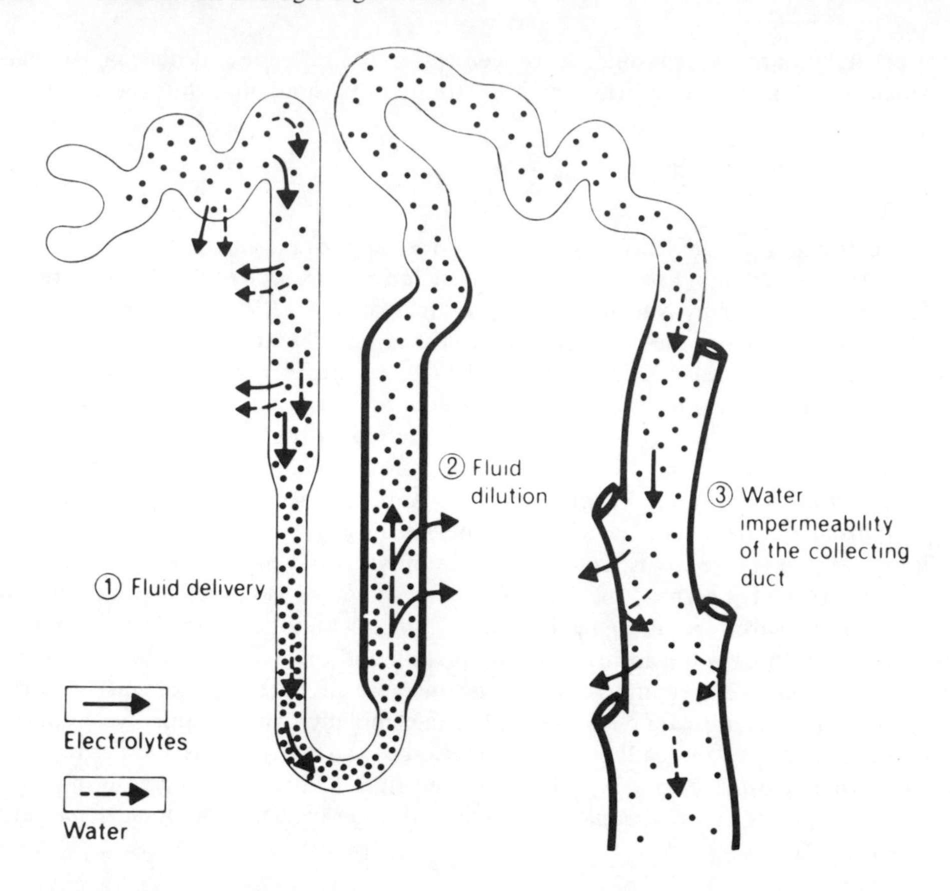

FIGURE 3. Schematic representation of the basic factors involved in renal excretion of water. (From Berl, T. and Schrier, R. W., in *Sodium and Water Homeostasis,* Brenner, B. M. and Stein, J. H., Eds., Churchill-Livingstone, New York, 1978, 3. With permission.)

diminution of the amount of fluid delivered to the distal tubule. The most common clinical setting leading to such abnormalities is a decrease in plasma volume, i.e., dehydration, hemorrhage.

Thus, water retention occurs in this condition despite the fact that other components of the renal diluting system may be functionally normal.

(2) Dilution of the tubular fluid takes place in the ascending limb of Henle. This segment of the nephron is water impermeable and is characterized by the active reabsorption of chloride which results in the selective reabsorption of solutes in excess of water and converts the hypertonic tubular fluid coming from the descending loop of Henle into hypotonic fluid.[6-8] By the time the tubular fluid reaches the distal convoluted tubule, it is hypotonic. Any abnormality or process disturbing the function of the ascending loop of Henle will eventually, by impairing the active resorption of solute, diminish the rate of renal water excretion which, coupled with increased water intake, will result in hyponatremia; an example is the administration of furosemide, a loop diuretic which inhibits active chloride reabsorption.

(3) The permeability to water of the collecting duct epithelium is dependent upon the presence or absence of ADH.[9] In the absence of ADH, the luminal segment of cell membrane is impermeable to water and there is no water transport across the collecting duct epithelium into the surrounding capillary beds. At the same time, the tubular fluid is made progressively more hypotonic by a continued active solute resorption. The end result is a hypotonic tubular fluid which reaches the lowest osmolality observed throughout the nephron between 15 and 20 mOsm/ℓ. This sequence is observed

in situations of maximal water diuresis, i.e., in patients with diabetes insipidus or compulsive water diuresis. In the presence of ADH through binding to cell membrane receptor and formation of cyclic adenosine monophosphate, there is an increased permeability of luminal plasma membrane allowing back diffusion of water along established concentration gradients into capillary beds in the renal medullary tissue.[10,11] The usual maximal concentration achieved in the urine in the hydropenic man is approximately 1000 mOsm/ℓ. Higher urine concentrating values reaching 1400 mOsm/ℓ may occur with prolonged water deficits. There are clinical conditions where the abnormal persistence of ADH in the circulation will increase free water reabsorption and interfere with renal water excretion, leading to hyponatremia.

In summary, an intact renal diluting system is necessary so that large amounts of water may be ingested without inducing hyponatremia. However, most causes of hyponatremia are not associated with an increased water intake above the renal excretory capacity; rather they are due to a defect in the renal diluting system involving one or most of the factors mentioned above.

VIII. CLINICAL COURSES OF HYPONATREMIA

A useful classification of the hyponatremic states according to the status of total body Na^+, water, and clinical features has been suggested by Berl et al.[12] These include: (1) hypovolemic hyponatremia, associated with decreased extracellular fluid volume; (2) normovolemic hyponatremia associated with minimal expansion of ECFV and no edema formation; and (3) hypervolemic hyponatremia associated with expanded extracellular fluid volume and edema formation.

A. Hypovolemic Hyponatremia

The common denominator of this group of disorders is the presence of volume contraction which leads to increased proximal tubular reabsorption, followed by decreased distal tubular delivery, which in turn results in impaired water excretion. In addition, circulating volume is an important nonosmolar factor controlling the release of ADH. Changes in blood volume are sensed by the intrathoracic receptors located in the carotid sinus, aortic arch, left atrium, and great pulmonary veins;[13,14] they respond to tension developed in the walls of the receptor organ and via the glossopharyngeal afferents; the impulses travel along the medulla and reticular formation of midbrain to reach the diencephalus and the supraoptic nuclei in the hypothalamus, resulting in increased ADH release. Afferent parasympathetic fiber also provide a link between these baroreceptors and the hypothalamus.[15] In the presence of volume contraction and a decrease in blood pressure, the baroreceptors are activated, resulting in an increased release of ADH. Although not firmly established, the renin-angiotensin system may also be involved in the increased release of ADH due to volume contraction.[16] Finally, hypovolemia stimulates thirst, which in turn, by increasing water intake, aggravates the hyponatremia.

The most common causes of volume contraction leading to hyponatremia are listed in Table 1. Approximately 30% of all hyponatremic patients belong to this group. A clinical history of fluid loss, the presence of decreased skin turgor, flat neck veins, dry mucous membranes, and hypotension are all suggestive of volume depletion. The two major sources of fluid loss are the kidneys and the gastrointestinal tract. Diarrhea and vomiting are the most frequent causes of extrarenal fluid loss; the kidney responds to such losses by increasing tubular Na^+ reabsorption which is reflected in a urinary Na^+ concentration of less than 10 mEq/ℓ and a high urinary osmolality. In the presence of renal disease or diuretics, this may not apply because the kidney can lose the capacity to conserve Na^+.

Table 1
ETIOLOGY OF HYPOVOLEMIC HYPONATREMIA

Renal causes (Urinary Na^+ concentration 20 mEq/ℓ)	Extrarenal causes (Urinary Na^+ concentration 20 mEq/ℓ)
Diuretic abuse	Gastrointestinal losses
Osmotic diuresis	Diarrhea, vomiting, "third space", intestinal obstruction
Adrenal insufficiency	
Salt losing nephritis	Skin losses: burns
Renal tubular acidosis	Blood loss

The single most common cause of hyponatremia may be diuretic therapy; many diuretics such as furosemide and ethacrynic acid act directly by inhibiting chloride reabsorption in the ascending loop of Henle which impairs the rate of water excretion.[17] The urine Na^+ concentration is usually greater than 20 mEq/ℓ despite the presence of hypovolemia. In addition to the stimulation of ADH release by the hypovolemia, hypokalemia is also induced and plays an important role in the generation of the hyponatremia; the intracellular concentration of K^+ is diminished and induces shifts of Na^+ into the cell; thus, these patients may have a marked decrease in exchangeable K^+ which also may exaggerate the volume receptor mediated release of ADH.[5] Potassium replacement in these patients may result in the correction of the hyponatremia even in the absence of water loss and a positive sodium balance.[5]

An infrequent cause of volume depletion leading to hyponatremia is adrenal insufficiency. The major factor seems to be an ADH-independent increase in water permeability of the collecting duct;[8] however, diarrhea, vomiting, and excessive losses of sodium via the kidney also play a role.[19] In addition, plasma levels of ADH have been found to be high.[20] The administration of cortisol without necessarily replacing volume usually induces a marked increase in the ability to excrete free water.[18] This response is rapid and occurs even in patients with diabetes insipidus and no ADH present, suggesting a direct effect of hydrocortisone on the collecting duct epithelium.[21] This response has important clinical implications because an acute clinical situation can be quickly corrected by the oral or intravenous administration of glucocorticoids in contrast to the other causes of hyponatremia, which do not respond to steroid administration.[22]

Salt-losing nephritis is another cause of hypovolemic hyponatremia. Though medullary cystic disease, polycystic disease, and chronic pyelonephritis are among the listed causes of salt-losing nephritis, advanced chronic renal failure is the most common. Patients with advanced renal insufficiency on a low Na^+ diet cannot achieve an appropriate minimal urinary Na^+ concentration. They usually have an obligatory loss of 10 to 15 mEq/ℓ of Na^+. As long as these patients are maintained on a normal Na^+ diet, they do well; however, once they are placed on a restricted Na^+ diet, they may become volume depleted.[23]

Renal tubular acidosis, especially those of the proximal type, also have renal wastage of Na^+ and K^+. These patients have a defect in the reclaiming of the bicarbonate in the proximal tubule which induces a marked bicarbonaturia, forcing the excretion of Na^+.[24]

B. Normovolemic Hyponatremia

This group represents at least 50% of hyponatremic patients seen in the hospital. It is characterized by a mild increase in ECFV, primarily due to water retention, no Na^+

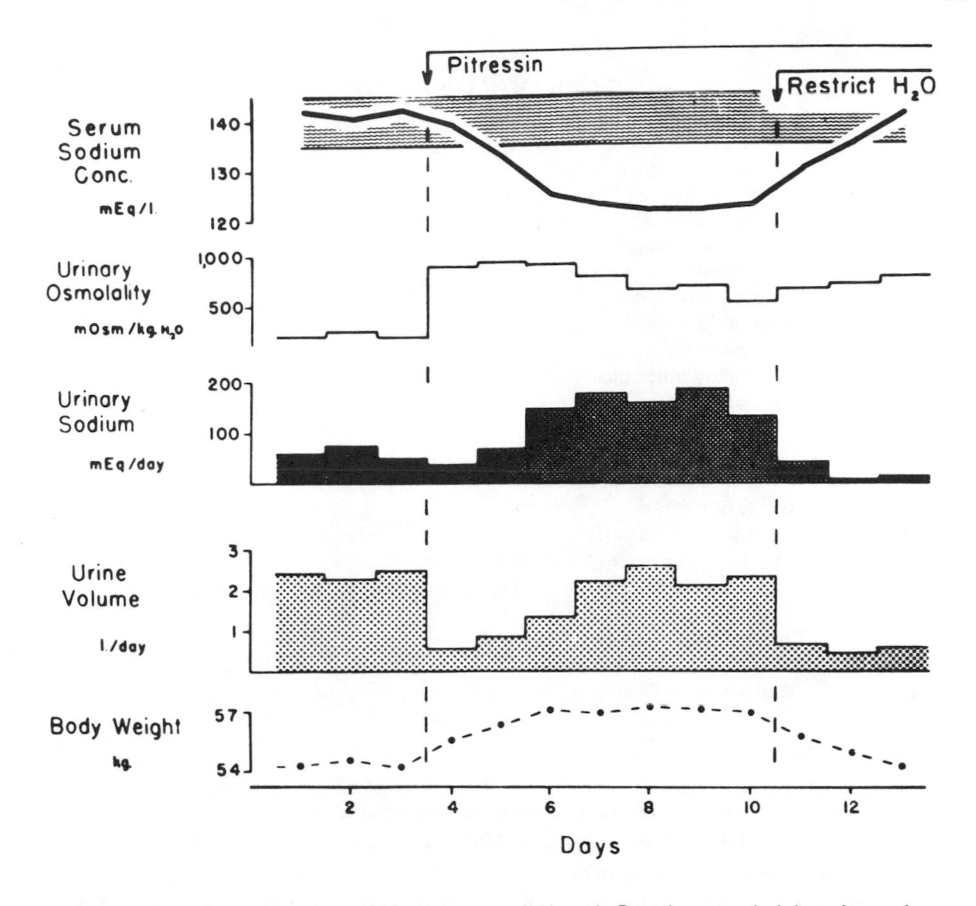

FIGURE 4. Schematic summary of the effects of Pitressin® and water administration and restriction to normal subjects. (From Kleeman, C. R., in *Clinical Disorders of Fluid and Electrolyte Metabolism,* 2nd ed., Maxwell, M. H. and Kleeman, C. R., Eds., McGraw-Hill, New York, 1962, 286. With permission.)

retention and the absence of edema formation. In general, clinical signs of fluid overload cannot be detected, urinary Na^+ excretion is usually greater than 20 mEq/ℓ, which in general, reflects dietary Na^+. If these patients are placed on a low Na^+ diet, they respond appropriately by conserving Na^+ and the urinary Na^+ excretion is low. Appropriate or inappropriate high levels of ADH are almost always present in these patients.

The clinical conditions associated with this group are given below.

1. The Syndrome of Inappropriate ADH Secretion (SIADH)

This syndrome was clearly defined by Schwartz and Bartter in 1957;[25] they reported two patients with bronchogenic carcinoma presenting with hyponatremia and a defect in urinary dilution. They concluded from indirect evidence that this symdrome was due to inappropriate secretion of ADH. The fundamental studies on which these authors based their clinical conclusions were performed earlier on the effect of ADH in normal subjects.[26] Thus the exogenous administration of ADH resulted in hyponatremia due to dilution caused by water retention and excessive urinary Na^+ excretion.[27] These changes were absent if water intake was restricted (Figure 4).

The common denominator is the presence of high ADH levels in the absence of normal physiologic stimulus, such as hyperosmolality or hypovolemia. The hormonal effect of ADH is to enhance renal water reabsorption, ingested water is retained, resulting in dilutional hyponatremia and hypo-osmolality with expansion of body fluids.

Table 2
ETIOLOGY OF SIADH

- Malignancies
  - Bronchogenic carcinoma
  - Adenocarcinoma of the pancreas
  - Lymphosarcoma and leukemias
  - Adenocarcinoma of the duodenum
  - Carcinoma of the thymus
  - Prostatic carcinoma
- Pulmonary disorders
  - Bronchogenic carcinoma
  - Pulmonary tuberculosis
  - Pulmonary abscess
  - Pneumonia: viral or bacterial
  - Chronic obstructive pulmonary disease
  - Ventilator-assisted ventilation
  - Cystic fibrosis
- Neuropsychiatric disorders
  - Infections: meningitis, encephalitis, abscess
  - Vascular: thrombosis, subarachnoid hemorrhages, subdural hematoma
  - Tumors: primary or metastatic
  - Trauma: skull fracture, CNS bleeding
  - Miscellaneous: Guillain-Barré syndrome, acute porphyria, acute psychosis, seizure disorders, central pontine myelinolysis
  - Drugs: (See Table 3)
  - Idiopathic

The increase in intravascular volume due to water retention decreases tubular reabsorption of Na and will increase urinary Na^+ excretion. Thus, in general, urinary Na^+ concentration is greater than 20 mEq/ℓ.

Although water retention decreases plasma Na^+ concentration by dilution it does not reduce the plasma HCO_3 concentration or plasma K^+ concentration. It appears that in hypotonic volume expansion, plasma HCO_3 concentration is maintained at normal levels by the movement of hydrogen ions into cells as well as by increased renal hydrogen ion excretion.[28] The intracellular movement of hydrogen ion is coupled with movement of K^+ out of the cell, which, in turn, maintains plasma K^+ concentration within normal range. In the presence of severe hypo-osmolality and hyponatremia, renal K^+ excretion is increased and mild hypokalemia may be present.[29] Recently, it has been shown that due to the volume expansion, uric acid excretion is usually increased and the plasma uric acid concentration is low (2.9 ± 0.4 ng/dℓ) in patients with this syndrome where other hyponatremic patients had normal levels.[30]

The various causes associated with SIADH are listed in Table 2. Hyponatremia occurs with a variety of malignant tumors, especially bronchogenic carcinoma.[25] The mechanism by which tumors can produce inappropriate antidiuresis may result from two distinct alterations in physiology. The first to be recognized in these patients was the production by the tumor cell of an aberrant peptide closely resembling ADH.[31,32] Early, it was thought that most cases of SIADH associated with malignancy were due to ectopic ADH production. However, 50% of these tumors have failed to show secretion of detectable amounts of antidiuretic substances.[31] The second mechanism may result from the endogenous secretion of ADH from the supraoptic hypophyseal system in quantities inappropriate for the normal physiological control; whether this is due to hypovolemia, invasion of the parasympathetic nerves, metastasis to the hypothalamus, or ectopic production of a neurotransmitter for ADH is still unknown.

Pulmonary disorders have frequently been associated with SIADH. Chronic pulmonary tuberculosis was referred to in the earlier literature as pulmonary salt wast-

Table 3
DRUGS CAUSING THE SIADH

ADH analogues
Lysine vasopressin (LVP)
1-diamino-8-D arginine vasopressin (DDAVP)
Oxytocin
Nicotine
Narcotics
Hypoglycemic agents
Chlorpropamide
Phenformin
Clofibrate
Tricyclic Agents
Carbamazepine (Tegretol®)
Amitriptyline (Elavil®)
Thiothixene (Navane®)
Fluphenazine
Thioridazine (Mellaril)
Cytotoxic agents
Vincristine
Cyclophosphamide
Analgesics
Acetaminophen
Indomethacin
Aspirin
Ibuprofen

ing.[25] The presence of ADH has been demonstrated in tuberculous lung tissue in a patient with pulmonary tuberculosis and SIADH.[33] However, ectopic production by the tissue was not demonstrated and the presence of ADH may have represented nonspecific hormone adsorption. Chronic obstructive pulmonary disease, acute lobular pneumonia, and empyema have all been associated with the SIADH. In some of these cases, the hypothesis has been advanced that stress on the pulmonary veins and/or left atrium receptors may have accounted for the high levels of ADH.

A variety of central nervous system and neuropsychiatric disorders have been reported to be associated with SIADH. Various types of malignancies, meningitis, encephalitis, cerebrovascular accidents, and trauma have all been associated with hyponatremia and SIADH.[25] Most of the data from these patients suggesting an increased ADH secretion is indirect and comes from the ability of ethanol, an inhibitor of ADH secretion, to lower urine osmolality and induce water diuresis;[34] in a few cases, the demonstration of ADH in the plasma or in the urine has been reported.[35]

Over the last few years, it has become clear that there are a number of drugs which can impair renal water excretion, leading to hyponatremia and mimicking the SIADH.[36] These drugs can impair water excretion by stimulation of ADH release from the neurohypophysis, by potentiation of the action of ADH on the renal medulla or by a nonmediated ADH alteration in renal function.[37]

The various drugs capable of inducing hyponatremia are listed in Table 3. ADH analogues have been developed and used primarily for the treatment of diabetes insipidus. Lysine vasopressin was one of the first to be developed;[38] recently 1-diamino-8-D-arginine vasopressin (DDAVP) has been synthesized. It has marked antidiuretic activity and, when administered as a nasal spray to patients with diabetes insipidus, its effects last 12 to 22 hr.[39] Oxytoxin is another product which, when administered in large amounts, may cause water retention. The intravenous administration of this hormone to stimulate labor in pregnant women has been reported to cause water retention and severe hyponatremia.[40]

Nicotine has long been recognized as a potent antidiuretic agent and is used as a clinical stimulation test of ADH secretion.[41] An increase in ADH levels in normal nonsmokers occurs after smoking two cigarettes over a period of 5 min.[42] It appears that such an effect on ADH release is mediated by a stimulation of the baroreceptors in the carotid sinus.[43]

Narcotics such as morphine have also been shown in the experimental animal to cause ADH release from the hypophysis and therefore may be implicated in the hyponatremia developing in the postoperative patient.[44]

Chlorpropamide, an oral hypoglycemic agent widely used in the diabetic patient, has been reported to cause hyponatremia in 4% of treated patients.[45] It can produce symptomatic hyponatremia even in patients with diabetes insipidus. It acts by potentiating the peripheral effect of ADH as well as by increasing ADH secretion.[44] Several other sulfonylurea agents, glyburide, tolazamide, and acetohexamine have diuretic action in hydrated normal subjects; this diuretic action most likely is due to inhibition of tubular reabsorption and is independent of ADH. These agents may be useful in the diabetic patient developing hyponatremia while on chlorpropamide.

Tricyclic agents such as carbamazepine (Tegretol®) has been used in the treatment of diabetes insipidus and may induce hyponatremia resembling the SIADH.[47] Its antidiuretic actions seem to be mediated via endogenous ADH release. Other tricyclic agents such as amitriptyline (Elavil®), thiothixene (Navane®) and fluphenazine, have been associated with hyponatremia.[48-49] Clofibrate has antidiuretic action in patients with diabetes insipidus; it appears to exert its antidiuretic action via the release of ADH from the hypophysis and it is ineffective in patients with nephrogenic diabetes insipidus.[50] Vincristine, an agent widely used in the treatment of leukemias, lymphomas, and other tumors is not an infrequent cause of hyponatremia.[51] A toxic mechanism in the neurohypophysis increasing ADH release is the most likely explanation.[52] Cyclophosphamide, a widely used immunosuppressive agent, has also been reported to cause hyponatremia, especially if the medication is given intravenously.[53]

Although, in the majority of patients, the cause of SIADH is immediately apparent, sometimes there may not be obvious causes and the diagnosis of an idiopathic form of SIADH is considered. Although isolated SIADH cases of long duration without apparent cause have been reported,[54] one must be cautious about the diagnosis of idiopathic SIADH. Sometimes, a malignant tumor may become apparent much later; there have been cases of this syndrome preceding overt malignant disease by 6 months.[55]

The criteria for the diagnosis of SIADH is (1) hyponatremia and hyperosmolality; (2) presence of inappropriately elevated urine osmolality, usually greater than 100 mOsm/ℓ, (3) a high urine Na^+ (greater than 20 mEq/ℓ), (4) the presence of clinical normovolemia, (5) normal renal and adrenal function, and (6) normal acid base and K^+ balance. It is also understood that any of these features may be absent and their presence may not be necessary for this diagnosis; thus, the hyperosmolality and hyponatremia respond to fluid restriction and, if the patient is not taking enough water, hyponatremia may not be present. Likewise, if the patient is on a low salt or water intake, urinary Na^+ may be low. Also, the presence of renal and adrenal disease may make the diagnosis more difficult.

A more direct evidence for the diagnosis of this syndrome comes from the recent development of radioimmunoassay which allows accurate measurement for ADH. The plasma ADH levels in patients with SIADH are not different from those of normal subjects; however, if the levels are correlated with plasma osmolality, they are abnormally high (Figure 5).

Figure 6 displays four patterns of response to hypertonic saline infusion which have been identified in patients with SIADH:[35] Type A pattern shows erratic fluctuations

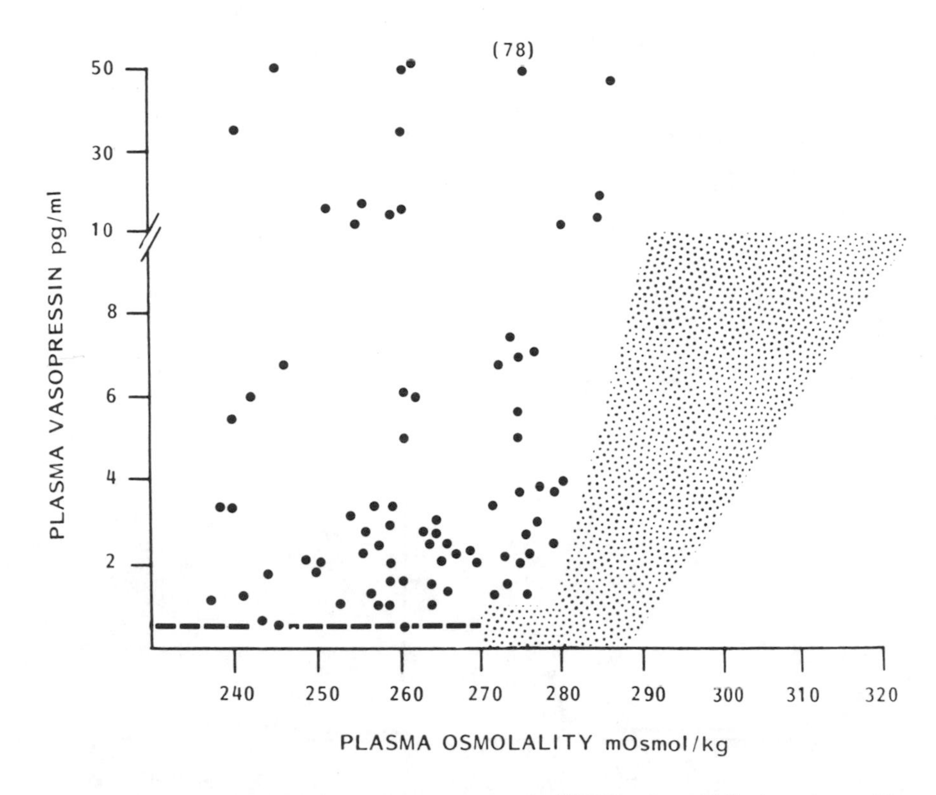

FIGURE 5. Relationship between plasma vasopressin (ADH) and osmolality in patients with SIADH. Shaded area represents the range of normal values. (From Zerbe, R., Stropes, L., and Robertson, G., *Annu. Rev. Med.*, 31, 317, 1980. With permission.)

of ADH levels which appear to be completely independent of osmotic control. This is a frequent pattern, occurring in 37% of patients. Type B shows a prompt and progressive increase in plasma ADH in response to hyperosmolality; it appears that the osmotic regulation of ADH is preserved but has been reset at a lower level. The term, essential hyponatremia or reset osmostat, has been used to describe this syndrome.[56] The mechanism by which this occurs is unknown; however, it should be noted that these patients behave quite differently from other forms of SIADH since they can produce a maximally diluted urine provided that they are made sufficiently hyponatremic. Approximately 33% of patients with SIADH have this pattern, including some patients with malignancies. The Type C pattern is characterized by a constant nonsuppressible leak of ADH and normal osmoreceptor function. At the normal range of plasma osmolality, ADH is released appropriately but below 278 mOsm/kg, there is still a constant secretion of ADH which is not suppressed by the hypo-osmolality. This pattern has been seen in 16% of patients with SIADH, especially in those with meningitis and skull fractures. Type D is characterized by low levels of ADH which are not increased by hyperosmolality. These patients continue to have a concentrated urine despite the presence of plasma hypo-osmolality and low ADH levels suggesting that antidiuretic factors other than ADH may be present. This has been observed in 10% of patients.

A useful feature of the diagnosis of SIADH is the presence of hypouricemia (Figure 7). It has been recently reported that these patients had substantial hypouricemia where those with hyponatremia due to other causes had normal levels of serum uric acid. The mechanism for such hypouricemia seems to be volume expansion and increased renal clearances of uric acid; water restriction in these patients did correct not only the hyponatremia but also the hypouricemia (Figure 8).

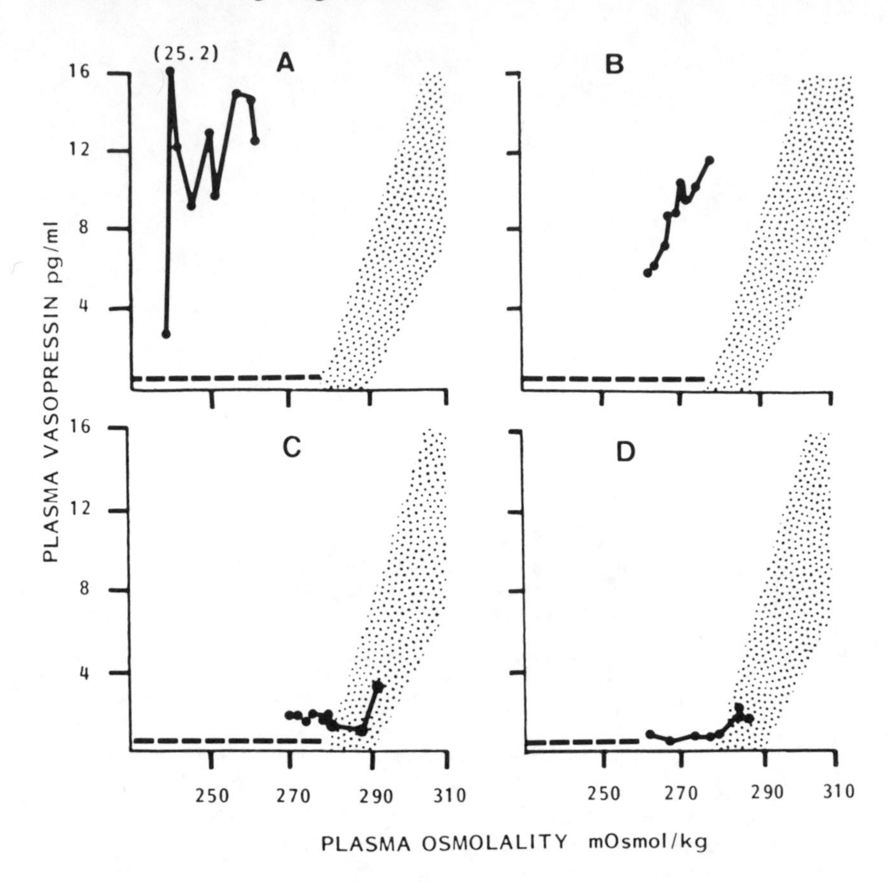

FIGURE 6. Patterns of vasopressin (ADH) response during hypertonic saline in patients with SIADH. (From Zerbe, R., Stropes, L., and Robertson, G., *Annu. Rev. Med.*, 31, 319, 1980. With permission.)

2. Hypothyroidism

Hyponatremia is infrequent in patients with myxedema. It has been reported only in a few cases with myxedema coma.[57] The hypothyroid patient cannot excrete a water load normally.[58] The mechanism of this defect in water excretion by the kidney is not clear. Studies on hypothyroid rats with congenital absence of ADH have shown that a decrease in glomerular filtration rate and distal sodium chloride delivery may be the most important pathogenic factor. Measurements of ADH in the hyponatremic hypothyroid patient are needed in order to solve this question. In general, the diagnosis is not a problem and should be considered only in the comatose patient. The administration of thyroid hormone rapidly corrects the clinical picture and the hyponatremia.[58]

3. Glucocorticoid Deficiency

Patients with Addison's disease are unable to excrete a water load normally and may develop hyponatremia.[59] In general, these patients present with prerenal azotemia due to volume depletion, hyperkalemia, acidosis, and hypotension. For this reason, it has been discussed in the group of hypovolemic hyponatremia; however, sometimes the clinical presentation may resemble SIADH. As previously discussed, the hyponatremia of glucocorticoid deficiency can be quickly corrected by the oral or intravenous administration of glucocorticoids, in contrast with the SIADH which will not be corrected by steroid administration.[22]

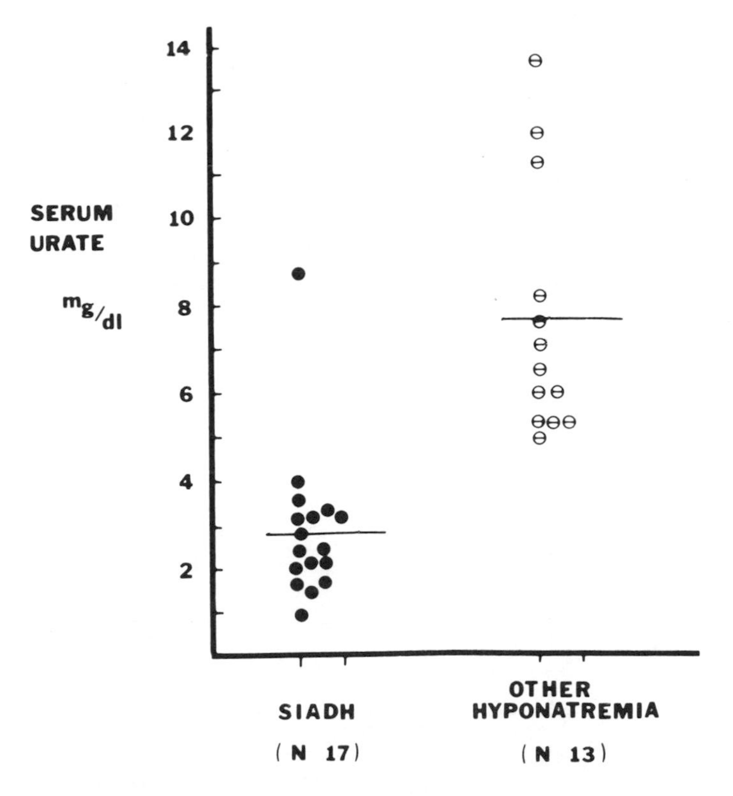

FIGURE 7. Serum urate concentration during hyponatremia in patients with SIADH as compared with other hyponatremic patients. (From Beck, L. H., *N. Engl. J. Med.*, 301, 528, 1979. With permission.)

4. Stress

Physical pain and surgical stress and emotional stimulus have been shown to alter renal water excretion and may lead to hyponatremia, especially if a high water intake is maintained. A variant of this group is the hyponatremia of the postoperative patients in which seemingly inappropriate ADH secretion may persist for several days. An increase in parasympathetic tone via afferent nerve caused by the stress may increase ADH release.[61] Other factors may also play a role: the administration of analgesics which may stimulate ADH release; manipulation of organs transmitting stress signals to the hypothalamus via afferents; and administration of intravenous hypotonic fluids may all contribute to the hyponatremia.

5. Psychogenic Polydipsia

This syndrome has been described in middle-aged women who ingested as much as 10 to 20 ℓ of water per day, leading to polyuria and on rare occasions hyponatremia. In general, these patients are not severely hyponatremic; the plasma Na^+ is at the lower end of the normal range; however, in rare instances, water intake may exceed 15 to 20 ℓ/day, overwhelming the renal capacity to excrete water and resulting in hyponatremia as low as 84 mEq/ℓ.[62] In such cases, urine osmolality is low and maximally diluted, reflecting appropriate inhibition of ADH secretion.[62] In some other cases, the hyponatremia has been associated with high urine osmolality, suggesting a defect in water excretion.

An interesting variety of this disorder is the hyponatremia attributed to consumption of excessive quantities of beer. These cases are complicated by the fact that liver and

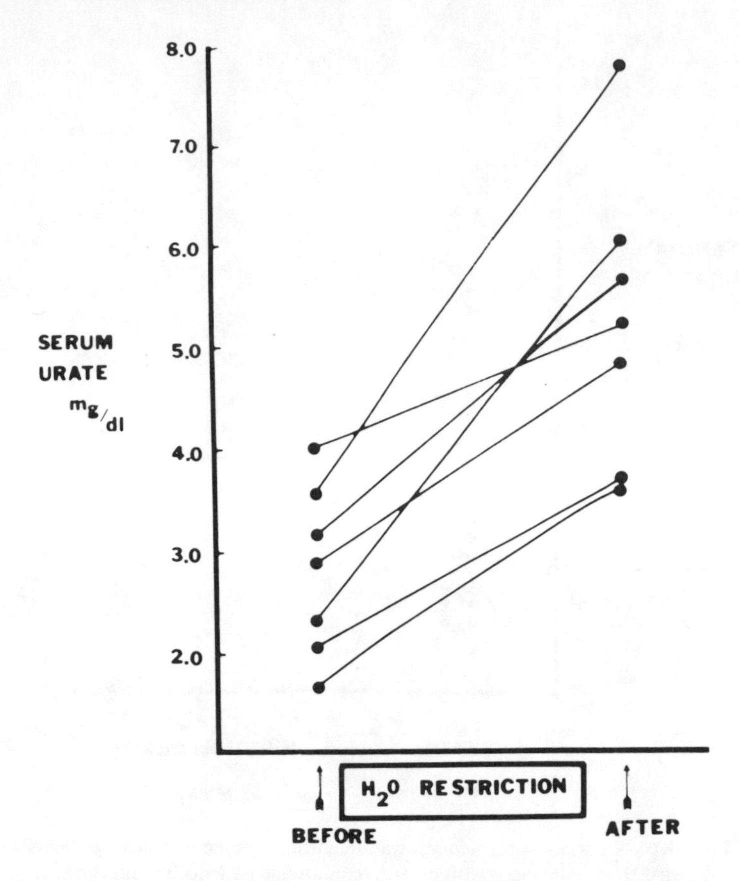

FIGURE 8. Serum urate concentration in patients with SIADH during hyponatremia before and after fluid restriction. (From Arieff, A. I., Llach, F., and Massry, S. G., *Medicine*, 55, 121, 1976. With permission.)

heart disease may be present. A poor dietary intake resulting in a decrease in the ingestion of total solutes coupled with excessive amounts of water intake have been advocated as major factors causing the hyponatremia. A hypothetical subject on a normal diet is able to ingest as much as 15 ℓ of water per day without developing hyponatremia. A normal subject ingesting 1 g of protein per kilogram of body weight (converted to urea) and a regular sodium diet will excrete approximately 900 mOsm/day. If the normal subject can lower urinary osmolality to 60 mOsm/kg, then the maximum daily urine volume will be 15 ℓ (900/60 = 15). However, if the dietary intake of solutes is markedly restricted, the daily solute excretion may decrease to 300 mOsm/kg or less. The maximum daily urine volume will be only 5 ℓ/day (300/60 = 5). A patient drinking an excessive amount of beer has a limited dietary intake of protein and the ingestion of 5 ℓ or more of fluid per day may induce hyponatremia. The majority of patients ingesting an excessive amount of beer have been shown to have low urinary Na^+ and osmolality;[63] however, there have been isolated cases of excessive ingestion of beer and hyponatremia with high urine osmolality.[64]

C. Hypervolemic Hyponatremia

This group of disorders is characterized by marked fluid retention, expanded ECFV, and the presence of edema. There is an increase in total body Na^+ but with a proportionately greater increase in TBW, the circumstances designated dilutional hyponatremia. Congestive heart failure is an example of dilutional hyponatremia and edema formation. The mechanism by which these patients develop hyponatremia is not completely understood. A decrease in cardiac output leading to a decrease in effective circulatory volume is sensed by the baroreceptors and results in increased ADH release.[65] In addition, a decrease in renal blood flow and glomerular filtration rate is present in these patients. This will result in increased proximal tubular reabsorption of Na^+, limiting the delivery of tubular fluid to the diluted segment of the nephron and further impairing renal water excretion.[66] This hypothesis is supported by the observation that when agents that increase distal fluid delivery, such as mannitol or furosemide, are used, they result in the conversion of a hypertonic into a hypotonic urine.[67]

Hepatic failure is also commonly associated with hyponatremia. A decrease in effective circulatory volume may be an important factor in the high ADH levels reported in these patients. As in congestive heart failure, a decrease in blood pressure may stimulate baroreceptors, resulting in the release of ADH;[65] a decrease in glomerular filtration rate and increased proximal tubular reabsorption has been observed in these patients and they may also respond to the administration of furosemide or mannitol with a decrease in urine osmolality. Portal hypertension and hypoalbuminemia may play an important role.

Nephrotic syndrome is also associated with hyponatremia and severe edema formation which occurs despite depressed proximal tubular reabsorption. The major factor is probably the hypoalbuminemia leading to volume contraction and increased tubular Na^+ reabsorption in the distal part of the nephron.[68]

The common denominator of all these edematous states associated with hypervolemic hyponatremia is the presence of low urinary Na^+, usually less than 20 mEq/ℓ. In general, the diagnosis in this group is not difficult because of the multiple symptoms and signs associated with the primary disorder.

Hyponatremia associated with expanded intravascular volume and edema may also occur in acute and chronic renal failure. Because of renal injury, these patients are unable to excrete water and therefore develop hyponatremia. In contrast with the previous group of edematous patients, the urinary Na^+ concentration in these patients is usually greater than 20 mEq/ℓ.

IX. CLINICAL MANIFESTATIONS OF HYPONATREMIA

The symptoms and signs associated with hyponatremia are often vague and nonspecific. The severity of the clinical manifestations depends on the degree of the hyponatremia as well as on the rapidity with which it develops. In general, symptoms do not develop until the plasma Na^+ concentration is less than 125 mEq/ℓ. With mild hyponatremia, the patient usually experiences loss of taste, nausea, vomiting, and abdominal discomfort. Weakness and muscle cramps may develop later. With more severe hyponatremia, neurological manifestations and signs such as depressed deep tendon reflexes, Cheyne-Stokes respirations, hypothermia, pathological reflexes, and seizure may develop. This may progress to lethargy and finally, coma. In an attempt to evaluate the pathophysiology of these symptoms, studies were carried out in 66 hyponatremic patients;[69] the degree of hyponatremia correlated well with the severity of the symptoms (Figure 9); however, there was a substantial overlap and some patients with a plasma Na^+ concentration of 120 mEq/ℓ were in coma while others with lower serum Na^+ concentration remained asymptomatic. Patients with acute hyponatremia (less than 12 hr duration) were all symptomatic. Symptomatic patients with hyponatremia of at least 3 days' duration had a lower plasma Na^+ concentration (115 $\pm$ 1 mEq/ℓ) than asymptomatic patients (122 $\pm$ 1 mEq/ℓ). Further studies were done in animals after induction of acute and chronic hyponatremia. Acute hyponatremic animals had a greater degree of brain edema than chronic hyponatremic animals (Figure 10). The

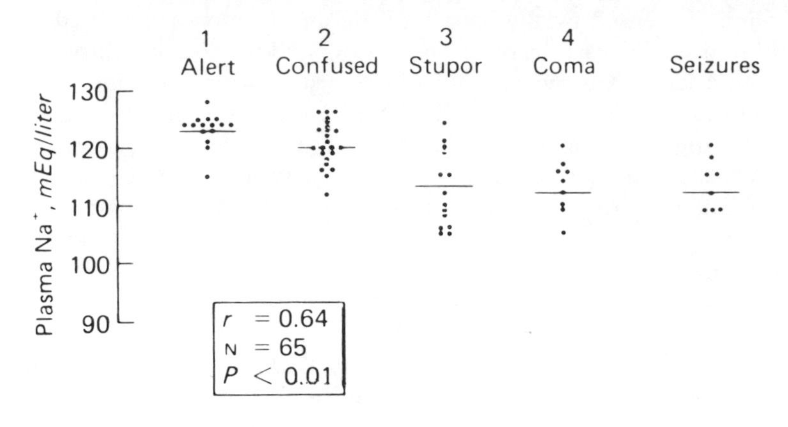

FIGURE 9. Relationship between plasma Na^+ concentration in 65 patients with plasma Na of 128 mEq/ℓ or less. (From Arieff, A. I., Llach, F., and Massry, S. G., *Medicine*, 55, 124, 1976. With permission.)

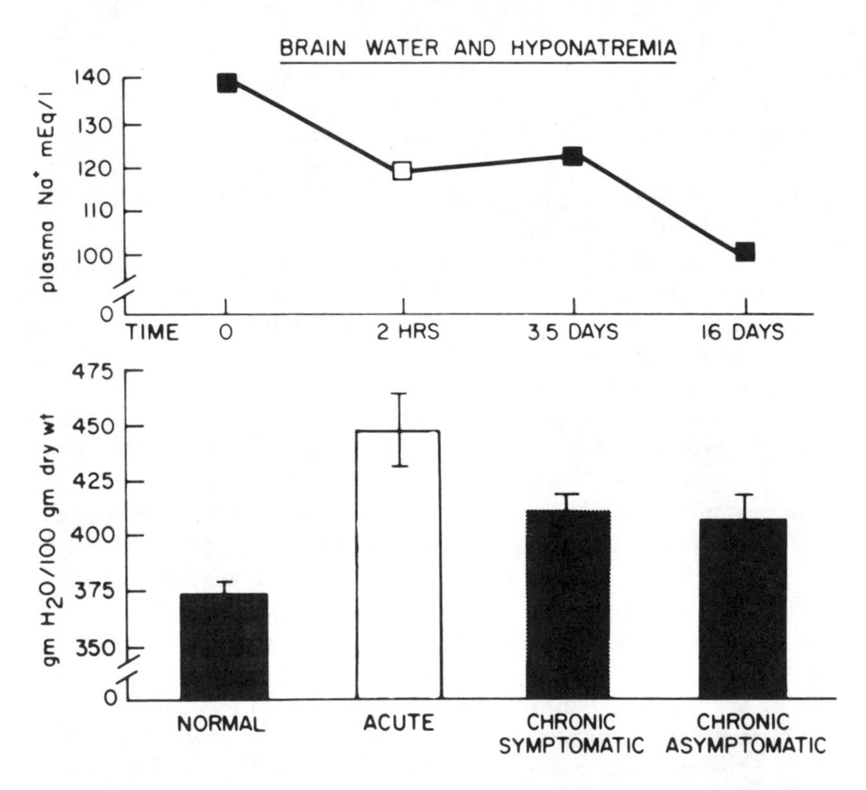

FIGURE 10. Brain water in normal and three groups of hyponatremic rabbits. In the animals with plasma Na lowered, in 2 hr the magnitude of brain water content is greater than when plasma Na is lowered more gradually over either 3½ or 16 days. (From Corey, C. M. and Arieff, A. I., in *Sodium and Water Homeostasis*, Brenner, B. M. and Stein, J. H., Eds., Churchill-Livingstone, New York, 1978, 231. With permission.)

brain content of Na^+, K^+, Cl^- and other osmols were lower in chronic hyponatremia. It appears that in chronic hyponatremia, the brain is able to compensate and decrease the osmotic gradient by losing osmols, minimizing the development of brain edema. Although most symptomatology of hyponatremia is related to brain edema, the loss of brain electrolytes observed in chronic hyponatremia may be of clinical significance; thus, Na^+ is an important ion in neuron excitability as well as appropriate neurotransmitter function. Therefore, it is conceivable that a low brain Na^+ content may also alter cerebral function.

X. DIAGNOSIS

The first step in the diagnosis of hyponatremia is to confirm the presence of hypoosmolality in order to exclude the possibility of pseudohyponatremia; that is, a condition in which plasma Na^+ concentration is low in the presence of normal or increased plasma osmolality. This underlies the importance of measuring plasma osmolality in patients with hyponatremia. The various conditions associated with pseudohyponatremia can be divided into two categories: (1) normo-osmolar hyponatremia; that is, hyponatremia associated with a normal plasma osmolality, and (2) hyperosmolar hyponatremia; that is, hyponatremia associated with an elevated plasma osmolality.

Normo-osmolar hyponatremia — A liter of normal plasma contains 930 mℓ of water. Plasma proteins and lipids account for 70 mℓ of each liter of plasma; in situations of severe hyperlipidemia and/or hyperlipoproteinemia, the amount of water contained in the plasma may be as low as 720 mℓ/ℓ of plasma. Plasma water exists in a different phase than protein and lipids and the osmometer measures only the activity of plasma water; consequently, the measurement of plasma osmolality will be unaffected. On the other hand, plasma Na^+ concentration is measured per liter of plasma, which will be reduced to 110 mEq/ℓ (153 mEq Na/ℓ of plasma water × 0.72 ℓ of plasma water per liter of plasma). Patients with normo-osmolar hyponatremia are asymptomatic. Normo-osmolar hyponatremia can be induced by the acute administration of isotonic fluid. In patients undergoing prostatectomy, a flushing solution of isotonic glycine may be used and a variable amount of the solution is reabsorbed into the circulation inducing a significant decrease in plasma Na^+ concentration without changes in plasma osmolality.

Hyperosmolar hyponatremia — Any solutes which penetrate the cells slowly such as glucose or mannitol will, by creating an osmotic gradient, shift water from the cells into the extracellular space (Figure 2*e*) leading to hyponatremia and hyperosmolality. An example is the administration of mannitol in patients with acute renal failure. An osmolar gap is usually seen in these conditions; that is, a difference between the measured and the calculated plasma osmolality (2 × plasma Na + glucose/18 + BUN/2.8). An osmolar gap can also be seen in hyperlipidemia, hypolipoproteinemia, and in drug overdoses such as ethanol, methanol, and ethylene glycol. The measurement of some of these substances may not be available in hospitals and, by calculating the osmolar gap, coupled with the specific clinical setting, the diagnosis of drug overdose may be apparent.

Once the diagnosis, hyponatremia, has been confirmed, the differential diagnosis frequently narrows down to volume contraction or the SIADH. The medical history in regard to vomiting, diarrhea, diuretic therapy, and physical examination, looking for signs of volume depletion such as decreased skin turgor, flat neck veins, dry mucous membranes, polyuria, and orthostatic hypotension are all helpful in establishing the presence of volume depletion.

The measurement of urinary Na^+ concentration is helpful; a low urinary Na^+, less than 10 mEq/ℓ is suggestive of volume contraction, whereas a urinary Na^+ greater than 20 mEq/ℓ suggests the SIADH. A useful diagnostic maneuver is a therapeutic trial of Na^+ replacement which will raise plasma Na^+ concentration in volume-depleted states but does not do so in the SIADH.[25]

The diagnosis of reset osmostat is based on the demonstration of hyponatremia

Table 4
CALCULATING SODIUM REPLACEMENT

When Na^+ deficit is the predominant factor:
Na^+ deficit = volume of distribution of plasma Na × plasma Na deficit
Na^+ deficit = (TBW) × (140 − plasma Na^+) = 0.6 × body weight × (140 − plasma Na)
For example, in a 60-kg man with a plasma Na concentration of 120 mEq/ℓ:
Na deficit = (0.6 × 60) (140 − 120) = 720 mEq.

which does not change with variations in Na^+ and water intake.[56] The diagnosis of hypervolemic hyponatremia is usually not a problem; the presence of ascites, jaundice, spider angiomata, peripheral edema, ventricular gallop, basal rales, and severe proteinuria usually make the diagnosis of cirrhosis or heart failure obvious.

A history of vomiting, diarrhea, and weakness associated with hyperkalemia and high urinary Na concentrations is suggestive of adrenal insufficiency. Hyperpigmentation may be present and the diagnosis is usually made by the measurement of plasma cortisol.

Measurement of other electrolytes as well as acid base parameters are usually helpful in the differential diagnosis of hyponatremia. Thus, the presence of metabolic alkalosis, hypochloremia, and hypokalemia are suggestive of vomiting or diuretic therapy, conditions associated with volume depletion and hyponatremia.

XI. TREATMENT

The approach to the hyponatremic patient is different, depending on whether it is an acute or a chronic hyponatremic state. That is, in the presence of neurological manifestations such as obtundation, seizures, and coma, the treatment is a medical emergency. Significant brain edema is present and no adaptive mechanism has taken effect yet. The immediate aim of therapy is to correct the severe cellular overhydration of the brain by rapidly increasing plasma osmolality with hypertonic saline. There are no objective data in regard to how rapidly plasma Na^+ concentration can safely be increased in acute hyponatremia. In addition, it is not known if such a rapid correction may result in a better neurological recovery than a more gradual treatment, such as water restriction would achieve. On the other hand, the correction of hyponatremia has to be done cautiously because of the dangers of inducing circulatory congestion and vascular hemorrhages. Nevertheless, in view of the high morbidity and mortality of acute hyponatremia,[69] an acute active therapeutic approach seems indicated until more objective data are available. A practical approach to calculating Na^+ requirement in the hyponatremic patient is shown in Table 4. Once the requirement has been calculated, half of the Na^+ dose is given over the first 8 hr. Then the patient's clinical status as well as the plasma Na^+ concentration is checked. The hypertonic saline infusion is continued until the neurological symptoms abate or the plasma Na^+ concentration exceeds 125 mEq/ℓ. Three percent NaCl (513 m*M*) or preferably 5% (855 m*M*) NaCl can be used. In general, the infusion of 855 m*M* NaCl at 70 mEq/hr has been shown to raise plasma Na^+ by 3 mEq/hr in most hyponatremic patients.[69] With the restoration of plasma Na^+ concentration to normal, neurological manifestations usually improve remarkably in a matter of hours. However, full recovery may take as long as 48 hr.

In a minority of patients with acute hyponatremia, especially those with SIADH, large amounts of Na^+ are lost in the urine and because of the expanded intravascular volumes of these patients, any given Na^+ load is rapidly excreted in the urine. Thus, Na requirement in this group of patients is usually underestimated and the correction of hyponatremia may be difficult. Furthermore, this syndrome occurs in the older

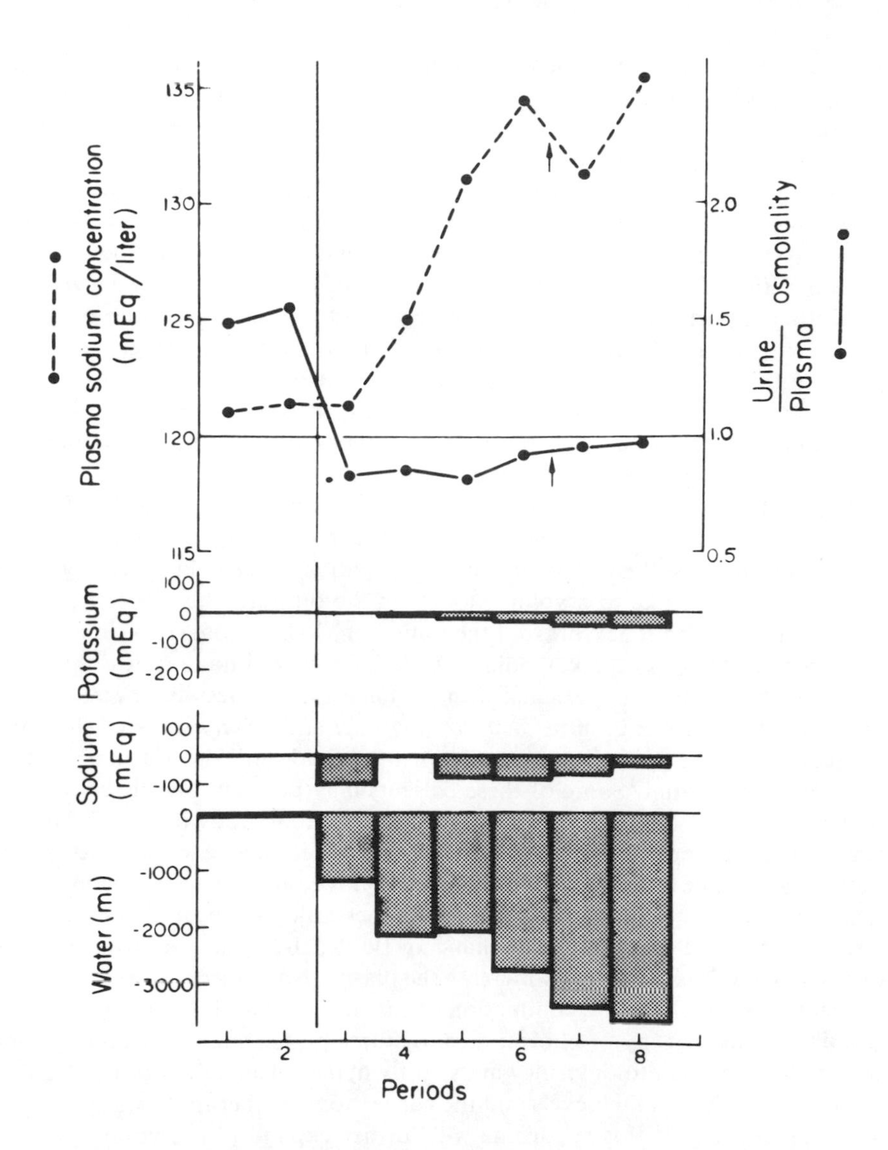

FIGURE 11. Correction of hyponatremia by inducing negative water balance with furosemide administered at 2½ periods (vertical line) and at 6½ periods (arrow). Observe the negative water balance (bottom) and the absence of change in K^+ and Na^+ balance accounting for the rise in plasma Na^+ concentration. (From Berl, T. and Schrier, R. W., in *Sodium and Water Homeostasis,* Brenner, B. M. and Stein, J. H., Eds., Churchill-Livingstone, New York, 1978, 17. With permission.)

patient at an age when hypertonic saline and rapid expansion of extracellular volume may precipitate acute congestive heart failure. For all these reasons, an alternative approach to the treatment of acute hyponatremia has been suggested. It involves the administration of a potent diuretic, such as furosemide, concomitant with hypertonic saline administration.[70] The diuretic is given to create a negative water balance and hypertonic saline to replace the urinary Na^+ losses. This approach is illustrated in Figure 11. With this approach, restoration of severe hyponatremia to normal has been achieved in 6 to 8 hr and with the advantages of not inducing circulatory congestion. Complications of such approach may be the rapid development of hypokalemia and hypomagnesemia.

Hypertonic mannitol has been advocated as an alternative approach to hypertonic

NaCl; however, it does not offer any specific advantage as compared to hypertonic saline and it has been proven to be not as effective in preventing neurological symptoms in the experimental animal.[71] Likewise, corticosteroids are not effective in preventing the neurological symptoms in the experimental animal and has not been shown to be useful in the treatment of the acute hyponatremic patients.[72]

The treatment of patients with chronic hyponatremia should be that of the underlying pathophysiological disorder. Thus, in patients with a decrease in the extracellular fluid volume, the administration of isotonic NaCl increases the distal delivery of Na^+ to the diluting segment, inhibits ADH release and allows renal excretion of the excess water. In general, one third of the estimated deficit of Na^+ may be given as isotonic saline in the first 6 hr and the remainder over the next 24 hr. In those situations of volume depletion secondary to vomiting, diarrhea or diuretic therapy, hypokalemia may be an important factor contributing to the hyponatremia and the administration of K may, in itself, increase plasma Na^+ concentration.

The hyponatremia of the edematous patients is essentially a problem of excess of water and any Na^+ therapy will aggravate the patient's underlying condition. The aim of therapy is to remove the excess of water. A negative water balance can be achieved by restricting water intake to a volume less than the output. The amount of water of daily food intake is about 850 mℓ and the obligatory losses from the skin and respiratory tract is 900 mℓ. Thus the key point is to restrict water intake to less than output. However, in the edematous patients, urine volume may be less than 800 mℓ/day so that the fluid restriction of 700 to 800 mℓ/day may still be in excess of the patient's needs. In such patients, water restriction has to be rigid until the plasma Na^+ concentration returns to normal. Some of these patients may be symptomatic because of the hyponatremia and, in addition to fluid restriction, diuretics may be beneficial. Hypertonic saline is contraindicated in these patients because it may aggravate the circulatory congestion. Sometimes, the only alternative to remove the excess water may be peritoneal dialysis. In cirrhotic patients, the use of hypertonic mannitol may be effective in increasing renal water excretion by enhancing fluid delivery to the distal part of the nephron and such a maneuver may increase the plasma Na^+ concentration.

In patients with SIADH, the elimination of the source of antidiuretic material whenever possible is the therapy of choice; most of these patients have chronic hyponatremia with mild symptomatology; they are slightly hypervolemic due to the water retention caused by the high ADH levels and the cornerstone of therapy is water restriction. Saline administration in this syndrome will further expand blood volumes, resulting in increase in renal Na^+ excretion and will not correct the hyponatremia.[25] As was mentioned before, this lack of response to saline is a useful diagnostic maneuver to differentiate patients with this syndrome from those with decreased intravascular volume. In general, these patients respond to fluid restriction within a few days and the plasma Na^+ concentration became an index of the magnitude of water intake. Once plasma Na^+ concentration returns to normal, water intake may be increased slowly and plasma Na^+ concentration monitored. A decrease in plasma Na^+ concentration with an increase in urine osmolality is suggestive of high ADH levels and requires the need for further water restriction.

The correction of chronic hyponatremia in patients with the SIADH may be difficult to achieve. In addition, the patient may not follow the water restriction regimen resulting in repeated admission to the hospital because of hyponatremia. Several drugs have been used in the treatment of chronic hyponatremia.

Ethanol, a well-known inhibitor of vasopressin release has been proposed as a therapy for this syndrome; however, its effects are short-lived, inconsistent, and require intoxicating levels in order to be effective. For all these reasons, the use of ethanol has become impractical and not safe. Diphenylhydantoin has been reported to be effective in some patients only when given intravenously.[73] Narcotics have been long-known to cause antidiuresis most likely due to an increase in ADH release. Consequently, several narcotic antagonists have been recently shown to have a diuretic effect which is mediated via the inhibition of ADH release.[74] These agents may have, in the future, therapeutic applications in the chronic treatment of SIADH.

The most widely used drugs in the treatment of SIADH are lithium and demeclocycline. Lithium is an agent which has been extensively used in neuropsychiatric patients. It does induce polyuria in normal subjects and has been used in the treatment of SIADH.[75] Lithium increases renal water excretion by both inhibition at the proximal and distal cyclic AMP formation.[76] However, the response to lithium is not predictable; these agents cause nephrogenic diabetes insipidus in about 20% of treated patients.[75] In addition, some of these patients may develop various toxic side effects which will include sluggishness, tremor, cardiotoxicity, and thyroid and renal dysfunction.[77,78]

Demeclocycline, an antibiotic of the tetracycline group appears to be a safer and a more effective agent in the treatment of SIADH.[79] It consistently induces nephrogenic diabetes insipidus in humans, by inhibiting the peripheral action of ADH through alterations in both the generation and action of cycline-adenosine-monophosphate.[80] A dose of 1200 mg/day is necessary in order to achieve the desired effect. Lower dose may produce only a partial nephrogenic diabetes insipidus. Demeclocycline is relatively safe but potential adverse effects may include nausea, photosensitivity, azotemia, and superinfection with resistant bacteria.[81,82] The nephrogenic diabetes insipidus induced by a dose of 1200 mg/day is reversible upon cessation of therapy.

The hyponatremia of adrenal insufficiency responds rather promptly to the administration of cortisol and if volume depletion is present, isotonic saline will ameliorate the hyponatremia. The use of mineralocorticoid alone such as 9-alpha-fluorohydrocortisone (Fluorinef) will not correct the hyponatremia.

REFERENCES

1. **Maffly, R. H. and Leaf, A.,** The potential of water in mammalian tissue, *J. Gen. Physiol.,* 42, 1257, 1959.
2. **Katz, A. and Epstein, F. H.,** Physiologic role of sodium-potassium-activated adenosine triphosphatase in the transport of cation across biologic membranes, *N. Engl. J. Med.,* 278, 253, 1968.
3. **Edelman, I. S., Leibman, J., O'Meara, M. P., and Birkenfeld, L. W.,** Interrelations between serum sodium concentration, serum osmolality and total exchangeable sodium, total exchangeable potassium and total body water, *J. Clin. Invest.,* 37, 1236, 1958.
4. **Laragh, J. H.,** The effect of potassium chloride on hyponatremia, *J. Clin. Invest.,* 33, 807, 1954.
5. **Fichman, M. P., Vorherr, H., Kleeman, C. R., and Telfer, N.,** Diuretic-induced hyponatremia, *Ann. Intern. Med.,* 75, 853, 1971.
6. **Rocha, A. S. and Kokko, J. P.,** Sodium chloride and water transport in the medullary thick ascending limb of Henle, *J. Clin. Invest.,* 52, 612, 1973.
7. **Burg, M. and Green, N.,** Function of the thick ascending limb of Henle's loop, *Am. J. Physiol.,* 224, 659, 1973.
8. **Jamison, R. L. and Maffly, R. H.,** The renal concentrating mechanism, *N. Engl. J. Med.,* 295, 1059, 1976.
9. **Kokko, J. P. and Tisher, C. C.,** Water movement across nephron segments involved with the countercurrent multiplication system, *Kidney Int.,* 10, 64, 1976.
10. **Jamison, R. L., Buerkert, J., and Levy, F. B. A.,** A micropuncture study of collecting duct function in rats with hereditary diabetes insipidus, *J. Clin. Invest.,* 50, 2444, 1971.

11. **Dousa, T. P. and Valtin, H.,** Cellular actions of vasopressin in the mammalian kidney, *Kidney Int.,* 10, 55, 1976.
12. **Berl, T., Anderson, R. J., McDonald, K., and Schrier, R. W.,** Clinical disorders of water metabolism, *Kidney Int.,* 10, 117, 1976.
13. **Henry, J. P., Gauer, O. H., and Reeves, J. L.,** Evidence of the atrial location of receptors influencing urine flow, *Circ. Res.,* 4, 85, 1956.
14. **Share, L.,** Vasopressin, its bioassay and the physiological control of its release, *Am. J. Med.,* 42, 701, 1968.
15. **Berl, T., Cadnapaphornchai, P., Harbottle J. A., and Schrier, R. W.,** Mechanism of stimulation of vasopressin release during beta adrenergic stimulation with isoproterenol, *J. Clin. Invest.,* 53, 857, 1974.
16. **Bonjour, J. P. and Malvin, R. L.,** Stimulation of ADH release by the renin angiotension system, *Am. J. Physiol.,* 218, 1555, 1970.
17. **Burg, M. B.,** Tubular chloride transport and the mode of action of some diuretics, *Kidney Int.,* 9, 189, 1976.
18. **Kleeman, C. R., Maxwell, M. H., and Rockney, R. E.,** Mechanisms of impaired water excretion in adrenal and pituitary insufficiency. I. The role of altered glomerular filtration rate and solute excretion, *J. Clin. Invest.,* 37, 1799, 1958.
19. **Gill, J. R., Jr., Gann, D. S., and Bartter, F. C.,** Restoration of water diuresis in Addisonian patients by expansion of the volume of extracellular fluid, *J. Clin. Invest.,* 41, 1078, 1962.
20. **Ahmed, A. B., George, B. C., Conyalez-Auvert, C., and Dingman, J. F.,** Increased plasma arginine vasopressin in clinical adrenocortical insufficiency and its inhibition by glucosteroids, *J. Clin. Invest.,* 46, 111, 1967.
21. **Green, H. H., Harrington, A. R., and Valtin, H.,** On the role of antidiuretic hormone in the inhibition of acute water diuresis in adrenal insufficiency and the effects of gluco- and mineralocorticoids in reversing the inhibition, *J. Clin. Invest.,* 49, 1724, 1970.
22. **Fichman, M. P. and Bethune, J. E.,** The role of adrenocorticoids in the inappropriate antidiuretic hormone syndrome, *Ann. Intern. Med.,* 68, 806, 1968.
23. **Schrier, R. W. and Regal, E. M.,** Influence of aldosterone on sodium, water, and potassium metabolism in chronic renal disease, *Kidney Int.,* 1, 156, 1972.
24. **Sebastian, A., McSherry, E., and Morris, R. C., Jr.,** On the mechanism of potassium wasting in renal tubular acidosis associated with the Fanconi Syndrome (Type 2 RTA), *J. Clin. Invest.,* 50, 231, 1971.
25. **Bartter, F. C. and Schwartz, W. B.,** The syndrome of inappropriate secretion of antidiuretic hormone, *Am. J. Med.,* 42, 790, 1967.
26. **Leaf, A., Bartter, F. C., Santos, R. F., and Wrong, O.,** Evidence in man that urinary electrolyte loss induced by Pitressin® is a function of water retention, *J. Clin. Invest.,* 32, 868, 1953.
27. **Jaenike, J. R. and Waterhouse, C.,** The renal response to sustained administration of vasopressin and water in man, *J. Clin. Endocrinol. Metab.,* 21, 231, 1961.
28. **Garella, S., Tzamaloukas, A. H., and Chazan, J. A.,** Effect of isotonic volume expansion on extracellular bicarbonate stores in normal dogs, *Am. J. Physiol.,* 225, 628, 1973.
29. **Lowance, D. C., Garfinkel, H. B., Mattern, W. D., and Schwartz, W. B.,** The effect of chronic hypotonic volume expansion on the renal regulation of acid-base equilibrium, *J. Clin. Invest.,* 51, 2928, 1972.
30. **Beck, L. H.,** Hypouricemia in the syndrome of inappropriate antidiuretic hormone secretion, *N. Engl. J. Med.,* 301, 528, 1979.
31. **Vorherr, H., Massry, S. B., Utiger, R. D., and Kleeman, C. R.,** Antidiuretic principle in malignant tumor extracts from patients with inappropriate ADH syndrome, *J. Clin. Endocrinol. Metab.,* 28, 162, 1968.
32. **George, J. M., Caper, C. C., and Phillips A. S.,** Biosynthesis of vasopressin in vitro and ultrastructure of a bronchogenic carcinoma, *J. Clin. Invest.,* 51, 141, 1972.
33. **Vorherr, H., Massry, S. G., Fallet, R., Kaplan, I., and Kleeman, C. R.,** Antidiuretic principle in tuberculosis and hyponatremia, *Ann. Intern. Med.,* 72, 383, 1970.
34. **Kleeman, C. R., Rubini, M. E., Lamdin, E., and Epstein, F. H.,** Studies on alcohol diuresis. II. The evaluation of ethyl alcohol as an inhibitor of the neurohypophysis, *J. Clin. Invest.,* 34, 448, 1955.
35. **Robertson, G. L.,** Vasopressin in osmotic regulation in man, *Ann. Rev. Med.,* 25, 315, 1974.
36. **Moses, A. M. and Miller, M.,** Drug induced dilutional hyponatremia, *N. Engl. J. Med.,* 291, 1234, 1974.
37. **Forrest, J. N., Jr. and Singer, I.,** Drug-induced interference with action of antidiuretic hormone, in *Disturbances in Body Fluid Osmolality,* Andreoli, T., Grantham, J. J., and Rector, F. C., Jr., Eds., American Physiological Society, Baltimore, 1977, 309.
38. **Moses, A. M.,** Synthetic lysine vasopressin nasal spray in the treatment of diabetes insipidus, *Clin. Pharmacol. Ther.,* 5, 422, 1964.

39. **Vavra, I., Machova, A., Holecek, V., Cort, J. H., Azoral, M., and Sorm, F.,** Effect of a synthetic analogue of vasopressin in animals and in patients with diabetes insipidus, *Lancet,* 1, 948, 1968.
40. **Pittman, J. G.,** Water intoxication due to oxytocin, *N. Engl. J. Med.,* 268, 481, 1963.
41. **Dingman, J. F., Benirschke, K., and Thorn, G. W.,** Studies of neurohypophyseal function in man, *Am. J. Med.,* 23, 226, 1957.
42. **Husain, M. K., Fernando, N., Shapiro, M., Kagan, A., and Glick, S. M.,** Radioimmunoassay of arginine vasopressin in human plasma, *J. Clin. Endocrinol. Metab.,* 37, 616, 1973.
43. **Cadnapaphornchai, P., Boykin, J. L., Berl, T., McDonald, K. M., and Schrier, R. W.,** Mechanism of effect of nicotine on renal water excretion, *Am. J. Physiol.,* 227, 1216, 1974.
44. **de Bodo, R. C.,** The antidiuretic action of morphine and its mechanism, *J. Pharmacol. Exp. Ther.,* 82, 74, 1944.
45. **Weissman, P., Shenkman, L., and Gregerman, R. I.,** Chlorpropamide hyponatremia, *N. Engl. J. Med.,* 284, 65, 1971.
46. **Liberman, B., Borges, R., and Wajchenberg, B. L.,** Evidence for a role of antidiuretic hormone (ADH) in the antidiuretic action of chlorpropamide, *J. Clin. Endocrinol. Metab.,* 36, 894, 1972.
47. **Rado, J. P.,** Water intoxication during carbamazepine treatment, *Br. Med. J.,* 3, 479, 1973.
48. **Luzecky, M. H., Burman, K. D., and Schultz, E. R.,** The syndrome of inappropriate secretion of antidiuretic hormone associated with amitriptyline administration, *South. Med. J.,* 67, 495, 1974.
49. **Ajlouni, K., Kern, M. W., Tures, J. F., Theil, G. B., and Hagen, T. C.,** Thiothixene-induced hyponatremia, *Arch. Intern. Med.,* 134, 1103, 1974.
50. **Moses, A. M., Howanitz, J., Van Gemert, M., and Miller, M.,** Clofibrate-induced antidiuresis, *J. Clin. Invest.,* 52, 535, 1973.
51. **Robertson, G. L., Boopalam, N., and Zelkowitz, L. J.,** Vincristine neurotoxicity and abnormal secretion of antidiuretic hormone, *Arch. Intern. Med.,* 132, 717, 1973.
52. **Rufener, C., Nordmann, J., and Rouiller, C.,** Effect of vincristine on the rat posterior pituitary in vitro, *Neuro Chir.,* 18, 137, 1972.
53. **De Fronzo, R. A., Braine, J., Colvin, O. M., and David, P. J.,** Water intoxication in man after cyclophosphamide therapy. Time course and relation to drug activation, *Ann. Intern. Med.,* 78, 861, 1973.
54. **Forrest, J. N., Jr., Cox, M., Hong, C., Morrison, G., Bia, M., and Singer, I.,** Superiority of demeclocycline over lithium in the treatment of chronic syndrome of inappropriate secretion of antidiuretic hormone, *N. Engl. J. Med.,* 298, 173, 1978.
55. **Martinez-Maldonado, M.,** Inappropriate antidiuretic hormone secretion of unknown origin (Nephrology Forum), *Kidney Int.,* 17, 554, 1980.
56. **De Fronzo, R. A., Goldberg, M., and Agus, Z. S.,** Normal diluting capacity in hyponatremia patients: reset osmostat or a variant of SIADH, *Ann. Intern. Med.,* 84, 538, 1976.
57. **Goldberg, M. and Reivich, M.,** Studies on the mechanism of hyponatremia and impaired water excretion in myxedema, *Ann. Intern. Med.,* 56, 120, 1962.
58. **Discala, V. A. and Kinney, M. J.,** Effects of myxedema on the renal diluting and concentrating mechanism, *Am. J. Med.,* 50, 325, 1971.
59. **Emmanouel, D. S., Lindheimer, M. D., and Katz, A. I.,** Mechanism of impaired water excretion in the hypothyroid rat, *J. Clin. Invest.,* 54, 926, 1974.
60. **Kleeman, C. R., Czaczker, J. W., and Cutler, R.,** Mechanisms of impaired water excretion in adrenal and pituitary insufficiency. IV. Antidiuretic hormone in primary and secondary adrenal insufficiency, *J. Clin. Invest.,* 43, 1641, 1964.
61. **Ukai, M., Moran, W. H., and Zimmerman, B.,** The role of visceral afferent pathways on vasopressin secretion and urinary excretory patterns during surgical stress, *Ann. Surg.,* 168, 16, 1968.
62. **Langgard, H. and Smith, W. O.,** Self-induced water intoxication without predisposing illness, *N. Engl. J. Med.,* 266, 378, 1962.
63. **Hilden, T. and Svendsen, T. L.,** Electrolyte disturbances in beer drinkers, a specific ''hypo-osmolality syndrome'', *Lancet,* 2, 245, 1975.
64. **Hobson, J. A. and English, J. T.,** Self-induced water intoxication, *Ann. Intern. Med.,* 58, 324, 1963.
65. **Stein, M., Schwartz, R., and Mersky, I. A.,** The antidiuretic activity of plasma of patients with hepatic cirrhosis, congestive heart failure, hypertension and other clinical disorders, *J. Clin. Invest.,* 33, 77, 1954.
66. **Anderson, R. J., Cadnapaphornchai, P., Harbottle, J. A., McDonald, K. M., and Schrier, R. W.,** Mechanism of effect of thoracic inferior vena cava constriction on renal water excretion, *J. Clin. Invest.,* 54, 1473, 1974.
67. **Schrier, R. W., Lehman, D., Zacherle, B., and Earley, L. E.,** Effect of furosemide on free water excretion in edematous patients with hyponatremia, *Kidney Int.,* 3, 30, 1974.
68. **Grausz, H., Lieberman, R., and Earley, L. E.,** Effect of plasma albumin on sodium reabsorption in patients with nephrotic syndrome, *Kidney Int.,* 1, 47, 1972.
69. **Arieff, A. I., Llach, F., and Massry, S. G.,** Neurological manifestations and morbidity of hyponatremia: correlation with brain water and electrolytes, *Medicine,* 55, 121, 1976.

70. **Hantman, D., Rossier, B., and Schrier, R.,** Rapid correction of hyponatremia in the syndrome of inappropriate secretion of antidiuretic hormone, *Ann. Intern. Med.,* 78, 870, 1973.
71. **Wakim, K. G.,** The pathophysiology of dialysis disequilibrium syndrome, *Mayo Clin. Proc.,* 44, 406, 1969.
72. **Rymer, M. M. and Fishman, R. A.,** Protective adaptation of brain to water intoxication, *Arch. Neurol. (Chicago),* 28, 49, 1973.
73. **Tanay, A., Yust, I., Pereselenschi, G., Ambramov, A. L., and Aviram, A.,** Long term treatment of the syndrome of inappropriate antidiuretic hormone secretion with phenyltoin, *Ann. Intern. Med.,* 90, 50, 1979.
74. **Miller, M.,** Institution of ADH release in the rat by narcotic antagonists, *Neuroendocrinology,* 19, 241, 1975.
75. **White, M. G. and Fetner, C. D.,** Treatment of the syndrome of inappropriate secretion of antidiuretic hormone with lithium carbonate, *N. Engl. J. Med.,* 282, 390, 1975.
76. **Forrest, J. N., Jr., Cohen, A. D., Torretti, J., Himmelhock, J. M., and Epstein, F. H.,** On the mechanism of lithium-induced diabetes insipidus in man and the rat, *J. Clin. Invest.,* 53, 1115, 1974.
77. **Tangedahl, T. N. and Gan, G. T.,** Myocardial irritability associated with lithium carbonate therapy, *N. Engl. J. Med.,* 287, 867, 1972.
78. **Hestbech, J., Hansen, H. E., Amidsen, A., and Olsen, S.,** Chronic renal lesions following long-term treatment with lithium, *Kidney Int.,* 12, 205, 1977.
79. **Forrest, J. N., Cox, M., Hon, J. C., Morrison, G., Bia, M., and Singer, I.,** Superiority of demeclocycline over lithium in the treatment of chronic syndrome of inappropriate antidiuretic hormone, *N. Engl. J. Med.,* 298, 173, 1978.
80. **Dousa, T. P. and Wilson, D. M.,** Effects of demethychlortetracycline on cellular action of antidiuretic hormone in vitro, *Kidney Int.,* 5, 279, 1974.
81. **Oster, J. R., Epstein, M., and Ulano, H. B.,** Deterioration of renal function with demeclocycline administration, *Curr. Ther. Res.,* 20, 794, 1976.
82. **Carrilho, F., Bosch, J., Arroyo, V., Mas, J., Viver, J., and Rades, J.,** Renal failure associated with demeclocycline in cirrhosis, *Ann. Intern. Med.,* 87, 195, 1977.
83. **Burton, D. R.,** *Clinical Physiology of Acid-Base and Electrolyte Disorders,* McGraw-Hill, New York, 1977.
84. **Berl, T. and Schrier, R. W.,** Water metabolism and the hypo-osmolar syndrome, in *Sodium and Water Homeostasis,* Brenner, B. M. and Stein, J. H., Eds., Churchill-Livingstone, New York, 1978.
85. **Kleeman, C. R.,** Water metabolism, in *Clinical Disorders of Fluid and Electrolyte Metabolism,* 2nd ed., Maxwell, M. H. and Kleeman, C. R., Eds., McGraw-Hill, New York, 1962.
86. **Zerbe, R., Stropes, L., and Robertson, G.,** Vasopressin function in the syndrome of inappropriate antidiuresis, *Annu. Rev. Med.,* 31, 317, 1980.
87. **Corey, C. M. and Arieff, A. I.,** Disorders of sodium and water metabolism and their effects on the central nervous system, in *Sodium and Water Homeostasis,* Brenner, B. M. and Stein, J. H., Eds., Churchill-Livingstone, New York, 1978, 231.

Chapter 12

HYPERNATREMIA

Laurence Finberg

TABLE OF CONTENTS

I. INTRODUCTION

The role of sodium in life processes is so pervasive that alterations in the concentrations of this ion produce a very wide range of disturbances. Among these, the hypernatremic states have a special place because they were probably not experienced as a disturbance at early levels of evolution. Although ultimately many other effects of high sodium levels may be uncovered, the characteristic of sodium that dominates our present understanding is its function as the cation half of an electrolyte pair that determines the partitioning of body water in higher animals. Presumably this arrangement exists because as very primitive life forms developed, they took some of the primeval sea as a bathing medium for cell functions to go on in relative tranquility. With this, an integument had to be added to protect the *milieu interieur* otherwise known as the extracellular fluid (ECF).

Hypernatremic disturbances are not readily corrected by homeostatic systems since all body secretions have a lower concentration of sodium than does ECF. The roles of such regulators as ADH, aldosterone, and prolactin are too little understood in hypernatremia to touch on in this brief chapter. Their physiologic roles in general have been discussed earlier and, of course, elsewhere.

Because virtually every hypernatremic state is also a dehydrated state — special experimental circumstances excepted — hypernatremic dehydration is a constant focus. This may occur with a body sodium content that is increased, normal, or decreased. All have been described; decrease is the most common. The discussion to follow will define the necessary concepts, give some clinical historical background, touch on etiologic factors, and describe clinical presentations, before moving to a more detailed discussion of pathophysiology and pathology. Finally, a brief discussion of therapeutic approaches is appended for interested clinicians.

II. DEFINITION

Hypernatremia is a physiologic disturbance defined by a concentration of sodium in serum of 150 mEq/ℓ or greater. This arbitrary definition is useful to clinicians because the measurement is one performed by clinical laboratories. This definition also has been accepted generally by authors, journals, and texts. Certain limitations of the definition for both physiologist and clinician need to be identified, however, for expert evaluation. The first and less important of these arises because the measurement of sodium concentrations in serum is not a measure of the chemically and biologically active sodium ion which would be the concentration of sodium in serum water. The nonwater portion of serum consists mostly of protein; since protein levels are usually similar, serum concentrations of sodium are comparable to one another even if the numerical value understates the active concentration. The second problem with using an arbitrary serum level stems from the fact that some of the important physiologic changes result from the imposition of a sudden gradient with the higher concentration on the extracellular (and vascular fluid) side. The rapid adjustment results from the difference in concentrations, not the level. Thus, if a subject has an adaptational sodium concentration of 125 mEq/ℓ and it is changed rapidly to 140 mEq/ℓ, some of the induced disturbances will be the same as if the change had been from 145 to 160.

A change in sodium concentration means, in effect, a change in body fluid osmolality because the sodium ion, along with a negatively charged ion or ions, is the principal solute of the ECF, and because sodium is relatively and actively excluded from body cell water. Thus, sodium salts act as the partitioner of body water between cells and ECF. One central part of the disturbance of hypernatremia is a shrinking of cell water with an expansion of extracellular fluid volume (ECFV). This may come about because of water loss or sodium gain or both. At steady state the subject may be dehydrated (most common), normally hydrated, or theoretically can be over-hydrated. Either of these last two is apt to happen only under circumstances of manipulation such as dialysis or faulty intravenous infusions. The second most important part of the hypernatremic disturbance affects the central nervous system uniquely when the gradient change is rapid (e.g., over hours, not days). The blood vessel changes in the CNS will not occur if hypernatremia develops slowly. The reasons for this will be discussed later.

III. HISTORICAL BACKGROUND

When that pioneer chemical anatomist, Schmidt,[1] published the chemical composition of dehydrated tissue, he recognized two types of dehydration: one, more common, resulting from loss of water and salts, which became the definition of dehydration by physiologists and clinicians; and the other from loss of water alone which has come to be known as hypernatremic dehydration. Between 1915 and 1920 Weed[2] performed experiments on the effects of hypertonic infusions upon the nervous system, first describing the pressure changes. In the 1940s Rapoport and Dodd[3] described the "postacidotic syndrome", an entity resulting in part from generous infusions of sodium bicarbonate. Subsequently Rapoport[4] described patients with hypernatremia from several causes. Other individual case reports from neurosurgeons and other also appeared in the 1940s. In the early 1950s, after the widespread use of flame photometry, Finberg and Harrison[5] and Weil and Wallacc[6] dcscribcd large series of infants with hypernatremic dehydration, mostly secondary to enteric (diarrheal) diseases. These papers, first presented at the same meeting (American Pediatric Society, 1954), defined the clinical syndrome in infants and delineated the modern prevalence and importance of the entity. Since that time additional insight in pathophysiology, pathology, and management have emerged and are described in the following text.

IV. IMPORTANCE AND EPIDEMIOLOGY

The importance of hypernatremia in infants emerged from the papers[5,6] which showcd an increased mortality, approximately 8% or fourfold more than otherwise similar patients with enteric disease and dehydration but not hypernatremia. An increased morbidity involving the CNS was also appreciated although follow-up data were not available. A careful follow-up study in the 1960s was carried out in Great Britain showing residual brain damage in hypernatremic dehydration in excess of other physiologic dehydrating disturbances.[7]

The incidence of hypernatremic dehydration in infants has had to be measured in hospitalized patients as a fraction of all infants hospitalized for dehydration over a given period of time. While not a true incidence, it is a clinically useful index of prevalence and incidence. From 1950 to 1980 in the U.S. the proportion has varied from about 15 to 30% with most reports between 20 and 25%. Similar data have come from Nigeria,[8] but most of the underdeveloped world, where malnutrition is common, report little hypernatremia when it is defined as sodium concentration greater than 150 mEq/ℓ. Most writers do indicate a clear increase of the disturbance in the winter, even in the tropics. This has implications — specific and nonspecific — taken up below.

In Great Britain[9] and Spain[10] during the 1960s and early 1970s there was a sharp increase in incidence of hypernatremic dehydration in infants. This occurred because in both countries dry milk was made available at low cost. Inappropriate mixing (too much powder for the recommended volume of water) resulted in very high solute intakes and a virtual epidemic of hypernatremia. In Great Britain, the elimination of the dry milk and the introduction of low solute feedings has been followed by a much lower incidence in English infants.[11]

V. ETIOLOGIC FACTORS

The factors that produce a hypernatremic disturbance involve failure of intake or abnormal loss of water on the one hand, or excess salt or failure to excrete sodium on the other. These events may occur singly or in any combination. The two simplest of these are uncomplicated thirsting and salt poisoning. Thirsting with or without solute intake will in some species (e.g., man, dog) produce a hypernatremic state. There is an extensive literature on shipwreck victims describing the clinical and chemical events in such circumstances.[12] In other species (e.g., rats, mice) cellular breakdown releasing cell water prevents hypernatremia from simple total thirst. Despite these species differences, death occurs in approximately the same time, suggesting that gradual development of hypernatremia is not an important factor in human death from total thirsting. Salt poisoning has occurred in a variety of ways including suicide in adults, improper mixing of infant feeds usually by accidentally substituting salt for sucrose,[13,14] and it is the usual method for producing hypernatremic states in experimental animals.

A diet or a fluid intake for certain patients, too high in solute for the kidney to excrete, is a frequent factor in producing hypernatremia. The human kidney does not defend well against high solute intake; the thirst mechanisms are the homeostatic system that defends primarily. It is, therefore, to be expected that the vulnerable individuals would be those whose thirst mechanism was either faulty or could not operate because of a problem in access to water. Small infants,[5] unconscious patients,[15] psychotics,[15] and patients with damage to the thirst center[16] all fit this description. Factors which lead to high insensible water loss also increase vulnerability. High environmental temperature, low humidity, high body temperature, small size with a higher ratio of surface area to weight, and hyperventilation,[17] are the principal ones encountered. Those things that contribute to poor solute excretion such as diabetes insipidus (pituitary or renal), renal immaturity (as in small and, especially, prematurely born infants), or renal medullary impairment also predispose to hypernatremia. Finally, loss of water from the gastrointestinal tract through vomiting and diarrhea usually results in proportionally (in the physiologic context of the extracellular fluid) more water loss than sodium loss. This is not true of the toxigenic diarrheas such as cholera and certain *Escherichia coli* infections which act on cyclic AMP or cyclic GMP to produce a stool water with high sodium concentration. It is theoretically true of those infections such as are produced by rotavirus which induce inflammation, malabsorption, and usually, low sodium stools. Confirmation of this point in a clinical setting is difficult because of the many variables (most importantly, dietary intake) that affect stool composition. Rotaviruses are recovered from patients mostly during winter epidemics when hypernatremia is more prevalent.

In recapitulation, small size, consciousness, rationality, immaturity, inability to self-regulate water intake, lack of renal water regulation, fast breathing, and sweating all represent vulnerabilities from the patient side. Hot, dry atmosphere, high solute intake, limited water access, and enteric infection with organisms causing inflammatory malabsorption, comprise the main environmental predisposing events.

VI. CLINICAL SYMPTOMS AND SIGNS

Patients or animals with hypernatremic dehydration have less circulatory disturbance for any given amount of fluid loss than those with physiologically proportionate water and sodium loss, i.e., isonatremic dehydration. The ECFV including the plasma volume is relatively preserved; the cell water volume is relatively reduced. Clinicians accustomed to finding circulatory deficits, therefore, may not recognize this kind of *inapparent* dehydration. Should the dehydration become very severe, e.g., 12 to 18% of body weight loss in 24 to 36 hr, circulatory signs will appear in addition to those described next.

Signs indicating nervous system involvement are almost always present in acute hypernatremic states. Disturbance of consciousness is the most constant, thus being an effect as well as a cause. A peculiar state of lethargy or even semicoma when not stimulated, coupled with hyperirritability to virtually any stimulus, is the usual clinical state. This picture is perhaps appreciated more readily in infants than in adults or children. Hypertonicity of muscles often occurs as does hyperreflexia. Nuchal rigidity, as a part of hypertonia, may falsely suggest meningitis especially in infants. More overt CNS manifestations sometimes occur, including muscle twitchings and either focal or generalized convulsions. Less commonly, hemiparesis or evidence of focal neurologic damage may be seen. Chvostek's sign may occasionally be elicited.

Fever is usual, again a cause and an effect, and in infants, blood-tinged vomitus has often been seen. The skin of infants frequently feels thickened or doughy, though even more frequent is a velvety soft skin texture.[18] In those patients who are given access to water, avid thirst is the rule unless an underlying illness interferes. In salt poisoning, pulmonary edema may occur early on with resultant hyperventilation which further aggravates the hypernatremia.

In chronic hypernatremic states including all those in which the phenomenon has occurred gradually, the CNS signs may be absent until the disturbance is extremely severe in terms of sodium concentration or osmolality of plasma (Na $>$ 175 mEq/ℓ, Osm $>$ 360 mOsm/kg).[19] At these very high levels however, CNS symptoms do appear presumably because of cellular dessication.

VII. PATHOPHYSIOLOGY

The principal role of sodium and chloride in mammalian physiology is that of determining the partition of body fluid into that in cells (intracellular) and that outside cells (extracellular). This role is accomplished because most cells extrude sodium (and an accompanying anion) through a biochemical transport system known as the Na-K ATPase or sodium "pump". Except at great extremes of concentration, this "pump" maintains a low (6 to 8 mEq/ℓ) concentration of sodium in cells. In the muscle cell, the dominant cell by mass, chloride is virtually absent in the millequivalent order of magnitude. There is controversy as to whether the chloride distribution is passive as in classic theory[20,21] or whether other explanations need to be invoked.[22] At present the preponderant weight of evidence gives the pump explanation and a passive role for chloride the most likely explanation.[23]

The various disturbances of hypernatremia are mostly related to this property of sodium regardless of why the property exists. Water moves freely across virtually all the membrane interfaces of the body. Hence, except for some very specialized tissues, osmolality will be the same in all body fluids. When a solute gradient is created, water moves to adjust concentration almost instantly. Specifically, if a sodium concentration in the ECF rises, water leaves cells to equilibrate the osmolality. Thus, ECF expands and ICF constricts. The foregoing explains why the circulation (plasma volume) is relatively expanded in hypernatremic states.

An unexpected phenomenon has been discovered when mammalian animals are subjected to extreme hypernatremia. In a variety of experimental approaches,[24-27] it has been demonstrated that osmols are generated within cells, a phenomenon that will reduce the efflux of water. This has been called idiogenic osmol production.[24,25] A number of theoretical explanations have been offered, most of which may coexist. Water binding by protein (not at present thought likely), osmotic activation of K^+ and

Mg^{++} from binding sites, inorganic PO_4 generation from complex organic phosphates,[28] and a breakdown of both peptides and proteins to amino acids have all been suggested as the source of the idiogenic osmols. At present the likely source of most of these milliosmols is the breakdown of peptides or protein to release (primarily) taurine, aspartate, and glutamine, among other amino acids.[29]

Taurine in particular does not diffuse readily from cells thus increasing the total of intracellular fluid (ICF) osmols and preserving cell volume in the face of an osmotic gradient. There are comparative phylogeny data of considerable interest here. Both in invertebrates[30] and primitive vertebrates,[31] taurine is an osmoregulator for species who inhabit both fresh water and salt water at different times in their life cycle. This ancient life phenomenon of regulation involves mollusks, crabs, fish, amphibians, and perhaps has been retained by mammals either vestigially or as a useful protective mechanism.[32] It should be noted that it takes time to generate "idiogenic" osmols and, that once present, the taurine will remain a cell constituent past the time osmolality has been returned to normal by, for example, an intravenous infusion of water. The brain cells of animals appear to produce more idiogenic (taurine) osmols than other tissue.[25,29] Thus, during rapid correction, while all cells may show an increase in volume, the taurine present may cause this to be exaggerated for brain cells and result in an important increment of brain size.

As previously mentioned, the central nervous system (CNS) seems to be a particularly vulnerable tissue for the symptomatic ill effects of hypernatremic states. There are currently three mechanisms known that may account for this. They are first, hemorrhage and thrombosis; second, idiogenic osmol formation; and finally response to a change in calcium concentration in the ECF. Each of these needs comment.

Infants with hypernatremic dehydration following enteritis[33] and salt poisoning[13,14] have been shown to have hemorrhage and thrombosis in the CNS and not in other tissues. Experimental animals have had similar lesions produced by salt poisoning.[34] When the brains of either the patients[14] or the animals[25,34] are examined, dilated capillaries and venules are in evidence in addition to hemorrhage and (presumed) secondary thrombosis.

Further experiments[24] demonstrated that hemorrhage, at least in the animals, was related to CSF pressure changes. When hypertonic (1*M*, 2000 mOsm/kg) solutions are infused intraperitoneally[25] or intravenously[35] the CSF pressure falls predictably below atmospheric and stays low for up to 6 hr. If the pressure is kept above atmospheric in the experiments by external manipulation, the hemorrhages do not occur nor does the capillary dilation.[25,34]

These last phenomena are explainable by the anatomy and physiology of the blood-brain and blood-CSF barriers. It is known that both the brain capillary endothelial cells and the cells of the choroid plexes are structured with tight junctions between them[36-38] unlike endothelial cells and membrane cells in most tissues. These tight junctions appear to be the basis of the long known physiologic and pharmacologic phenomena of nondiffusion or slow diffusion from blood to brain or CSF, and from brain or CSF to blood. Even the small crystalloid ions, Na and Cl, diffuse slowly into and out of the CSF and the brain interstitium; water flows instantly to return any osmolal gradient to a uniform concentration. Thus, when Na concentration in ECF rises rapidly, water leaves the interstitium and CSF, and enters the circulation. This, in turn, drags water from brain cells. The entire brain shrinks away from the cranium and blood vessels under the pressure of the cardiac pump dilate and sometimes rupture. Possibly bridging veins tear.[34,39]

To understand this phenomenon better, it helps to compare what happens in muscle tissue with what happens in the brain when a sudden sodium concentration gradient is created, plasma more than ICF. In the muscle the sodium and chloride pass readily through the endothelial "loose" junction to the interstitial fluid. The ions do not, in net, enter muscle cells which instead lose their water to the interstitium. Thus cells shrink and interstitial fluid expands; the tissue volume is essentially the same after equilibrium. In the brain the boundary at which water molecules and sodium chloride ions diverge in their diffusion characteristics is the blood vessel wall. Thus the entire CNS (brain cell, brain interstitial fluid, and CSF) shrinks.

The brain is encased in a rigid skeletal structure, the cranium. Shrinking in volume of the brain creates a lower pressure outside of the tissue including the capillaries which dilate, sometimes rupture, and sometimes have secondary thrombosis with possible propagation and small or large infarcts.

The evidence for the genesis of idiogenic osmols has been already mentioned. The presence of these particles, perhaps largely taurine molecules, will, after generation, defend cell volume. More of these particles may be present in brain cells than muscle cells as judged from available data.[25] An explanation may be that there is less entry of sodium into brain cells under the pathologic circumstance of hypernatremia. It is known that the muscle cell water concentration of sodium may triple whereas the non-chloride space (not as surely identical to cell water in brain as it is in muscle) of brain may not have any increase in sodium. Alternative explanations are possible from known data but all of them embody some difference between muscle and brain adaptation.

The experimental animal manifests dramatic CNS symptoms without hemorrhage but only after idiogenic osmols are present. Thus the change in the colligative nature of the cell material is at least a marker for the electrical disturbances that occur, if indeed it is not the cause. As noted earlier, once the taurine is present, it will not be reincorporated to protein quickly, nor will it diffuse well from the cells. Should sufficient (glucose) water now be infused, the cells will swell back to a size greater than prior to the taurine generation. Cerebral swelling is about as risky and certainly just as likely to produce symptoms as cerebral shrinkage. This in turn poses a problem for rapid therapy with dilute solutions.

The final mechanism known to produce CNS symptoms or signs is one of several metabolic changes. When data from hypernatremic patients are analyzed, it is noted that calcium levels in serum are depressed.[5,40] Most of the time the change is 1 to 3 mg/dℓ lower, not enough to cause symptoms even if all the change is in the ionized fraction. Rarely the level falls to the range of 4 to 5 mg/dℓ of total calcium, and clinical tetany then occurs. This phenomenon has been reproduced in experimental animals.[41] The mechanism has not yet been elucidated very well. The lowering of calcium does not occur during the experiments on rats loaded with sodium salts if the animals are also prevented from having a simultaneous potassium deficiency.[41] These data remain unexplained but have clinical relevance. The evidence indicates that the site of the original disturbance is at the interface of the ECF with the skeleton; neither renal nor intestinal sites of calcium loss need be invoked. In the experimental situation, the lowering is maximal 24 to 48 hr after loading with salt and resolves soon thereafter. Possible vitamin D metabolites are involved in regulating the release from bone but this remains speculation.

There are two other important metabolic changes in hypernatremic states; changes in glucose homeostasis and in hydrogen ion metabolism. The first recorded study on glucose changes was by Levin and Geller-Bernstein,[42] who found frequent hyperglycemia. Subsequently a number of authors have confirmed this finding.[43-45] In as many as 25% of infants (data on adults have not been systematically sought to date) the blood glucose level is >200 mg/dℓ. In some patients levels of over 1000 mg/dℓ have been recorded.[44] In a few studies it has been suggested that insulin levels are well below what would be expected as a normal response to hyperglycemia.[45] Thus this appears to be a transient diabetic state — transient, because no long term consequence nor later abnormal carbohydrate metabolism has occurred.

The presence of the hyperglycemic and insulinopenic state is of pathophysiologic consequence in worsening brain shrinkage. Physiologic levels of glucose cross the brain vessel endothelium by active transport.[38,39] One of the functions of insulin is to facilitate glucose transport into cells and possibly into the brain. With pathologic high levels of glucose and inadequate insulin, glucose becomes an extracellular obligate osmol in the same way as sodium and thus adds its osmolal concentration increment to that of the sodium salts in causing maldistribution of body water.

In severe experimental hypernatremia, cell desiccation causes release of hydrogen with resultant acidemia.[46] In less severe but steady state circumstances pH will fall because of an obligatory drop in plasma bicarbonate.[47] In clinical practice virtually any hydrogen ion disturbance (acidemia to alkalemia) may be seen concomitantly with hypernatremia because the number of variables is so great.

In addition to direct metabolic effects, hypernatremia also modifies renal function. If there is an expanded blood volume, an osmotic diuresis ensues so that the sodium U/P ratio falls to unity or even slightly less.[48] If dehydration is present with a low blood volume, oliguria occurs so that little sodium is excreted. Either way renal function does not adapt well to removal of sodium from humans with hypernatremia. In other species, sodium excretion in the urine is vastly more efficient.

Very high sodium concentrations may cause appreciable hemolysis by destroying red cell membranes in shrinking. Coupled with the expanded plasma volume and contracted red cell, a low hematocrit is common. More recently rhabdomyolysis has been demonstrated in patients as well with attendant risk for renal injury.[49]

VIII. PATHOLOGY

There are three principal sites for gross pathologic change from the physiologic disturbance of hypernatremia. These are the CNS, the kidney, and the skin. Although dehydration is usually also present, the CNS and renal changes can be produced experimentally in animals who are not dehydrated.[48,50]

In the preceding discussion of pathophysiology in the CNS, hemorrhage and thrombosis were discussed as the source of symptoms. These changes are also the cause of mortality and late morbidity. About one third of patients dying during hypernatremic dehydration following enteritis have extensive hemorrhage or thrombosis. All victims of salt poisoning show similar findings.[14] The predilectional sites for the hemorrhage are along the sagittal sinus and over the cerebellum. The bleeding may be intracerebral, subarachnoid, subdural, and intradural.[25,34] In low birth weight infants subjected to rapidly infused highly concentrated $NaHCO_3$, intraventricular bleeding may be precipitated.[51,52] Similar changes (except intraventricular) are seen in adults poisoned[53] or overtreated with hypertonic $NaHCO_3$ during resuscitation. Small hemorrhages are seen following most deaths in hypernatremic states but these do not cause death. The mechanism for CNS hemorrhage has already been discussed in the section on pathophysiology. Since the brain is very rich in thromboplastic substances, it is not surprising that thrombosis follows hemorrhage.

Renal pathology is primarily in the medulla and involves all of the tubular portions of the nephron. The following have been commonly seen: vacuolization of the tubular cytoplasm, sometimes the vacuoles coalesce to lift the cell off the basement membrane with necrosis, eosinophilic debris in tubules (Tamm-Horsfall protein) and medullary necrosis.[39] The lesions can be experimentally produced in dogs without concomitant dehydration.[48] The necrosis appears to be reversible if the animals or patient survives. Renal vein thrombosis has occurred in young infants, hypernatremic secondary to high solute feedings.[10]

Changes in the skin are rare and when seen are part of circulatory gangrene of a distal extremity — purpura fulminans. The mechanism is not known but local endothelial damage in a dehydrated patient seems likely.[54] Hypernatremic rather than isotonic dehydration appears more likely to do this, somewhat surprisingly in view of the relatively expanded blood volume. Perhaps endothelial cell damage together with platelet interaction occurs.

IX. THERAPEUTIC MANAGEMENT

Because the pathophysiology of hypernatremia has distinct features separating this form of dehydration from other more common varieties, a consideration of principles of therapy may help in understanding the disturbance. Restoration of hydration follows a different path when moderate hypernatremia is present. The objective of treatment is to restore hydration, to restore water distribution, and to correct the complicating disturbances. Consideration of treatment which at first glance seems to be simple replacement of water, needs as a first step, careful exploration for content of solution and for rate of administration. Two other circumstances also require comment: the presence of oliguria influences decision, and finally, salt poisoning should be considered as a separate entity.

Most patients with hypernatremic dehydration are not severely oliguric owing to the relatively expanded plasma volume. This group may then be considered first. If one were to infuse plain 5% glucose water into these patients, the risk would be cerebral swelling — actually water intoxication. This is the other side of the CNS endothelial cell tight junction coin. Just as rapid infusion of hypertonic salt results in brain shrinkage, so does rapid infusion of isotonic glucose water cause brain swelling. Remember glucose crosses the blood—CNS barrier by active transport so that unlike the red cell, the brain does recognize glucose as an osmol, at least at physiologic levels of glucose. Brain swelling causes convulsions. For some years after hypernatremic dehydration was described, convulsions were commonly seen during therapy.

This circumstance led many centers to suggest adding 75 mEq/ℓ or more of sodium salts to initial therapy. This will reduce risk of convulsions but adds to the sodium burden, frequently while excessive insensible water losses are in progress aggravating hypernatremia. Such therapy also frequently produces visible edema in patients, leading to prolonged recovery periods. An alternative to increasing concentration is to slow the rate of infusion which will also avoid convulsions, but at risk of being too slow in repairing dehydration again with suboptimal outcome.

We have found a compromise resolution to these problems in part by considering that a high potassium intake would offset cerebral swelling and would in part enter depleted cells (mostly muscle) carrying water into them, at the same time delivering water at a slow even rate, provided the patient has no initial strong evidence of circulatory deficiency. The repair solution is constructed with consideration of volume for 48 hr, glucose content, sodium content, potassium content, anion distribution, calcium additive, and rate of administration. An analysis of each of these points for use in a patient with hypernatremic dehydration not in shock, and producing visible urine is provided in Table 1. Shock, oliguria, and salt poisoning will be considered separately.

If the patient has circulatory impairment (shock), first infuse 20 mℓ/kg of 5% albumin solution (single donor plasma, plasma without immunoglobulin, whole blood are all substitutable). Sodium content in this up to 140 mEq/ℓ is not important since nearly the whole volume will remain intravascular. If the patient is producing urine, proceed as in the general manner given above.

If the patient has no apparent urine, try an albumin rapid infusion as above. If urine enters bladder, proceed as before. If no urine present in bladder, give furosemide 1

Table 1
SCHEME FOR THERAPY OF HYPERNATREMIC DEHYDRATION

Considerations (In order)	Action
1. Volume	a. Estimate the patient's deficit by clinical means first in mℓ/kg and multiply by kg of weight for total sum
	b. Estimate 48 hr worth of maintenance water following usual clinical rules
	c. Add a + b for tentative volume of solution for 2 days
2. Glucose content	Use 2½ (2 to 3%) to reduce problem with hyperglycemia
3. Sodium content	Allow 80 to 100 mEq/ℓ for deficit fraction of fluid and none for the maintenance portion, the resultant is usually 20 to 35 mEq/ℓ; use this concentration of sodium or estimate at 25 mEq/ℓ
4. Potassium content	Generally the maximum safe amount for an i.v. infusion or about 40 mEq/ℓ
5. Anion content	There is now 60 to 75 mEq/ℓ of cation; distribute the anions between chloride and base in accordance with clinical judgment; if desired, start with more base and change to more chloride after 6 to 12 hr; do not use HCO_3 as base because of calcium to be added; use acetate or lactate along with chloride
6. Calcium content	One ampule of 10% calcium gluconate for every 500 mℓ of infusate
7. Rate of administration	1/48 of volume per hour for 48 hr; in an infant the usual volume will be 275 to 350 mℓ/kg/48 hr or 6 to 7 mℓ/kg/hr

mg/kg. If urine comes, proceed as above; if not, manage without potassium in the infusion and boost sodium to 50 mEq/ℓ, slow rate and limit volume by subtracting one half of maintenance allowance.

In the event of massive salt poisoning (plasma concentration of Na >200 mEq/ℓ). Use peritoneal dialysis. For dialysing solution use 8% glucose with no electrolyte 100 mℓ/kg two times within 1 hr, 1 hr apart. Simultaneously maintain an i.v. solution to deliver a volume and content as above. The induced hyperglycemia by this method offsets the removal of sodium preventing water intoxication. As the glucose is metabolized, water slowly enters cells.

Insulin is not advisable for any of these patients because rapid removal of glucose is the physiologic equivalent of rapid water infusion.

In summary, treatment seems best handled with a *slow* infusion relatively low both in glucose and sodium and high in potassium with added calcium. For the past 12 years this regimen has been highly successful without producing complicating convulsions.

The complex disturbances resulting from disproportionate changes in sodium and water are by no means yet fully elucidated. Resolution of some of the existing gaps in knowledge will undoubtedly afford new insights in electrolyte physiology.

REFERENCES

1. **Schmidt, C.,** *Charakteristik der epidemischen Cholera gegenüber verwandten Transsudationsanomalieen,* G. A. Rehyer, Leipzig, 1850.
2. **Weed, L. H. and McKibben, P. S.,** Pressure changes in the cerebrospinal fluid following intravenous injection of solutions of various concentrations, *Am. J. Physiol.,* 48, 512, 1919.
3. **Rapoport, S., Dodd, K., Clark, M., and Syllm, I.,** Postacidotic state of infantile diarrheal dehydration: symptoms and chemical data, *Am. J. Dis. Child.,* 73, 391, 1947.
4. **Rapoport, S.,** Hyperosmolarity and hyperelectrolytemia in pathologic conditions of childhood, *Am. J. Dis. Child.,* 74, 682, 1947.

5. **Finberg, L. and Harrison, H. E.,** Hypernatremia in infants, *Pediatrics,* 16, 1, 1955.
6. **Weil, W. B. and Wallance W. M.,** Hypertonic dehydration in infancy Pediatrics, 17, 171, 1956.
7. **Macaulay, D. and Watson, M.,** Hypernatremia in infants as a cause of brain damage, *Arch. Dis. Child.,* 42, 485, 1967.
8. **Ahmed, I. and Agusto-Odutola, T. B.,** Hypernatemia in diarrhoeal infants in Lagos, *Arch. Dis. Child.,* 45, 97, 1970.
9. **Taitz, L. S. and Byers, H. D.,** High calorie/osmolar feeding and hypertonic dehydration, *Arch. Dis. Child.,* 47, 257, 1972.
10. **Rodrigo, F. and Ruza, F. J.,** Deshidratacion hipertonica en la infancia: fisiopatologia y pautas de tratomiento, *Rev. Esp. Pediatr.,* 174, 865, 1973.
11. **Davies, D. P., Ansari, B. M. and Mandal, B. K.,** The declining incidence of infantile hypernatremic dehydration in Great Britain, *Am. J. Dis. Child.,* 133, 148, 1979.
12. **Wolf, A. V.,** *Thirst,* Charles C Thomas, Springfield, Ill., 1958, Part I, chap. V and Part II.
13. **Miller, N. and Finberg, L.,** Peritoneal dialysis for salt poisoning, *N. Engl. J. Med.,* 263, 1347, 1960.
14. **Finberg, L., Kiley, J., and Luttrell, C. N.,** Mass accidental salt poisoning in infancy. A study of a hospital disaster, *JAMA,* 184, 187, 1963.
15. **Zierler, K. L.,** Hyperosmolarity in adults: a critical review, *J. Chron. Dis.,* 7, 1, 1958.
16. **Segar, W. E.,** Chronic hyperosmolality, *Am. J. Dis. Child.,* 112, 318, 1966.
17. **Rapoport, S.,** The role of overventilation in diseases of infancy, *Ann. Paediatr.,* 176, 137, 1951.
18. **Harrison, H. E. and Finberg, L.,** Hypernatremic dehydration, *Pediatr. Clin. N. Am.,* 11, 955, 1964.
19. **Crigler, J. F. and Suh, S.,** Hyperosmolarity in patients following radical treatment of craniopharyngioma, *Am. J. Dis. Child.,* 102, 469, 1961.
20. **Ussing, H. H., Erlij, D., and Lassen, U.,** Transport pathways in biologic membranes, *Am. Rev. Phys.,* 36, 17, 1974.
21. **Sweadner, K. J. and Goldin, S. M.,** Active transport of sodium and potassium ions, *N. Engl. J. Med.,* 302, 777, 1980.
22. **Ling, G. and Ochsenfeld, M. M.,** Studies on ion accumulation in muscle cells, *J. Gen. Physiol.,* 49, 819, 1966.
23. **Gulati, J. and Palmer, L. G.,** Potassium accumulation in frog muscle: the association-induction hypothesis versus the membrane theory, *Science,* 198, 1281, 1977.
24. **McDowell, M. E., Wolff, A. V., and Steer, A.,** Osmotic volumes of distribution, *Am. J. Physiol.,* 180, 545, 1955.
25. **Finberg, L., Luttrell C., and Redd, H.,** Pathogenesis of lesions in the nervous system in hypernatremic states. II. Experimental studies of gross anatomic changes and alterations of chemical composition of the tissues. *Pediatrics,* 23, 46, 1959.
26. **Arieff, A. I. and Kleeman, C. R.,** Studies on mechanisms of cerebral edema in diabetic comas: effects of hyperglycemia and rapid lowering of plasma glucose in normal rabbits, *J. Clin. Invest.,* 52, 571, 1973.
27. **Holiday, M. A., Kalayci, M. N., and Harrah, J.,** Factors that limit brain volume changes in response to acute and sustained hyper- and hyponatremia, *J. Clin. Invest.,* 47, 1916, 1968.
28. **Nitowsky, H. M., Herz, F., and Geller, S.,** Induction of alkaline phosphatase in dispersed cell cultures by changes in osmolarity, *Biochem. Biophys. Res. Commun.,* 12, 293, 1963.
29. **Thurston, J. H., Hauhart, R. E., and Dirges, J. A.,** Taurine: a role in osmotic regulation of mammalian brain and possible clinical significance, *Life Sci.,* 26, 1561, 1980.
30. **Bedford, J. J. and Leader, J. P.,** Hyperosmotic readjustment of the crab, *Hemigrapsus edwodsi, J. Comp. Phys.,* 128, 147, 1978.
31. **Gordon, M. S.,** Intracellular osmoregulation in skeletal muscle during salinity adaptation in two species of toads, *Biol. Bull.,* 128, 218, 1965.
32. **Jacobsen, J. G. and Smith, L. H., Jr.,** Biochemistry and physiology of taurine and taurine derivatives, *Physiol. Rev.,* 48, 424, 1968.
33. **Finberg, L. and Harrison, H. E.,** Hypernatremia in infants. An evaluation of the clinical and biochemical findings accompanying this state, *Pediatrics,* 16, 1, 1955.
34. **Luttrell, C. N., Finberg, L., and Drawdy, L.,** Hemorrhagic encephalopathy induced by hypernatremia. II. Experimental observations of hyperosmolarity in cats, *A.M.A. Arch. Neurol. Psychiatry,* 1, 153, 1959.
35. **Kravath, R. E., Aharon, A. S., Abal, G., and Finberg, L.,** Clinically significant physiologic danger from rapidly administered hypertonic solutions: acute osmol poisoning, *Pediatrics,* 46, 267, 1970.
36. **Reese, T. S. and Karnovsky, M. J.,** Fine structural localization of a blood-brain barrier to exogenous peroxidase, *J. Cell. Biol.,* 34, 207, 1967.
37. **Spector, R.,** Vitamin homeostasis in the central nervous system, *N. Engl. J. Med.,* 296, 1393, 1977.
38. **Goldstein, G. W.,** Pathogenesis of brain edema and hemorrhage: role of the brain capillary, *Pediatrics,* 64, 357, 1979.

39. **Finberg, L.**, Hypernatremic dehydration, *Adv. Pediatr.*, 16, 325, 1969.
40. **Finberg, L.**, Pathogenesi of lesions in the nervous system in hypernatremic states, *Pediatrics*, 23, 40, 1959.
41. **Finberg, L.**, Experimental studies of the mechanisms producing hypocalcemia in hypernatremic states, *J. Clin. Invest.*, 36, 434, 1957.
42. **Levin, S. and Geller-Bernstein, C.**, Alimentary hyperglycemia simulating diabetic ketosis, *Lancet*, 2, 595, 1964.
43. **Stevenson, R. E. and Bowyer, F. P.**, Hyperglycemia with hyperosmolal dehydration in nondiabetic infants, *J. Pediatr.*, 77, 818, 1970.
44. **Finberg, L.**, Current concepts: hypernatremic (hypertonic) dehydration in infants, *N. Engl. J. Med.*, 289, 196, 1973.
45. **Mandell, F. and Fellers, F. X.**, Hyperglycemia in hypernatremic dehydration, *Clin. Pediatr.*, 13, 367, 1974.
46. **Sotos, J. F., Dodge, P. R., and Talbot, N. B.**, Studies in experimental hypertonicity. II. Hypertonicity of body fluids as a cause of acidosis, *Pediatrics*, 30, 180, 1962.
47. **Winters, R. W., Scaglione, P. R., Nahas, G. G., and Verosky, M.**, The mechanism of acidosis produced by hyperosmotic infusions, *J. Clin. Invest.*, 43, 647, 1964.
48. **Finberg, L., Rush, B. F., and Cheung, C. S.**, Experimental observations of the renal excretion of sodium during hypernatremia, *Am. J. Dis. Child.*, 107, 483, 1964.
49. **Opas, L. M., Adler, R., Robinson, R., and Lieberman, E.**, Rhabdomyolysis with severe hypernatremia, *J. Pediatr.*, 90, 713, 1977.
50. **Rush, B. F., Finberg, L., Daviglus, G. F., and Cheung, C. S.**, Pathologic lesions in experimental hypernatremia induced by extracorporeal dialysis, *Surgery*, 50, 359, 1961.
51. **Papile, L., Burstein, J., Burstein, R., Koffler, H., and Koops, B.**, Relationship of intravenous sodium bicarbonate infusions and cerebral intraventricular hemorrhage, *J. Pediatr.*, 93, 834, 1978.
52. **Volpe, J. J.**, Intracranial hemorrhage in the newborn: current understanding and dilemmas, *Neurology*, 29, 632, 1979.
53. **Johnston, J. G. and Robertson, W. O.**, Fatal ingestion of table salt by an adult, *West. J. Med.*, 126, 141, 1977.
54. **Comay, S. C. and Karabus, C. D.**, Peripheral gangrene in hypernatremic dehydration of infancy, *Arch. Dis. Child.*, 50, 646, 1975.

INDEX

A

B

D

E

F

G

H

I

J

K

L

M

Q

R

S

V

W

X

Y

Z